Boolean Structures

Combinatorics, Codification, Representation

Boolean Structures

Combinatorics, Codification, Representation

Gennaro Auletta

University of Cassino and Southern Lazio, Italy
& Pontifical Gregorian University, Italy

World Scientific

NEW JERSEY · LONDON · SINGAPORE · BEIJING · SHANGHAI · HONG KONG · TAIPEI · CHENNAI · TOKYO

Published by

World Scientific Publishing Europe Ltd.

57 Shelton Street, Covent Garden, London WC2H 9HE

Head office: 5 Toh Tuck Link, Singapore 596224

USA office: 27 Warren Street, Suite 401-402, Hackensack, NJ 07601

Library of Congress Control Number: 2021012131

British Library Cataloguing-in-Publication Data
A catalogue record for this book is available from the British Library.

BOOLEAN STRUCTURES
Combinatorics, Codification, Representation

ISBN 978-1-80061-008-8 (hardcover)
ISBN 978-1-80061-009-5 (ebook for institutions)
ISBN 978-1-80061-010-1 (ebook for individuals)

For any available supplementary material, please visit
https://www.worldscientific.com/worldscibooks/10.1142/Q0296#t=suppl

Typeset by Stallion Press
Email: enquiries@stallionpress.com

I desire to thank both Doctor Mattia Manfredonia and Professor Massimo Stanzione for their helpful comments and suggestions.

Contents

List of Tables

List of Figures

Chapter 1

Introduction

1.1. Logic is the Only Self–grounding Knowledge

I think that the best definition of logic has been provided by Ludwig Wittgenstein, who told that logic must provide for itself,[1] meaning that it is the only discipline that does not need the contributions of other fields (and is also fully independent from experience[2]), although it can use mathematics to cast synthetically certain aspects. For this reason, to Wittgenstein logic is a science of the possible: whatever possible is, is also allowed. In other words, logic does not deal with empirical statements of any kind nor with assumptions about whatever material or formal entity. I add only that the connections among those possible combinations are necessary. Moreover, logic is a pure formal science. According to Wittgenstein, it occurs only in the field of logic that sign and what it denotes are identical.[3] This symmetry is evident by considering (as I shall show in the following) that logic deals with schemas of classes, and therefore the symbols for such schemes coincide with the latter.

---

[1] "Die Logik muss für such selber sorgen": [WITTGENSTEIN 1921, 5.473].
[2] [WITTGENSTEIN 1921, 5.552].
[3] [WITTGENSTEIN 1914–16, p. 91].

1.2. Is Logic Arbitrary?

Many logicians in the 20th century have assumed that whatever
statement is or can be is an admitted logical proposition, and this
assumption is the foundation of the so–called propositional logic.[4]
In other words, statements like "The sun is hot", "Today is a nice
day", "Napoleon won at Waterloo" are considered basic elements of
this logical calculus. If so, it is evident that logic would deal with
relations among "propositions" that are fully arbitrary. This arbitrariness shows that the notion of propositions has been taken here
from ordinary language and not from the language of science, where
proposition means proved statement, thus being a theorem. However,
natural languages are tools for communicating and, although instantiate some logical connectives, we cannot transpose their structures
as such to logic. Moreover, this raises a further question: how are
the logical connections themselves among these arbitrary statements
defined? To explore this problem, let us focus on the significance of
the material implication, a logical tool that was introduced in that
context.

Material implication appears to have no logical ground. In fact, if
expressions like $X \to Y$ were admitted (where X and Y could be here
any arbitrary statements or variables), we would count as a logical
proposition a connection between two logically unrelated variables.
Indeed, we cannot interpret material implication in terms of sufficient
condition when the antecedent (here X) is no condition whatsoever
of the consequent (here Y). The consequences of this step are well
known: we would run into the so–called logical paradoxes like "If
Napoleon won at Waterloo, I was the emperor of China" and similar absurd expressions. I think that to say that there are logical
paradoxes is the only true paradox here. At the opposite, I am con-
vinced that paradoxes simply do not exist either in logic or in any
other scientific discipline. "Paradox" may only be a useful term for
denoting difficult problems that have not been completely solved at

---

[4][WHITEHEAD/RUSSELL 1910].

certain stages of our progress in knowledge, like the "Schrödinger's cat paradox" and similar ones.[5]

Now, if material implication has no logical ground, it could still be considered as an important tool for dealing with inferences about concrete problems of empirical sciences or of everyday life. However, if we admit material implication (without the antecedent being a condition of the consequent) as scientific instrument, we are forced to also admit the so–called Transposition rule, i.e. $X \to Y$ would be equivalent to $Y' \to X'$ (where X' is the negation of X). Here, we already meet a problem: if they were only materially equivalent, to say this would be of no great help, but on the other side, as I shall show, they cannot be logically equivalent. This notion will be explored below. By now, suppose a material implication like "If Giorgio is not in Milan, he is at the University "La Sapienza"". This appears to be correct, or at least reasonable, given certain pieces of information (e.g. we know for sure that Giorgio would have gone that day to the University "La Sapienza" in Rome). However, the transposed statement tells us that "If Giorgio is not at the University "La Sapienza", he is in Milan". However, it is hard to see how these two hypothetical expressions should be logically equivalent on the basis of the same piece of information. In fact, although Giorgio is not at the University "La Sapienza", he could be still in Rome, for instance drinking a coffee in some cafeteria nearby. We should need some *additional* information for excluding such a possibility. However, if we admit this, we risk running in an endless regress.

One of the main misunderstandings here is that material implication is considered fundamental for reasoning in empirical sciences and practical matters. So, we suppose that "If it rains, there is traffic in Rome", and because we know that it rains, we can deduce that there is traffic in Rome. The hidden information here is represented by the fact if it rains today or in such and such context (as a matter of fact) that the whole inference would take the form of being $X \to Y$ and XZ, we can deduce XY, that is, that there is traffic.

---

[5][AULETTA 2019, Sec. 3.1.4].

Now, instead of speaking of material implication, we can obtain the same result but incur no paradoxes by dealing with relations among classes of objects X and Y and therefore interpret $X' + Y$ (where $+$ is the logical sum) as meaning that there are no objects X that are not Y. Why do we not incur in paradox here? Because we admit in advance that we are dealing with a *contingent* relation among *possible* classes of objects and not with a presumed logical relation that allows us to infer a certain consequence.

Then, the correct inference about rain in Rome has the following structure (X meaning raining, Y meaning traffic, and Z meaning the current context), where from now on I adopt the symbol $\rightarrow$ for logical implication *only*:

$$(X' + Y)XZ \rightarrow XY, \tag{1.1}$$

which is fully correct and is indeed a tautology. I stress here that *all* inferences of whatever kind need to be represented by a tautology with a logical implication connecting at least two premises and a consequence: this is the canonical syllogistic structure attributed to Aristotle,[6] whose archetypical structure $\forall X, Y, Z$ is

$$(X' + Y)(Z' + X) \rightarrow (Z' + Y). \tag{1.2}$$

In fact, the definition of logical implication is rigorous. As we shall see, it connects two nodes of Boolean algebra such that the logical ID (the binary sequence identifying the node) of the consequent needs to preserve *all* 1s of the antecedent in the *same* position and eventually increase their number. In the case in which the number does not increase, antecedent and consequent coincide according to the reflexive property of the logical implication. For instance, $\forall X, Y, Y \rightarrow (X' + Y)$ is a logical proposition telling us that if there are classes of objects Y, then there are also classes that cannot be X being not Y.

At the opposite, I admit the importance of material equivalence in mathematics and other sciences. Note that the mathematician first

---

[6][AULETTA 2013a, AULETTA 2017].

proves or tries to prove that something is the *condition* of something else as well as the condition in turn of the former, and *then* declares that the two items are equivalent. However, I distinguish material equivalence like $X \sim Y$ from logical equivalence $=$, e.g. $X(Y + Z) = XY + XZ$. Both satisfy the three fundamental requirements of equivalence (reflexivity, symmetry, and transitivity), but only the latter displays the *same* logical ID on the left and right side (it is an identity function), identifying therefore the *same* node of the Boolean algebra. Of course, material equivalence (in the binary case) is defined as $X \sim Y = XY + X'Y'$, while material contravalence is defined as $X \not\sim Y = XY' + X'Y$.

1.3. The Issue of Quantification

Since logic started as a propositional calculus in the way considered above, it was quite natural predicate calculus developed in order to ensure the correct quantification of the "logical" relations. However, logic, dealing with schemes of possible classes of objects and their relations is not concerned at all with the membership of those classes. It is true that it is concerned with the problem of knowing whether the relation between, say, classes X and Y is universal or particular. However, this quantification is already inscribed in the logical connectives, where logical sum denotes universal relations and logical product particular relations.

Quantification is very helpful in mathematics, especially when quantifying properties (second order) and not individuals. However, although strictly related (and in fact my approach to Boolean algebra is based on combinatorial calculus), logic and mathematics are two quite different disciplines and confusion between the two cannot really help us in the progress of knowledge.

1.4. Traditional Approach to Boolean Algebra

Let us consider the usual definition of Boolean algebra. Fix, once and for all, a set S, and denote by $\mathscr{B}(S)$ the collection of all subsets of S.

Next, let f be a Boolean function, i.e. mapping from $\mathscr{B}(S)$ to the set $\{0, 1\}$ of truth values, such that, $\forall X, Y \in \mathscr{B}(S)$, following Boolean operations are allowed[7]:

- *Intersection*, denoted by $X \times Y$ or XY, which, $\forall X, Y \in \mathscr{B}$, satisfies

$$f(XY) = 1 \text{ iff } f(X) = f(Y) = 1; \quad \text{otherwise } f(XY) = 0.$$
(1.3)

- *Union*, denoted by $X + Y$, which, $\forall X, Y \in \mathscr{B}$, satisfies

$$f(X + Y) = 0 \text{ iff } f(X) = f(Y) = 0; \quad \text{otherwise } f(X + Y) = 1,$$
(1.4)

  where iff denotes if and only if.
- *Complementation*, denoted by X', which, $\forall X \in \mathscr{B}$, satisfies

$$f(X') = 1 \text{ iff } f(X) = 0 \quad \text{and} \quad f(X') = 0 \text{ iff } f(X) = 1. \quad (1.5)$$

The map f comes therefore to represent truth–value assignment and the numbers 0 (false) and 1 (true) to represent truth values, which allows us to conceive sets and subsets as two–valued variables. Note that I am faithful to Boole's original symbols for the reason that these operations can also be understood in pure algebraic terms.[8]

In Boolean algebra $\mathscr{B}(S)$ the binary operations $\times, +$ (also called logical sum and product or inclusive disjunction and conjunction) satisfy the following properties:

1. *Idempotency*: $\forall X \in \mathscr{B}$, $X = XX = X + X$. From this it follows that, $\forall X \in \mathscr{B}$, either $f(X) = 1$ or $f(X) = 0$. In fact, suppose that $f(X)$ is not 1. Then, according to Eqs. (1.3), $f(XX) = 0$. However, from idempotency it follows that $f(X) = 0$. Similarly, suppose that, $f(X)$ is not 0. Then, from Equation (1.4), we have $f(X + X) = 1$. However, according to idempotency, it follows that $f(X) = 1$.

---

[7][BOOLE 1854]. For a canonical introduction to Boolean algebra see [GIVANT/HALMOS 2009].

[8]See [BOOLE 1854, p. 33].

2. *Commutative law:* $\forall X, Y \in \mathscr{B}$, $XY = YX$ and $X + Y = Y + X$.
3. *Associative law:* $\forall X, Y, Z \in \mathscr{B}$, $X(YZ) = (XY)Z$ and $X + (Y + Z) = (X + Y) + Z$.

Moreover, the two binary operations are *mutually distributive*, i.e., $\forall X, Y, Z \in \mathscr{B}$,

$$X(Y + Z) = XY + XZ \quad \text{and} \quad X + YZ = (X + Y)(X + Z).$$
$$(1.6)$$

In order for this collection $\mathscr{B}(S)$ of all subsets of S, together with the operations of union, intersection, and complementation, to be Boolean algebra, there are some requirements. In the literature, there is no general agreement about those requirements or about what are the primitive ones. Here, I introduce them in a rather loose way since I shall later present another approach that can bypass this problem by simply looking at the *structural* properties of Boolean algebra. A necessary requirement is that such a collection must be a Partially Ordered Set (POSet) by inclusion (*implication*), symbolised as $\rightarrow$, displaying

- *Reflexivity:* $\forall X \in \mathscr{B}$, $X \rightarrow X$;
- *Transitivity:* $\forall X, Y \in \mathscr{B}$, if $XY \rightarrow X$ and $X \rightarrow X + Y$, then $XY \rightarrow X + Y$;
- *Antisymmetry:* $\forall X, Y \in \mathscr{B}$, the two expressions $XY \rightarrow Y$ and $Y \rightarrow XY$, are logically different and neither implies or is equivalent to the other.

In other words, the different subsets are connected through logical implication.

As mentioned, in the present book I adopt another approach based on combinatorial calculus without basic principles or assumptions but by requiring only the basic definitions of logical connectives, logical operations, and logical propositions. Specifically, I will address three different but interrelated issues: the combinatorics of truth values (determining the nodes), the structural characters and patterns of the numerical sequences (the IDs of the nodes) resulting from codification, and the logical representation of such numerical sequences.

1.5. About the Notation

In the book I use

- Indexed X's (like $X_{1.1}, X_{2.3.6}$, and so on) for denoting specific nodes of the Boolean algebra.
- The first letters of the alphabet (like A, B, C, and so on) for denoting logical variables that are related to each other through some kind of symmetry or replacement.
- Traditional forms (like X, Y, Z, and so on) for denoting arbitrary logical variables.

All of these symbols stand for sets and not for propositions, and this is the reason for using capital letters (in order to distinguish them for traditional letters, like $p, q, r, \ldots$, denoting what are ordinarily understood as propositions). Moreover, letters like x, y, z denote numerical variables or vectors.

Chapter 2

Preliminary Notions

2.1. Basic Operations

The basic logical operations are assumed to be complementation, logical sum, and logical product. We distinguish between the notions of class (and set) and the notion of logical proposition. Nevertheless, it is convenient to use a single symbol for the different operations: thus, logical sum and product apply both to classes (and sets) and logical propositions, so that they turn out to be union and intersection for the first case and conjunction and disjunction for the second case, respectively. Table 2.1 shows how operations act on truth values (complementation simply flips the truth value). The fact that we can use the same symbols for both cases simplifies a lot. Using the same symbols also allows us to speak of the truth value of a set or a class, meaning that when we say, for example, that a set is true, it is meant that such a set exists (or that the considered possible object, whether set or class, shows that aspect), and when it is said that it is not true it is meant that its complement does exist (or the object does not display that aspect).

All sets are denoted by capital letters, like X, Y. Since in logic we only deal with schemes of sets (and not with concrete examples), $X, Y, \ldots$ are variables. Moreover, the Boolean algebra considered here

Table 2.1. The four basic logical operations, as well as logical subtraction (corresponding to AND NOT) and logical division (corresponding to OR NOT).

Sum	$0+0=0$	$0+1=1$	$1+0=1$	$1+1=1$
Product	$0 \times 0 = 0$	$0 \times 1 = 0$	$1 \times 0 = 0$	$1 \times 1 = 1$
Subtraction	$0 - 0 = 0$	$0 - 1 = 0$	$1 - 0 = 1$	$1 - 1 = 0$
Division	$0 : 0 = 1$	$0 : 1 = 0$	$1 : 0 = 1$	$1 : 1 = 1$

comprehends universal bounds: $\mathbf{1}$ (the universal set or *supremum*, given by the union of all elements) and $\mathbf{0}$ (the empty set or *infimum*, given by the intersection of all elements), which $\forall X$ satisfies the following properties:

$$\mathbf{0}X = \mathbf{0}, \quad \mathbf{1} + X = \mathbf{1}, \quad \mathbf{0} + X = X, \quad \mathbf{1}X = X, \qquad (2.1)$$

so that they are the identity elements for product and sum, respectively. Since both logical sum and product are associative (see Section 1.4), Boolean algebra is a monoid[1] under both sum and product.

As mentioned, we need to sharply distinguish between propositions and relations among classes. I stress again that a logical proposition needs to be a tautology, while a relation between sets (like $X + Y$) is contingent and has no logical significance apart from representing a node of a Boolean network. Thus, as anticipated, in the following statement the only propositions admitted are those constituted by a logical implication (satisfying reflexivity, anti-symmetry and transitivity) or combinations of logical implications, like

$$XY \to X, \quad X \to (X + Y), \quad (XY \to X)(YZ \to Z). \qquad (2.2)$$

In other words, X or $X \to Y$ are not propositions. As mentioned, the so-called material implication $X \to Y$ turns out to be simply the logical sum $X' + Y$.

---

[1] [SPIVAK 2013].

Let us now define two new operations that are crucial for the following:

- The *reversal* operation, which reverses the binary ID (the sequence of truth values identifying a node, whose formal definition will be provided below); e.g. $(10001101)^{-1} = 10110001$. Such an operation transforms XY into $X'Y'$ as well as $X + Y$ into $X' + Y'$, and $\mathbf{0}$ into $\mathbf{0}$. These expressions are also called contradual of each other. Note that in the case of complex expressions we have $\forall X, Y, Z$

$$(XY + X'Z)^{-1} = (XY)^{-1} + (X'Z)^{-1} = X'Y' + XZ'. \quad (2.3)$$

Mathematically, we can effect the reversal operation thanks to the reversal or exchange matrix (with all 1s located on the counter-diagonal). There is an important rule to be observed here: we cannot drop parentheses if there are $\mathbf{0}$s or $\mathbf{1}$s inside unless the whole expression between parentheses is $\mathbf{0}$ or $\mathbf{1}$.

- The *neg–reversal* operation (it does not matter the order of negation and reversal): $(10001100)^{\dagger} = 11001110$. In other words, with $\forall X$ we have

$$X^{\dagger} = (X')^{-1} = (X^{-1})', \quad \text{which implies} \quad (X^{\dagger})^{-1} = (X^{-1})^{\dagger} = X'.$$
$$(2.4)$$

Note that we can understand complementation or negation as a combination of neg–reversal and reversal. In other words, as for complementation, both two applications of reversal and neg–reversal operations bring the proposition back to the initial element (and so their compositions). They correspond therefore to the inversion in group theory. Note that we cannot represent the neg–reversal operation with a single matrix because we cannot find a general matrix for negation (it works only in the 2D case, e.g. through the X matrix in quantum computation; however, in such a case it coincides with the reversal matrix).

The neg–reversal operation transforms XY into $X + Y$ and vice versa, as well as $\mathbf{0}$ into $\mathbf{1}$ and vice versa. These expressions are also called the dual of each other. Moreover, with $\forall X, Y, Z$ we have

$$(XY + X'Z)^\dagger = (XY)^\dagger(X'Z)^\dagger = (X + Y)(X' + Z) = X'Y + XZ,$$
$$(2.5)$$

since

$$\begin{aligned}
X'Y + XZ + YZ &= X'Y + XZ + 1YZ \\
&= X'Y + XZ + (X' + X)YZ \\
&= X'Y + XZ + X'YZ + XYZ \\
&= X'Y(1 + Z) + XZ(1 + Y) \\
&= X'Y + XZ.
\end{aligned} \qquad (2.6)$$

In the case of triplets, the computation is more complex: with $\forall X, Y, Z$ we have:

$$\begin{aligned}
(XY' + X'YZ)^\dagger &= (X + Y')(X' + Y + Z) \\
&= X'Y' + (XY + Y'Z + XZ) \\
&= X'Y' + XY + Y'Z,
\end{aligned} \qquad (2.7)$$

and therefore, $\forall X, Y, Z, W$,

$$\begin{aligned}
(XY + X'Y'Z + YZ'W)^\dagger &= (X + Y)(X' + Y' + Z)(Y + Z' + W) \\
&= (XY' + XZ + X'Y + YZ)(Y + Z' + W) \\
&= X'Y + YZ + X(Y'Z' + Y'W + ZW) \\
&= X'Y + YZ + XY'Z' + XZW,
\end{aligned} \qquad (2.8)$$

where in the last step I use Equation (2.6). Thus, while logical sum and product are connectives taking two or more expressions and transforming them into a single one, there are four operations or transformations mapping the set of Boolean expressions one–to–one onto itself: the identity function $=$ (taking each expression to itself), the complement function $'$ (taking each expression to its complement), the contradual function $^{-1}$ (bringing an expression to its reverse), and the dual function † (bringing an expression to its negative–reverse). Thus, all operations introduced so far map elements of the algebra to the same or other elements of the algebra.

2.2. Sets and Classes

For practical purposes, in this book I use following nomenclature: sets are collections of items that share a single basic property (all objects being X), while classes are collections of items sharing several properties and therefore displaying several relations among the basic sets themselves. *Sets* express characters of objects, or better (in the language of category theory) *aspects*,[2] while classes are objects showing an internal articulation of different aspects. So an expression like XY means "The objects sharing aspects X and Y", and therefore the expression constitutes a (very elementary) class. For instance, the collection of all red items is a set, while mammals represent a class. Moreover, as said, the extension of sets and classes is irrelevant in the present examination.

How can we formally distinguish between these two kinds of objects? The neg–reversal operation will help us to this purpose. Indeed, for any algebra, there is a set of variables whose binary ID is divided in two equal halves with the first half the neg–reversal of the second. Examples are: 0000 1111, 0011 0011, 1001 0110. Let us denote the set collecting those variables as NR. This designation implies that such variables are the neg–reversal of themselves, meaning that with $\forall X \in NR$ we have $X' = X^{-1}$. We can call these variables also *self–dual variables*. Self–dual variables (SD variables) represent basic *sets*, while their logical combinations (which are no longer self–dual) represent *classes*.

2.3. Basic Numbers

In this book I we focus on what appears to be a particular class of Boolean algebra (with the set S being in fact the universal set **1**,

---

[2][SPIVAK 2013].

so that we write $\mathscr{B}(1)$) that is characterised by some fundamental numbers:

- n is the number of basic sets (SD variables) whose combinations can span the algebra and determine its dimension. With $\forall n \in \mathbb{N}$, we denote the nD algebra by $\mathscr{B}_n$ (we drop the dependence on the universal set since all algebra shares such a character). It is easy to see that with $\forall n \in \mathbb{N}$ all elements of $\mathscr{B}_n$ can be derived from some combination of the neg–reversal variables due to their self–duality.

- $m(n) = 2^n$ is the number of truth–value assignments that determine the truth–table: I have expressed the dependence of m on n. In any algebra, all objects (whether sets or classes) are represented by a sequence of m 0s and 1s representing falsity and truth, respectively. I call any of these Boolean bitstrings *the binary ID* of the object, ID in short.

- $k(n) = 2^m$ is the overall number of all sets and classes that can be generated in such algebra (represented by nodes of the Boolean network) through relations among the n basic sets. In particular, the generation of all classes (being sets improper classes) for any n–dimensional algebra obeys following binomial series:

$$k(n) = \sum_{x=0}^{2^n} \binom{2^n}{x}, \tag{2.9}$$

where the binomial coefficient is notoriously given by

$$\binom{x}{y} = \frac{x!}{y!(x-y)!} \tag{2.10}$$

and $x! = 1 \cdot 2 \cdot 3 \cdots (x-1) \cdot x$ is called the factorial. The binomial coefficient expresses the different ways to "put" y objects in x "boxes". Equation (2.9) is an instance of the general formula

$$\sum_{x=0}^{2^n} \binom{2^n}{x} = 2^{\sum_{y=0}^{n} \binom{n}{y}}. \tag{2.11}$$

- Each of the binomial coefficients in the series (Equation (2.9)) gives rise to all classes with the same number of 0s and 1s. These are called the *Levels* of the algebra. The number l of levels is always $l(n) = m(n) + 1$.
- The number of all basic sets (which for $n \geq 3$ is larger than the number of dimensions or the number of sets spanning the algebra) together with their complements is given by $s(n) = k(n - 1)$. Of course, if the complements need to be excluded, the equation is $s'(n) = k(n - 1)/2$.
- The Boolean algebra is in fact a Partially Ordered Set (POSet), satisfying reflexivity, transitivity, and antisymmetry. This set builds a network where several nodes are connected through logical implication. Many use the symbol $\leq$ for such relations. However, it is suitable to use the arrow $\rightarrow$ since it expresses in a better way the structure of a graph (with relative pathways) that a Boolean network is in fact. Thus, the number r of logical relations (logical implications) that can be constituted among all nodes of the network is given by

$$r(n) = \sum_{t=0}^{m-1} (m - t) \binom{m}{m - t}, \tag{2.12}$$

where t is the number of 1s (truths) present in the $m = 2^n$ ID. By putting $x = m - t$, the previous equation can be rewritten as

$$r(n) = \sum_{x=1}^{m} x \binom{m}{x}$$

$$= m2^{m-1} = 2^n 2^{m-1}$$

$$= 2^{m(n)+(n-1)}. \tag{2.13}$$

Note that the arrows departing from a node go as a decreasing series (according to the different levels $l(n)$)

$$m, m - 1, m - 2, \ldots, 0. \tag{2.14}$$

Of course, the series is inverted if we consider the arrows going to a node.

- The product of all logical relations gives the number $p(n) = m(n)!$ of possible paths going from the lowest to the highest node.

I prove now that, for each $\mathscr{B}_n$, the number of basic sets or SD variables (and their complements) is equal to the number of all objects (sets and classes) of $\mathscr{B}_{n-1}$. In fact, SD variables are identified by half the sequence of their binary ID. An immediate consequence is that, for each $\mathscr{B}_n$ ($n \in \mathbb{N}$), with $k(n) = 2^m(n)$ (where $m(n) = 2^n$) objects, the number of these variables is $k(n - 1)$, with $m(n - 1) = 2^{n-1}$.

Later on, I shall show that any other kind of Boolean algebra is in fact an extrapolation from the Boolean algebra treated in this book, so that they finally represent the general case.

2.4.   First Examples of Algebra

Let us consider now some basic examples of algebra. In the following, for reasons that described below, I use indexed variables for denoting basic sets (SD variables). Note that we are free to choose whatever symbol we like for such variables (for instance, α or λ). However, the choice of X as the classical unknown variable seems suitable. Moreover, I recall that the use of the capital letter is to stress that these variables denote sets. The indexes themselves are not conventional but "forced" upon us by the objective structure of the algebra. In fact, although we could use different sorts of symbols for indexing, we are forced to use natural numbers due to the exponential growth of the indexes themselves across the different dimensions. In general, most of the choices we make are not arbitrary at all.

The easiest case is represented by $\mathscr{B}_1$, which can be represented by a single SD variable X_0 and its complement X_0', as displayed in Figure 2.1. Clearly, such algebra has $m(1) = 2$ and thus generates $k(1) = 4$ combinations, $X_0, X_0', \mathbf{0}, \mathbf{1}$, to which the following IDs are respectively assigned: 01, 10, 00, 11. The number of logical relations is $r(1) = 4$, of levels is $l(1) = 3$, and of paths is $p(1) = 2$, giving rise to the network displayed in Figure 2.2. If we started with $\mathscr{B}_0$ (that I introduce here only for argumentation), which clearly has only 0 and 1 as nodes, we see that the IDs or nodes of $\mathscr{B}_1$ are generated by the

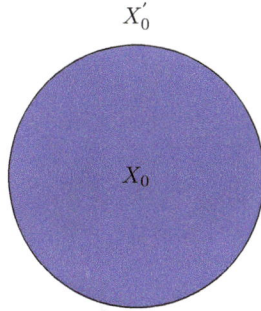

X_0'

X_0

Figure 2.1. The set X_0 and its complement X_0'. Clearly, the sum of the two covers the whole universe of all possible objects, i.e. $X_0 + X_0' = \mathbf{1}$. On the other hand, their product is $X_0 X_0' = \mathbf{0}$.

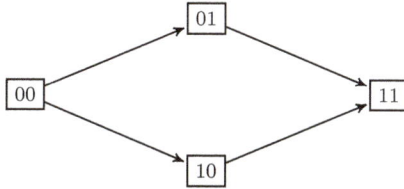

Figure 2.2. The $\mathscr{B}_1$ network. We have two pathways for going from 00 to 11. Note that we also have arrows going from each node back to itself (not shown in the diagram). Since the two nodes X_0 and X_0' have a least upper bound ($\mathbf{1}$) and a greatest lower bound ($\mathbf{0}$), it is a (complete) lattice. In fact, a lattice equipped with complementation and satisfying distributivity is Boolean algebra and vice versa.

Cartesian product between these two elements, and in the same way we build all higher–dimensional algebra.

It may be wondered what is the use of variables like X_0 if we already have the IDs. Clearly, IDs (and their sums and products) represent the main content of the Boolean algebra. Nevertheless, the use of variables is crucial for inspecting some important invariances and symmetries of the algebra that would otherwise be hidden. This will become clearer with higher–dimensional algebra. Moreover, the use of variables allows a straight connection with computational architectures.

For $\mathscr{B}_2$ (with $m(2) = 4$, $k(2) = 16$, $s(2) = 4$, $l(2) = 5$, $p(2) = 4! = 24$) we get the situation represented in Figure 2.3, which gives

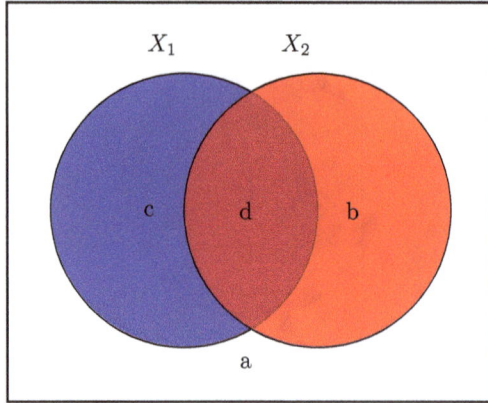

Figure 2.3.   The sets X_1 and X_2. X_1 is represented as the sum of areas c e d while X_2 as the union of areas b and d (see Table 2.2).

Table 2.2.   The IDs of X_1 and X_2 for $\mathscr{B}_2$.

Sets	X_1	0	0	1	1
	X_2	0	1	0	1
Areas			a	b	c	d

rise to the IDs for the SD variables X_1, X_2 displayed in Table 2.2. Note again that we are relatively free in denoting the different columns in Table 2.2. Of course, we could have reversed the order, putting what here is d at the place of a. Nevertheless, the relations that are established between X_1 and X_2 are invariant across all these possible choices.

Since we deal with SD variables (the half of whose IDs is the neg–reversal of the other half), these IDs can be generated from those of $\mathscr{B}_1$ as follows: we split the original IDs of X_0 and X_0' in two parts and put in between them the IDs of X_0 and X_0' so as to obtain the results shown in Table 2.3. Note that either the first two numbers or the latter two of the two variables and their complements cover all 4 classes of $\mathscr{B}_1$.

All 16 classes can be summarised as in Table 2.4. Here, we can remark that the so–called modulo-2 addition, whose properties are

$$0 \oplus 0 = 0, \quad 0 \oplus 1 = 1 \oplus 0 = 1, \quad 1 \oplus 1 = 0, \qquad (2.15)$$

Table 2.3. Generation of the IDs of X_1, X_2.

X_1	0	01	1	X_2	0	10	1
X_1'	1	10	0	X_2'	1	01	0

Table 2.4. The different classes generated in $\mathscr{B}_2$. The advantage to labelling the levels with fractional numbers is that we get invariant levels across all dimensions of the algebra. In fact, the middle level that is here $1/2$ is always $1/2$ for all other nD Boolean algebra, and in every one of such levels we will always find SD variables and binary material equivalences, as well as binary products for all levels $1/4$. Note that all levels but the intermediate one are the negation of each other: Levels $0/4$ and $4/4$ as well as Levels $1/4$ and $3/4$ (in both cases the expressions of a level can also be understood as the neg–reversal of the expressions of the complementary level); Level $2/4$ presents 6 expressions, the half of which is the neg–reversal of itself or of the other ones. All nodes are, in fact, the reversal of each other or of themselves. This is true for any $\mathscr{B}_n$ algebra.

Level	ID	Classes	Areas
4/4	1111	$X_1 + X_1' + X_2 + X_2'$	$a + b + c + d = (a'b'c'd')'$
3/4	1110	$X_1' + X_2'$	$a + b + c = d'$
	1101	$X_1' + X_2$	$a + b + d = c'$
	1011	$X_1 + X_2'$	$a + c + d = b'$
	0111	$X_1 + X_2$	$b + c + d = a'$
2/4	1100	X_1'	$a + b = c'd'$
	1010	X_2'	$a + c = b'd'$
	1001	$X_1 \sim X_2$	$a + d = b'c'$
	0110	$X_1 \approx X_2$	$b + c = a'd'$
	0101	X_2	$b + d = a'c'$
	0011	X_1	$c + d = a'b'$
1/4	1000	$X_1'X_2'$	$a = b'c'd'$
	0100	$X_1'X_2$	$b = a'c'd'$
	0010	X_1X_2'	$c = a'b'd'$
	0001	X_1X_2	$d = a'b'c'$
0/4	0000	$X_1X_1'X_2X_2'$	$a'b'c'd' = (a + b + c + d)'$

can be understood as a material contravalence if 0 and 1 stand for numbers or as a logical contravalence if they stand for truth values.

Note that the levels of the algebra are numbered according to the number of 1s present in the ID; labelling them with fractional numbers allows us to immediately individuate the dimension of the algebra. Note also that Levels $0/4 = 0, 2/4 = 1/2, 4/4 = 1$ are already present in $\mathscr{B}_1$. The number of nodes for each level follows the binomial series

$$\binom{4}{0}, \binom{4}{1}, \binom{4}{2}, \binom{4}{3}, \binom{4}{4}. \tag{2.16}$$

Each of the IDs generated by combining (either through sum or product) the SD variables are computed by using the IDs of the variables themselves. For example, we have

$$
\begin{array}{llll}
X_1 & 0011 & \times & \qquad X_1 \qquad 0011 \quad + \\
X_2 & 0101 & = & \qquad X_2 \qquad 0101 \quad = \\
X_1 X_2 & 0001 & & \qquad X_1 + X_2 \quad 0111
\end{array}
$$

The generation of the IDs, displayed in Table 2.5, according to the series of natural numbers, obeys the rule of Cartesian product. In fact, we write the general formula

$$k(n) = 2^{m(n)} = 2^{m(n-1)} \times 2^{m(n-1)} = k(n-1) \times k(n-1). \tag{2.17}$$

The number of logical relations is $r(2) = 4 + 12 + 12 + 4 = 32 = 2^{m(2)+1}$, and the relative 2D Boolean network is shown in Figure 2.4. Note that in any Boolean network, any node representing the product of n sets is the source of n arrows (or paths) leading from the product itself to these sets, while any node representing the sum of n sets is the target of n arrows (or paths) leading from those sets to the sum. This implies that, whenever we have a structure like that shown in Figure 2.5, we have a logical equivalence between the sum of the products and the product of the sums, which is represented by the

Table 2.5. The generation
of the IDs of $\mathscr{B}_1$ and $\mathscr{B}_2$. As
anticipated, the IDs of any
algebra $\mathscr{B}_n$ are generated
through Cartesian product
of the IDs of $\mathscr{B}_{n-1}$.

0	00		0000
			0001
			0010
			0011
	01		0100
			0101
			0110
			0111
1	10		1000
			1001
			1010
			1011
	11		1100
			1101
			1110
			1111

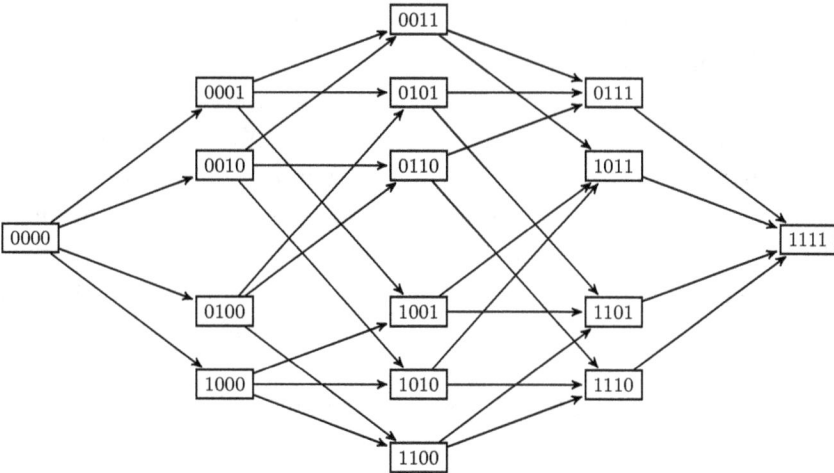

Figure 2.4. The $\mathscr{B}_2$ network. We have 24 pathways bringing us from 0000 to
1111.

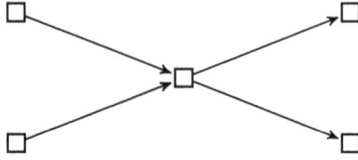

Figure 2.5.   Logical equivalence.

central node. For instance,

$$X_1 X_2 + X_1 X_2' = X_1 = (X_1 + X_2)(X_1 + X_2'), \qquad (2.18)$$

where the central node is X_1. Due to the transitivity of the logical implication this kind of structure can cut across many levels of the algebra.

Let us compare $\mathscr{B}_1$ and $\mathscr{B}_2$. It is evident that the latter algebra keeps the main characters of the levels of the former but adds two new levels. Level 1/2 can be understood as a selection of an element (either X_0 or X_0') not only out of 0/2 but also out of 2/2. With two SD variables in $\mathscr{B}_2$, Level 1/4 selects two elements (e.g. X_1, X_2 in the product $X_1 X_2$) each time out of Level 0/4 and Level 3/4 selects two elements (e.g. X_1, X_2 in the sum $X_1 + X_2$) out of Level 4/4. Level 2/4 maintains the same general character of Level 1/2 but with the novelty that any node of Level 2/4 is now the result of a sum of two binary products, e.g. $X_1 = X_1 X_2 + X_1 X_2'$ and $X_1 \sim X_2 = X_1 X_2 + X_1' X_2$. This means that we have summed two expressions of Level 1/4. Of course, we also have the reverse; that is, we can make the product of two expressions of Level 3/4 to get the same result: $X_1 = (X_1 + X_2)(X_1 + X_2')$ and $X_1 \sim X_2 = (X_1 + X_2')(X_1' + X_2)$. Such patterns are further developed when passing to algebra with higher dimensions.

I have said that only SD variables can span the whole of a complete Boolean algebra in the sense understood here. Let us consider a bit of this problem. Assume that it is possible to pick up two other nodes in $\mathscr{B}_2$ and use them to generate all other (14) nodes. It seems convenient to this purpose to choose what in my language are binary products. Let us choose, for instance, the nodes 0001 and 0010 and label them X_1 and X_2, respectively. Now, we can generate 0000, 0011 (which is $X_1 + X_2$), 1110 (X_1'), 1101 (X_2'), 1100 ($X_1'X_2'$), and 1111. However, there are 8 nodes that cannot be generated in this way (the problem is worse with higher dimensions). However, we could choose to take as basic all four binary products. In this way, we can certainly generate all 16 nodes. However, apart from the fact that it is less economic (and therefore be a solution not satisfying Occam's razor), there are also some weird consequences. Suppose that we label 0001 as X_1, 0010 as X_2, 0100 as X_3, and 1000 as X_4. Now, it happens that the node 0111 that results by summing the first three is also the complement of X_4, that is $X_1 + X_2 + X_3 = X_4'$. The resulting ambiguities increase with the number of dimensions.

What is important to understand is that Boolean algebra (with ID length given by $m = 2^n$) is a complete self–consistent system. This means that we have no principles, no axioms, no theorems to be proved. Whatever logical (in the sense of formal) expression is already contained in the Boolean network and suffices to simply follow the arrows (this is what I have meant with the structural properties of Boolean algebra). It could be objected that at least we presuppose the definitions listed in Table 2.1. However, as I shall show, everything can be accomplished through complementation, reversal, and neg–reversal (or any two of these operations), so that operations mapping the set of Boolean expressions one–to–one onto itself do suffice. However, one could again object that we need the definitions of such operations. But suppose that we build the nD algebra by using the Cartesian product of the IDs of $n - 1$ D algebra, according to what was said previously, and knowing nothing of the logical expressions. Then, a simple look at the algebra will convince us that all the nodes' IDs are simply the complements, the reversal, and the neg–reversal of other nodes (or of themselves). Thus, again the structural characters of the algebra fully suffice to define it, and the logical relations are a (important) super–structure that ultimately relies on that structure and naturally follow from those structural relations. A fundamental part of this structure is represented by logical implications, so that at least the definition of this connective seems presupposed. However, we may know nothing about implication and say simply that it represents a path from less truth to more truth. Too much focus on logical sum and product and on the logical expressions without taking into account the IDs, as in the traditional approach to Boolean algebra, obscures this state of affairs.

The same is true for all the following so-called logical laws. In general they pertain to two different classes: rules of substitution and rules of derivation. Let us consider the first class. A simple look at Table 2.4 makes us understand that each product is the negation of a corresponding logical sum, which implies that in the network are inscribed all De Morgan relations with two SD variables:

$$X_1 X_2 = (X_1' + X_2')', \quad X_1' + X_2' = (X_1 X_2)', \quad (2.19a)$$

$$X_1' X_2 = (X_1 + X_2')', \quad X_1' + X_2 = (X_1 X_2')', \quad (2.19b)$$

$$X_1 X_2' = (X_1' + X_2)', \quad X_1 + X_2' = (X_1' X_2)', \quad (2.19c)$$

$$X_1' X_2' = (X_1 + X_2)', \quad X_1 + X_2 = (X_1' X_2')'. \quad (2.19d)$$

Another way to get evidence of the previous relations is to inspect the Boolean network and realise that there exists no path leading from any of the products to its contraequivalent sum. In other words, given any two nodes not located at the same level of the algebra, either there is a logical arrow (path) connecting them, whose orientation depends on the number of 1s in the two nodes' IDs, and thus building a tautology, or they are complements and their product is $\mathbf{0}$.

Also important are the absorption rules:

$$X_1 X_1 = X_1, \quad X_1 + X_1 = X_1,$$

$$X_1 + X_1 X_2 = X_1(1 + X_2) = X_1 \mathbf{1} = X_1. \quad (2.20)$$

The first two logical equivalences are justified by the fact that expressions like $X_1 X_1$ and $X_1 + X_1$ need to be understood as arrows going from a node to the same node and therefore representing identity. For the third equivalence (which has a recursive structure), I have used the tautologies and contradictions (2.1). To have evidence of the latter ones (that I have previously introduced without justifying them), it suffices to consider that the only source node that is common to $\mathbf{0}$ and X_1 is $\mathbf{0}$ for $X_1\mathbf{0}$ and the only target node is X_1 for $X_1 + \mathbf{0}$, while the only target node that is common to $\mathbf{1}$ and X_1 is $\mathbf{1}$ for $X_1 + \mathbf{1}$ and the only source node is X_1 for $X_1\mathbf{1}$.

For the derivation rules, we need to follow the arrows of the network. A look at Figure 2.4 shows that we have both $X_1X_2 \to X_1$ and $X_1X_2 \to X_2$, so that whenever X_1X_2 is true we can derive X_1 or X_2 (this is called Simplification). On the other hand, since the product X_1X_2 is the source of both X_1 and X_2, this shows that whenever X_1 and X_2 are true, also their product is true. For the so–called rule of addition, we have $X_1 \to X_1 + X_2$, which shows that whenever X_1 is true we can always add an arbitrary X_2. On the other hand, suppose that $X_1 + X_2$ is true but X_1 is false. Following the arrows back, we see that the only source node of them is X_2. This rule is called disjunctive syllogism. I mention that there is a substitution rule that cannot be displayed in $\mathscr{B}_2$ but requires $\mathscr{B}_3$ (to be discussed below): distribution.

In the following section, I make use of logical rules only for explicitly proving results that are already inscribed in the structure of the Boolean algebra.

2.5. Algebraic Computation

The use of the AND NOT connective (Table 2.1) makes clear a character of the algebra that is quite relevant. Note first that we have

$$1 - (X_1X_2 + X_1X_2' + X_1'X_2 + X_1'X_2') = 1 - 1$$
$$= \mathbf{0}, \qquad (2.21)$$

that is, $\mathbf{0} = \mathbf{0}$. In other words, we have here the logical counterpart of an ordinary mathematical equation. It can be showed that all logical identities introduced in this book can be reduced to the form $\mathbf{0} = \mathbf{0}$. This gives us the opportunity to extend our formalism in order to make the parallelism with mathematical equations more evident. In fact, we can introduce the OR NOT connective that behaves like a division, and from the previous equations write

$$1 = \frac{X_1 X_2}{X_1 X_2' + X_1' X_2 + X_1' X_2'}, \tag{2.22}$$

which is another way to write

$$1 = X_1 X_2 + (X_1 X_2)'. \tag{2.23}$$

Note that, rewriting Equation (2.23) as

$$0 = X_1 X_2 + (X_1 X_2)' - 1, \tag{2.24}$$

we obtain the definition

$$\frac{0}{X_1 X_2} = (X_1 X_2)', \tag{2.25}$$

by dividing left and right sides by $X_1 X_2$ and taking into account that

$$\frac{1}{X_1 X_2} = 1 \quad \text{and} \quad \frac{(X_1 X_2)'}{X_1 X_2} = (X_1 X_2)'. \tag{2.26}$$

In Table 2.6 are some noticeable expressions.

Let us now shift to a more general symbolism and apply these resources. First, consider following logical identity $\forall X, Y, Z$:

$$X'Y' + X'Z + Y'Z + Z = X'Y' + Z. \tag{2.27}$$

In order to prove that the two sides of the equation are indeed equal, we need to reduce it to a form $X = X$. In the following equations, I consider the particular forms $0 = 0$ and $1 = 1$ because they allow us to apply subtraction and division, respectively. In the first case, we

Table 2.6. Noticeable expressions.

$0 + X = X$	$X + 0 = X$	$0 - X = 0$	$X - 0 = X$
$0 : X = X'$	$X : 0 = 1$	$0 \times X = 0$	$X \times 0 = 0$
$1 + X = 1$	$X + 1 = 1$	$1 - X = X'$	$X - 1 = 0$
$1 : X = 1$	$X : 1 = X$	$1 \times X = X$	$X \times 1 = X$

need to subtract to both sides the rhs (or the lhs):

$$(X'Y' + X'Z + Y'Z + Z) - (X'Y' + Z) = (X'Y' + Z) - (X'Y' + Z),$$
$$(X'Y' + X'Z + Y'Z + Z)(X'Y' + Z)' = 0,$$
$$(X'Y' + X'Z + Y'Z + Z)(X'Y')'Z' = 0,$$
$$X'Y'Z'(X'Y')' = 0,$$
$$X'Y'Z'(X + Y) = 0,$$
$$0 = 0. \tag{2.28}$$

Another example is the following:

$$X'YZ + X'Z + YZ' = X'Z + YZ',$$
$$(X'YZ + X'Z + YZ')(X'Z + YZ')' = 0,$$
$$(X'YZ + X'Z + YZ')(X + Z')(Y' + Z) = 0,$$
$$(XYZ' + YZ')(Y' + Z) = 0,$$
$$0 = 0. \tag{2.29}$$

Let us now use the form $\mathbf{1 = 1}$ by making use of OR NOT; indeed, we need to divide (OR NOT) both sides by one of the two sides. Let us apply this procedure to the example (2.27):

$$\frac{X'Y' + X'Z + Y'Z + Z}{X'Y' + Z} = \frac{X'Y' + Z}{X'Y' + Z}, \tag{2.30}$$

from which it follows:

$$X'Y' + X'Z + Y'Z + Z + (X'Y' + Z)' = 1$$
$$X'Y' + X'Z + Y'Z + Z + Z'(X'Y')' = 1$$
$$X'Y' + X'Z + Y'Z + Z + Z'(X + Y) = 1$$
$$Z(X' + Y' + 1) + X'Y' + Z'(X + Y) = 1$$
$$Z + X'Y' + Z'(X + Y) = 1$$
$$Z + (X + Y)' + Z'(X + Y) = 1$$
$$[Z + Z'(X + Y)] + (X + Y)' = 1$$

$$(Z + Z')[Z + (X + Y)] + (X + Y)' = 1$$
$$Z + (X + Y) + (X + Y)' = 1$$
$$1 = 1, \qquad (2.31)$$

where the passage from the fourth to the third last line is due to a distribution (the reader may substitute the variable Φ to $X + Y$ in order to simplify this step). Similarly, in the case of (2.29), we have:

$$\frac{X'YZ + X'Z + YZ'}{X'Z + YZ'} = \frac{X'Z + YZ'}{X'Z + YZ'}$$
$$X'YZ + X'Z + YZ' + (X'Z + YZ')' = 1$$
$$X'YZ + X'Z + YZ' + (X'Z)'(YZ')' = 1$$
$$X'YZ + X'Z + YZ' + (X + Z')(Y' + Z) = 1$$
$$X'YZ + X'Z + YZ' + XY' + XZ + Y'Z' = 1$$
$$X'YZ + XY' + Z(X + X') + Z'(Y + Y') = 1$$
$$X'YZ + XY' + Z + Z' = 1$$
$$1 = 1. \qquad (2.32)$$

Chapter 3

$\mathscr{B}_3$

3.1. General Considerations

3.1.1. *Main code alphabets*

Let us now consider $\mathscr{B}_3$. For $n = 3$, we have $m(3) = 8, k(3) = 256$, $s(3) = 16, l(3) = 9$,

$$r(3) = 8+56+168+280+280+168+56+8 = 1024 = 2^{10} = 2^{m(3)+2},$$
(3.1)

and

$$p(3) = 8! = 40,320. \tag{3.2}$$

The previous examples can be considered introductory ones since, for dimensions $D \leq 2$, the Boolean algebra does not display some of its basic features. The most distinguished feature for $\mathscr{B}_n$ with $n \geq 3$ is that the number of the SD variables is larger than the dimensions themselves. This allows us a distinction between the object (node) and its multiple logical expressions (and relative representations), which of course are all logically equivalent. This is due to the fact that only for $\mathscr{B}_1$ and $\mathscr{B}_2$ the number of SD variables together with their complements is equal to m. For the case under consideration, we have 8 SD variables (plus their 8 complements). A choice of the SD variables is displayed in Figure 3.1. Note that we generate the

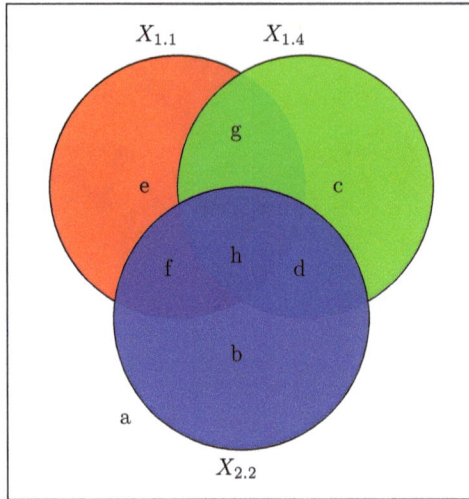

Figure 3.1.   Three SD variables of $\mathscr{B}_3$.

IDs of all those variables by reiterating the same procedure used for $\mathscr{B}_2$. Thus, we get the two columns of Table 3.1. Note again that we are relatively free to choose either this order or to reverse it, and, for example, denote with $X_{1.1}$ what is here $X_{1.4}$. Nevertheless, the possible combinations are these ones and need to be put in such a sequence (or its inverse), which is also the sequence of natural numbers.

There is no other convenient strategy for generating SD variables that do work. Of course, inverting the sequence, that is, putting first the elements with more 1s on the left, will simply invert the indexes and change nothing. Anyway, it is opportune to use criteria of clarity and simplicity in making a choice, and this strategy justifies the codification choice made in this book (introduced in Table 3.1). Of course also other methods can be imagined, like pairing an ID of a 2D SD variable with other similar IDs, but will not work since they produce IDs that are not self–dual.

The areas shown in Figure 3.1 can be summarised as in Table 3.2. As already happened for $\mathscr{B}_1$ and $\mathscr{B}_2$ and will be also true for any $\mathscr{B}_n$ algebra with $n > 3$, such tables display a first column with all 0s and a last column with all 1s and the first half that is the mirror–like

Table 3.1. The generation of the SD variables of $\mathscr{B}_3$ and their complements. Note that either the columns represented by the first four numbers or those represented by the latter 4 ones cover all 16 classes of $\mathscr{B}_2$. Parenthetically, this table explains why $\forall n$ the number of the SD variables and their complements of $\mathscr{B}_n$ has the size of $k(n-1)$.

$X_{1.1}$	00	0011	11	$X_{2.1}$	01	0011	01
$X_{1.2}$	00	0101	11	$X_{2.2}$	01	0101	01
$X_{1.3}$	00	1010	11	$X_{2.3}$	01	1010	01
$X_{1.4}$	00	1100	11	$X_{2.4}$	01	1100	01
$X'_{1.4}$	11	0011	00	$X'_{2.4}$	10	0011	10
$X'_{1.3}$	11	0101	00	$X'_{2.3}$	10	0101	10
$X'_{1.2}$	11	1010	00	$X'_{2.2}$	10	1010	10
$X'_{1.1}$	11	1100	00	$X'_{2.1}$	10	1100	10

Table 3.2. The 8 3D areas.

a	b	c	d	e	f	g	h
0	0	0	0	1	1	1	1
0	0	1	1	0	0	1	1
0	1	0	1	0	1	0	1

negation of the other. However, this arrangement does not mean that the relative logical expressions are the negation of each other. In order to be that, we need an ID that runs across all values from a to h. For instance, let us consider the columns c and f: according to Figure 3.1 we can express them as $X'_{1.1}X_{1.4}X'_{2.2}$ and $X_{1.1}X'_{1.4}X_{2.2}$, respectively. Their product gives **0** but their sum gives the material equivalence $X_{1.1} \sim X'_{1.4} \sim X_{2.2}$. In practice, we can represent the previous truth–value table as a tree structure, as displayed in Figure 3.2.

This structure means that we can have different choices of SD variables for giving rise to all of the 256 classes. In theory, we could have

$$\binom{8}{3} = 56 \tag{3.3}$$

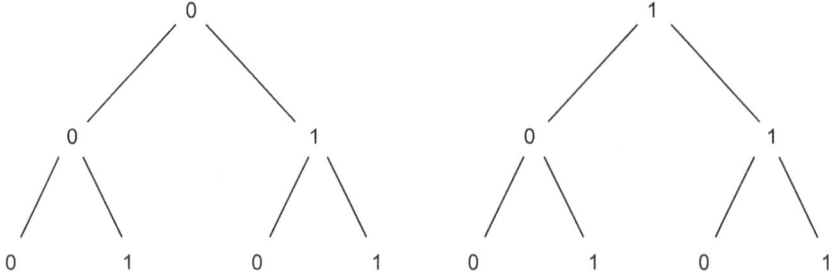

Figure 3.2. The 3D tree of truth values. Each double connection represents a column of Table 3.2, in the same order.

Table 3.3. The two main code alphabets of SD variables of $\mathscr{B}_3$. Here and in the following section, I divide any ID into segments (*chunks*) of 4 values or digits (in other words, I am using the term "chunk" in a quite narrow sense). This segmenting has not only practical reasons (the ID is more readable, especially when long), but it is also fundamental for catching certain regularities and patterns of Boolean algebra.

	abcd	efgh	Sets		abcd	efgh	Sets
	0000	1111	$X_{1.1}$		0001	0111	$X_{1.2}$
CA1	0011	0011	$X_{1.4}$	CA2	0010	1011	$X_{1.3}$
	0101	0101	$X_{2.2}$		0100	1101	$X_{2.1}$
	0110	1001	$X_{2.3}$		0111	0001	$X_{2.4}$

different choices. In practice, many combinations do not work and finally, as displayed in Table 3.3, we have two main code alphabets (CA1 and CA2)

$$\{X_{1.1}, X_{1.4}, X_{2.2}, X_{2.3}\} \quad \text{and} \quad \{X_{1.2}, X_{1.3}, X_{2.1}, X_{2.4}\}. \quad (3.4)$$

I call *main alphabet code* (main CA, or simply CA in short) whatever list of SD variables that allows the formation of the right combinations (here triplets) allowing the generation of all words (nodes) of the algebra.

I have said that not every combination will work. It is therefore important to stress the general rules (that, apart from the last two, are automatically satisfied for $\mathscr{B}_1$ and $\mathscr{B}_2$) which govern the generation of CAs 1 and 2 and similar ones for higher–dimensional algebra.

- Apart from the the first and last columns (that have all 0s and 1s, respectively), all columns need to have half 1s and half 0s.
- Moreover, apart from the second and the penultimate columns (which have all 1s under and above the horizontal central line, respectively), one fourth of the 1s (and of the 0s) need to be above the central horizontal line and half below. In fact, here only the 4 central columns are involved. This means that we have 4 0s in the first column, 4 1s in the last one, and two 1s in all the 6 central columns. The reason for these two requirements is that the SD variables need to cover all nodes of the algebra.
- Furthermore, we cannot have two or more replicates of the same column. The reason is that otherwise there will be ambiguities in the truth–value assignment.
- We need to cross the two main lists of variables. In particular, note that for each CA we have *doublets* (with the same root in a $(n-1)$D variable): e.g. $X_{1.1}$ and $X_{1.4}$. The *root variable*, e.g. X_1, gives rise to the doublet $X_{1.1}$ and $X_{1.4}$ in CA1 as well as to the doublet $X_{1.2}$ and $X_{1.3}$ in CA2. These doublets are such that the internal part of the ID of each element is the complement of the internal part of the other one: 0011 and 1100 for the first doublet and 0101 and 1010 for the second doublet. Such a structure makes that the sum of the last indexes of the two elements of a doublet is $5 = 4 + 1$ (where 4 is the number of variables generated by each root, i.e. X_1 and X_2). These characters are common to any nD algebra, with $n \geq 3$, so that in general the sum of the last index of a doublet needs to be equal to $s(n - 1) + 1$.
- Another relevant point is that in CA1 the internal and external parts of the IDs are coherent ($X_{1.1}$ and $X_{1.4}$ have X_1 as internal part and again X_1 or its complement as external part, and similarly for $X_{2.2}, X_{2.3}$ with X_2), while in CA2 these two parts are incoherent. This is not a specific character of CAs 1 and 2 in particular, but of all CAs for any nD algebra (with $n \geq 3$), thus displaying such a dichotomy.

Table 3.4. The two cycles (left and right of the vertical bar) of both $X_{1.x}$ and $X_{2.y}$: the first moves CW; the second CCW.

$X_{1.1}$-$X_{1.2}$	$X_{2.3}$-$X_{2.4}$
$X_{1.4}$-$X_{1.3}$	$X_{2.2}$-$X_{2.1}$

It is easy to verify that the only allowed combinations are precisely CAs 1 and 2 (at most we can get some permutation of the same rows).

Note that both the $X_{1.x}$s and the $X_{2.x}$s ($x = 1, 2, 3, 4$) perform a cycle of just 1 element each step according to Table 3.4. Such a cyclic property is clearer with higher dimensions.

We have the 16 chunks of SD variables and their complements displayed in Table 3.5. The chunks represent the 16 nodes of $\mathscr{B}_2$. The additional information is represented by the two CAs, which regroup all the IDs of Level 2/4 (plus **0** and **1**) on the one hand and all those of Levels 1/4 and 3/4, on the other (for any nD algebra the sum of all nodes already present in the levels of the $n-1$ algebra is equal to the sum of the nodes present in new levels). A simple calculation shows that any of these chunks can be paired with any of the other 15 ones as well as with itself so that we obtain $16 \times 16 = 256$ combinations; we can cover in this way all the nodes of $\mathscr{B}_3$. Thus, while 3D SD variables are generated by splitting the IDs of 2D SD variables, all 3D nodes are generated by coupling the chunks of such variables. In other words, here and in the following we are additionally employing a 16–element CA good for all dimensions $n \geq 3$. Thus, a node is generated by pure combinatorics of 0s and 1s and its ID by a combination (sum) of two (or more) chunks of the IDs of SD variables (which turn out to be the IDs of the nodes of the $n-1$ algebra), and is logically represented by a logical combination (complementation, sum, and product) of the relative logical expressions. For instance, we can symbolically represent the two chunks in the first row on the left as $X_{1.1.A}$ and $X_{1.1.B}$, and $X_{1.1}$ as $X_{1.1.A} + X_{1.1.B}$, where it is understood that the parts that fail in this and similar sums are filled by 0s: 0000 0000 + 0000 1111 = 0000 1111.

Table 3.5. Chunks of 3D SD variables (and their complements). Note that for CA1 we have repeats for the same 3D variable or its complement, while for CA2 we have repeats between different variables and their complements. Moreover, only in CA1 we have complete 2D SD variables and their complements. Note that any chunk of any variable and its complement is represented. Note also that all chunks of CA1 combine two couples of digits that are either elements of X_1 or of X_2 (or their complements), while the chunks of CA2 combine two digits of X_1 and two digits of X_2 (or their complements).

CA1			
$X_{1.1.A} = X'_{1.1.B}$	0000	$X_{1.1.B} = X'_{1.1.A}$	1111
$X_{1.4.A} = X_{1.4.B}$	0011	$X'_{1.4.B} = X'_{1.4.A}$	1100
$X_{2.2.A} = X_{2.2.B}$	0101	$X'_{2.2.B} = X'_{2.2.A}$	1010
$X_{2.3.A} = X'_{2.3.B}$	0110	$X_{2.3.B} = X'_{2.3.A}$	1001

CA2			
$X_{1.2.A} = X_{2.4.B}$	0001	$X'_{2.4.B} = X'_{1.2.A}$	1110
$X_{1.3.A} = X'_{2.1.B}$	0010	$X_{2.1.B} = X'_{1.3.A}$	1101
$X_{2.1.A} = X'_{1.3.B}$	0100	$X_{1.3.B} = X'_{2.1.A}$	1011
$X_{2.4.A} = X_{1.2.B}$	0111	$X'_{1.2.B} = X'_{2.4.A}$	1000

3.1.2. Subcode alphabets

Thus, CAs 1 and 2 represent sets out of which we can pick out subcode alphabets (SCAs) of three elements (for $\mathscr{B}_3$). Also each SCA needs to fulfil a combinatorial criterion. When we build a triplet, each column represents in fact the product among elements and their complements as displayed in Tables 3.7–3.8. Now, we have one column with all complements and one column with no complement (the first and last columns) and 6 columns displaying all possible combinations of one or two complements. In other words we have the series

$$\binom{3}{0}, \binom{3}{1}, \binom{3}{2}, \binom{3}{3}, \tag{3.5}$$

that is, 1, 3, 3, 1 columns for 0, 1, 2, 3 numbers of 1s for column, respectively. As said, we cannot have repeats from one column to the other, and this is the reason why we cannot have, for example, a SCA $\{X_{1.1}, X_{1.4}, X_{2.1}\}$. We can have exchange of columns when passing from either a CA or a SCA to another one, but we cannot change such a fundamental structure.

Thus, each of the two main CAs allows the formation of four SCAs:

$$\{\{X_{1.1}, X_{1.4}, X_{2.2}\}, \{X_{1.4}, X_{1.1}, X_{2.3}\}, \{X_{2.2}, X_{2.3}, X_{1.1}\},$$

$$\{X_{2.3}, X_{2.2}, X_{1.4}\}\}, \tag{3.6a}$$

$$\{\{X_{1.2}, X_{1.3}, X_{2.1}\}, \{X_{1.3}, X_{1.2}, X_{2.4}\}, \{X_{2.1}, X_{2.4}, X_{1.2}\},$$

$$\{X_{2.4}, X_{2.1}, X_{1.3}\}\}. \tag{3.6b}$$

It is evident that a single SCA out of the previous one is sufficient to generate any class of $\mathscr{B}_3$. Nevertheless, the use of all CAs and SCAs simultaneously is much more satisfactory from different points of view that need to be explored later on.

Note that in such a way in each nD algebra there are several subalgebras that are isomorphic to $\leq n$ algebra. For $\mathscr{B}_2$ there are 2 $\mathscr{B}_1$ subalgebras; for $\mathscr{B}_3$ there are 8 $\mathscr{B}_1$ subalgebras, $2 \times 3 \times 2 = 12$ $\mathscr{B}_2$ subalgebras, and 8 $\mathscr{B}_3$ subalgebras; and for $\mathscr{B}_4$ there are 128 $\mathscr{B}_1$ subalgebras, $28 \times 16 = 448$ $\mathscr{B}_2$ subalgebras, $56 \times 16 = 896$ $\mathscr{B}_3$ subalgebras, and $56 \times 16 = 896$ $\mathscr{B}_4$ subalgebras.

The best method is to align the two main CAs when there triplets, as displayed for Level 1/8 in Tables 3.7–3.8. I have considered as standard triplets (where complements are not considered for simplicity)

$$X_{1.1}X_{1.4}X_{2.2}, \ X_{1.4}X_{1.1}X_{2.3}, \ X_{2.2}X_{2.3}X_{1.1}, \ X_{2.3}X_{2.2}X_{1.4} \tag{3.7a}$$

for CA1 and the perfectly aligned triplets

$$X_{1.2}X_{1.3}X_{2.1}, \ X_{1.3}X_{1.2}X_{2.4}, \ X_{2.1}X_{2.4}X_{1.2}, \ X_{2.4}X_{2.1}X_{1.3} \tag{3.7b}$$

for CA2. Let us call the first 4 SCAs SCA1.a, 1.b, 1.c, and 1.d, and similarly for the second 4 SCAs, i.e. SCA2.a, 2.b, 2.c, and 2.d.[1]

---

[1]In [AULETTA 2013b], I used the SCA2.a only.

Table 3.6. The transformations for going from a SCA to another one. Note that on each row we have the 4 elements of CA1.

SCA1.a		SCA1.b		SCA1.c		SCA1.d	
0000	1111	0011	0011	0101	0101	0110	1001
0011	0011	0000	1111	0110	1001	0101	0101
0101	0101	0110	1001	0000	1111	0011	0011

Thus, the order of the elements in all SCAs is relevant. In fact, for both main CAs, we pass from the first triplet to the second one by exchanging the second index, from the first to the third one by exchanging the first index, and from the first to the fourth one by exchanging both the first and the second index. These changes are induced by exchange of columns as displayed in Table 3.6 for CA1: we pass from SCA1.a to SCA1.b by exchanging c-d with f-e, from SCA1.a to SCA1.c by exchanging b-d with g-e, and from SCA1.a to SCA1.d by exchanging b-c with g-f.

Note that the elements of CA1 have the same ID of the correspondent elements of CA2 but with the central columns (d and e) exchanged. This arrangement induces a change of the central rows in Tables 3.7–3.8. It is easy to verify that the exchange of lines (and the particular values of each row) between Tables 3.7 and 3.8 is produced just in the case in which the first element of each triplet has a value that is different from the other two. Nevertheless, the triplets are fully parallel for these two central rows.

About the SCAs (3.7), note that, when all values in the first triplet are equal (first and last rows), all SCAs are aligned for both CAs 1 and 2. For both CAs 1 and 2 and for all rows, when, in the first triplet, the first element of SCA1.b ($X_{1.4}$) has a value different from that of the first element of SCA1.a ($X_{1.1}$); all the values of the second column (in the order of SCA1.b) are reversed relative to those of the first column (in the order of SCA1.a). Something similar happens for the third column relative to the first one, when the values of $X_{2.2}$ and $X_{1.1}$ do not mach in the first column. Finally, when, in the second triplet, the values of $X_{2.3}$ do not match with those of $X_{1.4}$, the fourth column displays inverted value assignations relative to the

Table 3.7. The $\binom{8}{1} = 8$ classes of Level 1/8 using CA1. Note that the former four IDs are the reversal of the latter four. Here and in the following example, it is understood that different expression under the same logical ID represents in fact logical equivalences.

1	1000	0000	$X'_{1.1}X'_{1.4}X'_{2.2}$	$X'_{1.4}X'_{1.1}X'_{2.3}$	$X'_{2.2}X'_{2.3}X'_{1.1}$	$X'_{2.3}X'_{2.2}X'_{1.4}$
2	0100	0000	$X'_{1.1}X'_{1.4}X_{2.2}$	$X'_{1.4}X'_{1.1}X_{2.3}$	$X_{2.2}X_{2.3}X'_{1.1}$	$X_{2.3}X_{2.2}X'_{1.4}$
3	0010	0000	$X'_{1.1}X_{1.4}X'_{2.2}$	$X_{1.4}X'_{1.1}X_{2.3}$	$X'_{2.2}X_{2.3}X'_{1.1}$	$X_{2.3}X'_{2.2}X_{1.4}$
4	0001	0000	$X'_{1.1}X_{1.4}X_{2.2}$	$X_{1.4}X'_{1.1}X'_{2.3}$	$X_{2.2}X'_{2.3}X'_{1.1}$	$X'_{2.3}X_{2.2}X_{1.4}$
5	0000	1000	$X_{1.1}X'_{1.4}X'_{2.2}$	$X'_{1.4}X_{1.1}X_{2.3}$	$X'_{2.2}X_{2.3}X_{1.1}$	$X_{2.3}X'_{2.2}X'_{1.4}$
6	0000	0100	$X_{1.1}X'_{1.4}X_{2.2}$	$X'_{1.4}X_{1.1}X'_{2.3}$	$X_{2.2}X'_{2.3}X_{1.1}$	$X'_{2.3}X_{2.2}X'_{1.4}$
7	0000	0010	$X_{1.1}X_{1.4}X'_{2.2}$	$X_{1.4}X_{1.1}X'_{2.3}$	$X'_{2.2}X'_{2.3}X_{1.1}$	$X'_{2.3}X'_{2.2}X_{1.4}$
8	0000	0001	$X_{1.1}X_{1.4}X_{2.2}$	$X_{1.4}X_{1.1}X_{2.3}$	$X_{2.2}X_{2.3}X_{1.1}$	$X_{2.3}X_{2.2}X_{1.4}$

Table 3.8. The $\binom{8}{1} = 8$ classes of Level 1/8 for CA2. Also here the former four IDs are the reversal of the latter four.

1	1000	0000	$X'_{1.2}X'_{1.3}X'_{2.1}$	$X'_{1.3}X'_{1.2}X'_{2.4}$	$X'_{2.1}X'_{2.4}X'_{1.2}$	$X'_{2.4}X'_{2.1}X'_{1.3}$
2	0100	0000	$X'_{1.2}X'_{1.3}X_{2.1}$	$X'_{1.3}X'_{1.2}X_{2.4}$	$X_{2.1}X_{2.4}X'_{1.2}$	$X_{2.4}X_{2.1}X'_{1.3}$
3	0010	0000	$X'_{1.2}X_{1.3}X'_{2.1}$	$X_{1.3}X'_{1.2}X_{2.4}$	$X'_{2.1}X_{2.4}X'_{1.2}$	$X_{2.4}X'_{2.1}X_{1.3}$
4	0001	0000	$X_{1.2}X'_{1.3}X'_{2.1}$	$X'_{1.3}X_{1.2}X_{2.4}$	$X'_{2.1}X_{2.4}X_{1.2}$	$X_{2.4}X'_{2.1}X'_{1.3}$
5	0000	1000	$X'_{1.2}X_{1.3}X_{2.1}$	$X_{1.3}X'_{1.2}X'_{2.4}$	$X_{2.1}X'_{2.4}X'_{1.2}$	$X'_{2.4}X_{2.1}X_{1.3}$
6	0000	0100	$X_{1.2}X'_{1.3}X_{2.1}$	$X'_{1.3}X_{1.2}X'_{2.4}$	$X_{2.1}X'_{2.4}X_{1.2}$	$X'_{2.4}X_{2.1}X'_{1.3}$
7	0000	0010	$X_{1.2}X_{1.3}X'_{2.1}$	$X_{1.3}X_{1.2}X'_{2.4}$	$X'_{2.1}X'_{2.4}X_{1.2}$	$X'_{2.4}X'_{2.1}X_{1.3}$
8	0000	0001	$X_{1.2}X_{1.3}X_{2.1}$	$X_{1.3}X_{1.2}X_{2.4}$	$X_{2.1}X_{2.4}X_{1.2}$	$X_{2.4}X_{2.1}X_{1.3}$

first column. It can be further noted that in all rows apart from the first and the last, two of the expressions are complementary to the other two.

All of this can be explained as follows (see Table 3.9):

- When for SCA1.a the value of $X_{1.1}$ is different from that of the other two sets (row 4 in Table 3.7), we are in fact performing an exchange of the columns a-b-c-d with columns d-e-f-g-h. Indeed, this exchange leaves both $X_{1.4}$ and $X_{2.2}$ unchanged in the same SCA (and similarly for sets $X_{2.2}$ and $X_{1.4}$ for SCA1.d). However, for SCA1.b (as well as for SCA1.c) the same exchange of columns induces complementation of both $X_{1.1}$ and $X_{2.3}$. In other words, complementation in the two CAs is reversed.

Table 3.9. Transformations of SCAs: the exchanges are made on the standard forms of SD variables. First, second, and third sets refer to the different permutations characterising the SCAs. It is easy to pick up the sets and their complements: the former ones have a 0 in the a position, the latter ones a 1. Note that the values of the sets are invariant across all SCAs in the same transformation.

Transformation	SCA	First	set	Second	set	Third	set
Exchange of	1.a	1111	0000	0011	0011	0101	0101
a-b-c-d	1.b	0011	0011	1111	0000	1001	0110
with	1.c	0101	0101	1001	0110	1111	0000
e-f-g-h	1.d	1001	0110	0101	0101	0011	0011
Exchange of	1.a	0000	1111	1100	1100	0101	0101
a-b with c-d	1.b	1100	1100	0000	1111	1001	0110
and	1.c	0101	0101	1001	0110	0000	1111
e-f with g-h	1.d	1001	0110	0101	0101	1100	1100
Exchange of	1.a	0000	1111	0011	0011	1010	1010
a with b, c with d,	1.b	0011	0011	0000	1111	1001	0110
e with f,	1.c	1010	1010	1001	0110	0000	1111
and g with h	1.d	1001	0110	1010	1010	0011	0011

- When for SCA1.a the value of $X_{1.4}$ is different from the other two sets and similarly for the value of $X_{2.3}$ for SCA1.d (row 6 in Table 3.7), we are in fact performing an exchange of both columns a-b with c-d and e-f with g-h. Indeed, this exchange leaves both $X_{1.1}$ and $X_{2.2}$ unchanged in the same SCA. However, for SCA1.d this exchange of columns induces again an opposite complementation. SCAs 1.c and 1.b behave like SCAs 1.a and 1.d, respectively.
- Finally, when for SCA1.a (and 1.b) the value of $X_{2.2}$ ($X_{2.3}$) is different from that of the other two sets (row 7 in Table 3.7), we are in fact performing exchanges of column a with b, c with d, e with f, and g with h, inducing the same opposite complementation for SCA1.c (and 1.d). This exchange leaves both $X_{1.1}$ and $X_{1.4}$ unchanged.

Note that similar transformations hold for CA2. In fact, complementing only $X_{1.1}$ in SCA1.a (together with exchange of columns d and e) corresponds to SCA2.d; complementing only $X_{1.4}$ in SCA1.b (together with exchange of columns d and e) corresponds to SCA2.c; complementing only $X_{2.2}$ in SCA1.c (together with exchange of columns d and e) corresponds to SCA2.b; finally, complementing only $X_{2.3}$ in SCA1.d (together with exchange of columns d and e) corresponds to SCA2.a.

On this basis, it can easily be understood why no other code alphabet can work apart from the eight ones proposed here: they give rise to erroneous IDs generating ambiguity and incapability to generate certain nodes. For example, let us take $X_{1.1}, X_{1.2}$, and $X_{1.3}$ as constituting a code alphabet. Then, we shall have $X_{1.1}X_{1.2}X_{1.3} = 0000\ 0011$. Moreover, if we take $X_{1.1}, X_{1.4}$, and $X_{2.1}$, we get $X_{1.1}X'_{1.4}X_{2.1} = 0000\ 1100$.

3.1.3. *Transformation rules*

Before proceeding in the analysis of the levels, let us look at a general property. Considering CAs 1 and 2 in Table 3.3, we have the following transformations for CA1 (note that each SD variable is expressed in a different SCA from its own; for example, $X_{1.1}$ that pertains to SCAs 1.a, 1.b, and 1.c is expressed by means of SCA 1.d)

$$X_{1.1} = X_{1.4} \sim (X_{2.2} \sim X_{2.3}) = X_{2.2} \sim (X_{1.4} \sim X_{2.3})$$
$$= X_{2.3} \sim (X_{1.4} \sim X_{2.2}),$$
$$X_{1.4} = X_{1.1} \sim (X_{2.2} \sim X_{2.3}) = X_{2.2} \sim (X_{1.1} \sim X_{2.3})$$
$$= X_{2.3} \sim (X_{1.1} \sim X_{2.2}),$$
$$X_{2.2} = X_{1.1} \sim (X_{1.4} \sim X_{2.3}) = X_{1.4} \sim (X_{1.1} \sim X_{2.3})$$
$$= X_{2.3} \sim (X_{1.1} \sim X_{1.4}),$$
$$X_{2.3} = X_{1.1} \sim (X_{1.4} \sim X_{2.2}) = X_{1.4} \sim (X_{1.1} \sim X_{2.2})$$
$$= X_{2.2} \sim (X_{1.1} \sim X_{1.4}),$$

where, for example, $X_{1.4} \sim (X_{2.2} \sim X_{2.3})$ is not $X_{1.4} \sim X_{2.2} \sim X_{2.3}$ but $X_{1.4}(X_{2.2} \sim X_{2.3}) + X'_{1.4}(X_{2.2} \sim X_{2.3})'$. This expression shows that material equivalence is not associative (different from product and sum). The same transformation rules apply for CA2. The above expressions are another way to say that out of the 4 SD variables constituting a single CA, we can always drop one of them and reduce them to triplets (SCAs).

Of course, since the main CAs each represent a complete set of of SD variables, we also have:

$$X_{1.1} = (X_{1.2} + X_{1.3})(X_{2.1} + X'_{2.4}) = (X_{1.2} + X_{2.1})(X_{1.3} + X'_{2.4})$$
$$= (X_{1.2} + X'_{2.4})(X_{1.3} + X_{2.1}),$$

$$X_{1.4} = (X_{1.2} + X_{1.3})(X'_{2.1} + X_{2.4}) = (X_{1.2} + X'_{2.1})(X_{1.3} + X_{2.4})$$
$$= (X_{1.2} + X_{2.4})(X_{1.3} + X'_{2.1}),$$

$$X_{2.2} = (X_{1.2} + X'_{1.3})(X_{2.1} + X_{2.4}) = (X_{1.2} + X_{2.1})(X'_{1.3} + X_{2.4})$$
$$= (X_{1.2} + X_{2.4})(X'_{1.3} + X_{2.1}),$$

$$X_{2.3} = (X'_{1.2} + X_{1.3})(X_{2.1} + X_{2.4}) = (X'_{1.2} + X_{2.1})(X_{1.3} + X_{2.4})$$
$$= (X'_{1.2} + X_{2.4})(X_{1.3} + X_{2.1}),$$

$$X_{1.2} = (X_{1.1} + X_{1.4})(X_{2.2} + X'_{2.3}) = (X_{1.1} + X_{2.2})(X_{1.4} + X'_{2.3})$$
$$= (X_{1.1} + X'_{2.3})(X_{1.4} + X_{2.2}),$$

$$X_{1.3} = (X_{1.1} + X_{1.4})(X'_{2.2} + X_{2.3}) = (X_{1.1} + X'_{2.2})(X_{1.4} + X_{2.3})$$
$$= (X_{1.1} + X_{2.3})(X_{1.4} + X'_{2.2}),$$

$$X_{2.1} = (X_{1.1} + X'_{1.4})(X_{2.2} + X_{2.3}) = (X_{1.1} + X_{2.2})(X'_{1.4} + X_{2.3})$$
$$= (X_{1.1} + X_{2.3})(X'_{1.4} + X_{2.2}),$$

$$X_{2.4} = (X'_{1.1} + X_{1.4})(X_{2.2} + X_{2.3}) = (X'_{1.1} + X_{2.2})(X_{1.4} + X_{2.3})$$
$$= (X'_{1.1} + X_{2.3})(X_{1.4} + X_{2.2}).$$

The first 4 cases refer to the expressions of the elements of CA1 by means of those of CA2, the latter 4 cases display the inverse transformations. These equations show that all formulations of a SD variable in terms of the other ones is ultimately a resolution of the identity. Moreover, the previous set of equations shows that what does matter is to establish which is the complement variable (e.g. $X'_{2.4}$) giving rise to a certain variable of the other CA through any combination (having the character of a sum times another sum) with the other variables of its CA. In other words, $\forall A, B, C, D \in$ CA1 (or $\in$ CA2) we have the general structure $(A + B)(C + D') = X$ with all pairwise permutations of the four terms on the left giving the same result (same row). By displacing the prime from one variable to another we get different sets (different cases). Note that such invariances represent a logical character of SD variables although it comes out when we partition those variables in sets (3.4) and subsets (3.6). Moreover, the two sets of transformation rules are the same across any nD algebra (with $n \geq 3$). Let us call this important logical property the *permutation–invariance* of SD variables of a CA (or SCA) expressed by those of another CA (or SCA).

3.2. Different Levels

In the following, I use only the simplest logical expressions that are possible for each node, and, as anticipated, for this reason I use different CAs and SCAs simultaneously: our goal is to understand which are the typical logical forms determined by the 8 SD variables (and their complements). Moreover, I show only the differences among CAs and SCAs, so that, I use SCAs 1.a and 2.a, i.e. $\{X_{1.1}, X_{1.4}, X_{2.2}\}$ and $\{X_{1.2}, X_{1.3}, X_{2.1}\}$, when each is parallel to all other SCAs of its CA. Similarly, when the two main CAs are fully parallel, I use only CA1. This means that when I display both CAs, the listed combinations are the only allowed and analogously for SCAs. Note that CAs 1 and 2 mostly show parallel logical forms and the only difference concerns the way in which SD variables are combined inside a single CA for expressing the relative nodes of the algebra. Of course, there are 8 nodes representing the whole of the 3D SD variables. However, the previous transformations allow us to easily recover the meaning of those nodes in whatever CA or SCA.

3.2.1. *From Level 1/8 to Level 2/8*

Level 2/8 displays the ID combinations of Table 3.10. From a combinatorial point of view, we have (i) 12 cases with two 1s in a single chunk (2 times the combinations of Level 2/4, which are 6) and

Table 3.10. The $\binom{8}{2} = 28$ nodes of Level 2/8 as sums of pairs of Level 1/8. Note that the former 14 IDs are the reverse of the latter 14 IDs (not in the same order).

1	1100	0000	=	10000000+01000000
2	1010	0000	=	10000000+00100000
3	1001	0000	=	10000000+00010000
4	1000	1000	=	10000000+00001000
5	1000	0100	=	10000000+00000100
6	1000	0010	=	10000000+00000010
7	1000	0001	=	10000000+00000001
8	0110	0000	=	01000000+00100000
9	0101	0000	=	01000000+00010000
10	0100	1000	=	01000000+00001000
11	0100	0100	=	01000000+00000100
12	0100	0010	=	01000000+00000010
13	0100	0001	=	01000000+00000001
14	0011	0000	=	00100000+00010000
15	0010	1000	=	00100000+00001000
16	0010	0100	=	00100000+00000100
17	0010	0010	=	00100000+00000010
18	0010	0001	=	00100000+00000001
19	0001	1000	=	00010000+00001000
20	0001	0100	=	00010000+00000100
21	0001	0010	=	00010000+00000010
22	0001	0001	=	00010000+00000001
23	0000	1100	=	00001000+00000100
24	0000	1010	=	00001000+00000010
25	0000	1001	=	00001000+00000001
26	0000	0110	=	00000100+00000010
27	0000	0101	=	00000100+00000001
28	0000	0011	=	00000010+00000001

(ii) 16 cases with a 1 in each of the two chunks (so that each chunk displays all combinations of Level 1/4, which are 4). In other words, we can analyse the combinatorial patterns as

$$\binom{8}{2} = \left[\binom{4}{2} + \binom{4}{2} \right] + \binom{4}{1}\binom{4}{1}. \qquad (3.8)$$

(i) In the former cases, the ID is constituted for this half by all IDs of Level 2/4 plus another half that is simply the empty set 0000 (0/4),

while for Level 1/8 we sum a 2D product (Level 1/4) plus 0000.
(ii) In the latter case, we sum two chunks of Level 1/4: in fact, $1/4 +$
$1/4 = 2/8$, where, when passing from an algebra to another, we sum
both numerator and denominator. (The reader may wonder the fact
that we use 1/4 and 2/8 as different; however, in most cases keeping
the α/m form [with $0 \leq \alpha \leq m$] for the nD algebra is very helpful.)
Note that, according to Table 3.5, we sum either two chunks of CA1
or two of CA2.

Note that of all the 16 cases with a 1 in each chunk, 12 couple
the two 1s randomly (thus, generating products), while the latter
four (the ternary equivalences) couple them in a symmetric way (one
half is the reverse or mirror of the other half of the ID). Thus, if
we look at this character, we can better express the logical form by
saying that we have 24 asymmetric cases (the products) and 4 that
are symmetric (the material equivalences).

Following the previous rules, in Table 3.11 the 24 logical products
have a single simple expression for each CA while the ternary material equivalences are fully paralleled between the two main CAs and
among the SCAs of each main CA. In all cases in which we have vari-
ability of patterns across expressions of the same group (as it occurs
here for products), all cases are shown (also those that are paralleled
in order to avoid confusion). Each of these classes is obtained by
summing two appropriate classes of Level 1/8, such as:

$$X'_{1.1}X'_{1.4}X'_{2.2} + X'_{1.1}X'_{1.4}X_{2.2} = X'_{1.1}X'_{1.4} \times \mathbf{1} = X'_{1.1}X'_{1.4} \qquad (3.9)$$

$$X'_{1.1}X_{1.4}X_{2.2} + X_{1.1}X'_{1.4}X'_{2.2} = X'_{1.1} \sim X_{1.4} \sim X_{2.2}. \qquad (3.10)$$

In other words, we have $1/8 + 1/8 = 2/8$. Such expressions can
appear a kind of game. At the opposite, we discover that they give
us fundamental hints in the way in which logical expressions are generated. There is, however, a limitation: we can apply them only to
the Levels from $1/n$ to $\alpha/m = 1/2$. Nevertheless, since the subsequent levels express logical formula that are the negation of those
included in the previous levels, such a limitation does not hinder our
inquiry.

Table 3.11. The 28 classes of Level 2/8 for $\mathscr{B}_3$. Note that in each group of four nodes, two IDs are the reverse of the other two. The last four are invariant under reversal.

Binary ID		CA1			CA2
1100	0000	1.a, 1.b	$X'_{1.1}X'_{1.4}$	$X'_{1.2}X'_{1.3}$	2.a, 2.b
0011	0000	1.a, 1.b	$X'_{1.1}X_{1.4}$	$X'_{2.1}X_{2.4}$	2.c, 2.d
0000	1100	1.a, 1.b	$X_{1.1}X'_{1.4}$	$X_{2.1}X'_{2.4}$	2.c, 2.d
0000	0011	1.a, 1.b	$X_{1.1}X_{1.4}$	$X_{1.2}X_{1.3}$	2.a, 2.b
1010	0000	1.a, 1.c	$X'_{1.1}X'_{2.2}$	$X'_{1.2}X'_{2.1}$	2.a, 2.c
0101	0000	1.a, 1.c	$X'_{1.1}X_{2.2}$	$X'_{1.3}X_{2.4}$	2.b, 2.d
0000	1010	1.a, 1.c	$X_{1.1}X'_{2.2}$	$X_{1.3}X'_{2.4}$	2.b, 2.d
0000	0101	1.a, 1.c	$X_{1.1}X_{2.2}$	$X_{1.2}X_{2.1}$	2.a, 2.c
1001	0000	1.b, 1.c	$X'_{1.1}X'_{2.3}$	$X'_{1.3}X'_{2.1}$	2.a, 2.d
0110	0000	1.b, 1.c	$X'_{1.1}X_{2.3}$	$X'_{1.2}X_{2.4}$	2.b, 2.c
0000	0110	1.b, 1.c	$X_{1.1}X'_{2.3}$	$X_{1.2}X'_{2.4}$	2.b, 2.c
0000	1001	1.b, 1.c	$X_{1.1}X_{2.3}$	$X_{1.3}X_{2.1}$	2.a, 2.d
1000	1000	1.a, 1.d	$X'_{1.4}X'_{2.2}$	$X'_{1.2}X'_{2.4}$	2.b, 2.c
0100	0100	1.a, 1.d	$X'_{1.4}X_{2.2}$	$X'_{1.3}X_{2.1}$	2.a, 2.d
0010	0010	1.a, 1.d	$X_{1.4}X'_{2.2}$	$X_{1.3}X'_{2.1}$	2.a, 2.d
0001	0001	1.a, 1.d	$X_{1.4}X_{2.2}$	$X_{1.2}X_{2.4}$	2.b, 2.c
1000	0100	1.b, 1.d	$X'_{1.4}X'_{2.3}$	$X'_{1.3}X'_{2.4}$	2.b, 2.d
0100	1000	1.b, 1.d	$X'_{1.4}X_{2.3}$	$X'_{1.2}X_{2.1}$	2.a, 2.c
0001	0010	1.b, 1.d	$X_{1.4}X'_{2.3}$	$X_{1.2}X'_{2.1}$	2.a, 2.c
0010	0001	1.b, 1.d	$X_{1.4}X_{2.3}$	$X_{1.3}X_{2.4}$	2.b, 2.d
1000	0010	1.c, 1.d	$X'_{2.2}X'_{2.3}$	$X'_{2.1}X'_{2.4}$	2.c, 2.d
0010	1000	1.c, 1.d	$X'_{2.2}X_{2.3}$	$X'_{1.2}X_{1.3}$	2.a, 2.b
0001	0100	1.c, 1.d	$X_{2.2}X'_{2.3}$	$X_{1.2}X'_{1.3}$	2.a, 2.b
0100	0001	1.c, 1.d	$X_{2.2}X_{2.3}$	$X_{2.1}X_{2.4}$	2.c, 2.d
1000	0001		$X_{1.1} \sim X_{1.4} \sim X_{2.2}$		
0100	0010		$X_{1.1} \sim X_{1.4} \sim X'_{2.2}$		
0010	0100		$X_{1.1} \sim X'_{1.4} \sim X_{2.2}$		
0001	1000		$X'_{1.1} \sim X_{1.4} \sim X_{2.2}$		

Reciprocally, the ternary products of Level 1/8 can be understood as products of three expressions of Level 2/8, for example:

$$X_{1.1}X_{1.4}X_{2.2} = (X_{1.1}X_{1.4})(X_{1.1}X_{2.2})(X_{1.4}X_{2.2}). \qquad (3.11)$$

However, there is a difference here. Nodes of lower level can be understood as products of nodes of higher level. However, the proper logical expression of Level 1/8 is a product of three SD variables, and such sets are expresses by Level 4/8. Indeed, $4/8 \times 4/8 \times 4/8 = 1/8$. The same is true for $\mathscr{B}_2$, where the products of Level 1/4 can be computed according to $2/4 \times 2/4 = 1/4$.

Thus, both the 24 products and the 4 material equivalences are sums of ternary products. In the first case the products give rise to *reduced forms* (as well as, in the 2D case, e.g. X_1 is a reduced form of $X_1 X_2 + X_1 X_2'$), while the equivalences are simply symbolic shorthands for the relative (irreducible) sums. Then, we can take the sum of two ternary products as the *common form* of Level 2.8 (as well as the sum of two binary products, in which the term in a product is complemented in the other product is the common form of Level 2/4). By "common form" I mean that logical form to which all logical expressions of a given level can be led.

As anticipated (Table 2.4), levels are invariant across all dimensions, so that for level 2/8 (3D algebra) we find the same expressions (binary products) that we found for level 1/4 (2D algebra). Thus we can establish a tree structure between IDs of 2D binary products and those of 3D ones, as displayed in Table 3.12. In other words, we split the IDs of Level 1/4 with the same procedure used for SD variables. Note also that products of variables generated from the same root (i.e. $X_{1.1} X_{1.4}$ and similar ones as well as $X_{2.2} X_{2.3}$ and similar ones) do not instantiate such pattern. Indeed, when we have self–products (i.e. two variables generated by the same root variable of less dimension), the IDs explore the possible positions of a fixed pattern and its complement, as displayed by Table 3.13.

As for two dimensions, the sum of every variant (four as a whole) of the product between two variables is equal to identity. In this way we have six groups of products (each one pertaining to two SCAs; e.g. $X_{1.1} X_{1.4}$ pertains to both SCA1.a and SCA1.b). Moreover, according to Table 3.6, we have that the product $X_{1.1} X_{1.4}$ maps to the product $X_{2.2} X_{2.3}$ thanks to the exchange of columns b-d with

Table 3.12. Comparison between IDs of Level 1/4 and IDs of Level 2/8.

Level 1/4		Level 2/8		
$X_1'X_2'$	1000	$X_{1.1}'X_{2.2}'$	1010	0000
		$X_{1.1}'X_{2.3}'$	1001	0000
		$X_{1.4}'X_{2.2}'$	1000	1000
		$X_{1.4}'X_{2.3}'$	1000	0100
$X_1'X_2$	0100	$X_{1.1}'X_{2.2}$	0101	0000
		$X_{1.1}'X_{2.3}$	0110	0000
		$X_{1.4}'X_{2.2}$	0100	0100
		$X_{1.4}'X_{2.3}$	0100	1000
X_1X_2'	0010	$X_{1.1}X_{2.2}'$	0000	1010
		$X_{1.1}X_{2.3}'$	0000	0110
		$X_{1.4}X_{2.2}'$	0010	0010
		$X_{1.4}X_{2.3}'$	0001	0010
X_1X_2	0001	$X_{1.1}X_{2.2}$	0000	0101
		$X_{1.1}X_{2.3}$	0000	1001
		$X_{1.4}X_{2.2}$	0001	0001
		$X_{1.4}X_{2.3}$	0010	0001

Table 3.13. Further comparison between IDs of Level 1/4 and IDs of Level 2/8.

$X_{1.1}'X_{1.4}'$	1100	0000	1000	0010	$X_{2.2}'X_{2.3}'$
$X_{1.1}'X_{1.4}$	0011	0000	0010	1000	$X_{2.2}'X_{2.3}$
$X_{1.1}X_{1.4}'$	0000	1100	0001	0100	$X_{2.2}X_{2.3}'$
$X_{1.1}X_{1.4}$	0000	0011	0100	0001	$X_{2.2}X_{2.3}$

g-e, $X_{1.1}X_{2.2}$ maps to $X_{1.4}X_{2.3}$ by exchanging the column c-d with f-e, and $X_{1.1}X_{2.3}$ maps to $X_{1.4}X_{2.2}$ by exchanging again c-d with f-e. Of course, changing the order of the products will also affect the transformations (since we deal with different SCAs). The products $X_{1.1}X_{2.2}, X_{1.1}X_{2.3}$, and $X_{1.4}X_{2.2}$, relative to the product X_1X_2, are obtained by simply doubling segments of the ID (0001): 1/4, 1/2, and the whole of that ID, respectively.

As a consequence, products display characteristic patterns:

- The first 4 products are such that, when there is value mismatch between $X_{1.1}$ and $X_{1.4}$, for CA2 we have an exchange of the first index (between 1 and 2) relative to the alignment of Tables 3.7–3.8;
- For the second 4 products, when there is value mismatch between $X_{1.1}$ and $X_{2.2}$, for CA2 both the first and the second indexes are exchanged;
- For the third 4 products, when there is value matching between $X_{1.1}$ and $X_{2.3}$, for CA2 the second index (3 with 1) is exchanged;
- For the fourth 4 products when the values of $X_{1.4}$ and $X_{2.2}$ do match we have exchange of the second index (2 with 4);
- For the fifth 4 products, when there is value mismatch between $X_{1.4}$ and $X_{2.3}$, for CA2 both the first and the second indexes are exchanged;
- Finally, for the last 4 products, when there is value mismatch between $X_{2.2}$ and $X_{2.3}$, for CA2 there is exchange in the first index (1 and 2).

In other words, for couples $X_{1.1}$–$X_{1.4}$ and $X_{2.2}$–$X_{2.3}$, the CA2 counterparts behave the same when their values match, and similarly for couples $X_{1.1}$–$X_{2.2}$ and $X_{1.4}$–$X_{2.3}$. At the opposite, for couples $X_{1.1}$–$X_{2.3}$ and $X_{1.4}$–$X_{2.2}$, there is change in the indexes when their values match. Mismatches between CAs 1 and 2 happen when there is a 1 either in the fourth or in the fifth column (reflecting the exchange of the central rows in Tables 3.7–3.8). Since in the last ternary equivalence we have a 1 in both the fourth and fifth columns, there is compensation and therefore matching.

Let us give a closer look to ternary equivalences. Ternary material equivalences are defined as the product of three binary equivalences (or contravalences):

$$X_{1.1} \sim X_{1.4} \sim X'_{2.2} = X_{1.1}X_{1.4}X'_{2.2} + X'_{1.1}X'_{1.4}X_{2.2}$$
$$= (X_{1.1} \sim X_{1.4})(X_{1.1} \not\sim X_{2.2})(X_{1.4} \not\sim X_{2.2}).$$
$$(3.12)$$

Actually, the product of two binary equivalences does suffice, since

$$X_{1.1}X_{1.4}X_{2.2} + X'_{1.1}X'_{1.4}X'_{2.2}$$
$$= (X_{1.1}X_{1.4} + X'_{1.1}X'_{1.4})(X_{1.1}X_{2.2} + X'_{1.1}X'_{2.2})$$
$$= (X_{1.1}X_{1.4} + X'_{1.1}X'_{1.4})(X_{1.4}X_{2.2} + X'_{1.4}X'_{2.2})$$
$$= (X_{1.1}X_{2.2} + X'_{1.1}X'_{2.2})(X_{1.4}X_{2.2} + X'_{1.4}X'_{2.2}).$$

In other words, the relations between a first and a second term and between a first and third term suffice to also determine the relation between a second and third term.

It is easy to see that we can get the same ternary equivalence by exchanging each term with its complement: $X_{1.1} \sim X_{1.4} \sim X'_{2.2} = X'_{1.1} \sim X'_{1.4} \sim X_{2.2}$. Note that the ID of the ternary equivalence $X_{1.1} \sim X_{1.4} \sim X_{2.2}$ has exactly the same structure of the ID of the binary equivalence $X_1 \sim X_2$ (two 1s at the extreme points of the ID). Moreover, for the 2D algebra we cannot have equivalences with complements since $X_1 \sim X'_2 = X_1 \nsim X_2$.

Finally, note that in the ternary equivalence we have the values 0100 0010 and 0010 0100 (which express discordance as far as 0010 and 0100 are the two values for which the truth values X_1 and X_2 are opposite) when either $X_{1.4}$ is true and $X_{2.2}$ false or vice versa, which means that these two variables are parallel. Since we also have

$$X_{1.1} \sim X_{1.4} \sim X'_{2.3} = 0100 \ 0010 \quad \text{and}$$

$$X'_{1.1} \sim X_{1.4} \sim X_{2.3} = 0010 \ 0100, \tag{3.13}$$

then also $X_{1.1}$ and $X_{2.3}$ are parallel. A similar analysis also shows that the pairs $X_{1.3} - X_{2.1}$ and $X_{1.2} - X_{2.4}$ are parallel, respectively. The couples so formed share, in fact, fundamental properties. The two chunks of both $X_{1.1}$ and $X_{2.3}$ are the negation of each other, while the two chunks of both $X_{1.4}$ and $X_{2.2}$ repeat the same pattern. On the other hand, the two chunks of $X_{1.2}$ are cross–identical with those of $X_{2.4}$, while the two chunks of $X_{1.3}$ are cross–negative relative to those of $X_{2.1}$.

3.2.2. *From Level 2/8 to Level 3/8*

The IDs of Level 3/8 are displayed in Table 3.14, while relative classes are shown in Table 3.15. From a combinatorial point of view, we have (i) 8 cases (the first ones in Table 3.15) with three 1s in a single chunk and (ii) 48 different combinations of two 1s in a chunk and another in the other chunk. In other words, we have either a chunk of Level 3/4 plus 0/4 or a chunk of Level 1/4 and another of Level 2/4. Summarising, we have

$$\binom{8}{3} = \left[\binom{4}{3} + \binom{4}{3} \right] + \left[\binom{4}{2}\binom{4}{1} + \binom{4}{2}\binom{4}{1} \right]. \quad (3.14)$$

Level 3/8 is new (it is not a doubling of a level already present in $\mathscr{B}_2$), and therefore its IDs are a patchwork of two elements. All of the 56 IDs are constituted of two halves of 3D SD variables, one coming from CA1, the other from CA2 (Table 3.5). Each chunk of e.g. CA1 in the first 32 IDs is combined with 8 chunks of CA2. For instance, the first chunk of $X_{1.1}$ is combined with the second chunk of $X_{1.2}, X_{1.3}, X_{2.1}, X'_{2.4}$ and the second chunk of $X'_{1.1}$ is combined with the first chunk of $X'_{1.2}, X'_{1.3}, X'_{2.1}, X_{2.4}$: this is evident by looking at the first 8 IDs. This pattern is followed by all of the 32 first expressions. The other 24 also show similar combinations but, for obvious reasons, $X_{1.1}$ (and $X'_{1.1}$) is not present.

About the relation between ID and logical expression, we need to distinguish several cases:

- There are 8 cases (the first ones in Table 3.15) represented by a chunk being a sum of Level 3/4 and 0000: they are all a product of a term and a binary sum. Note that there is agreement between the sums occurring here and those occurring in Level 3/4. For instance, the IDs 1110 0000 and 0000 1110 have both sums $X'_{1.4} + X'_{2.2}$ that correspond to the sum $X'_1 + X'_2$ representing the ID 1110.
- All the other IDs have a chunk with two 1s and a chunk with a single 1. We need to distinguish several cases. If the chunk with two 1s represents either X_1 or X_2 (or their complements), we need to see whether there is agreement with the single 1 in the other chunk

Table 3.14. The $\binom{8}{3} = 56$ nodes of Level 3/8. Note that the former 28 IDs are the reverse of the latter 28 (not in the same order).

1	1110	0000	=	11000000+10100000+01100000
2	1101	0000	=	11000000+10010000+01010000
3	1100	1000	=	11000000+10001000+01001000
4	1100	0100	=	11000000+10000100+01000100
5	1100	0010	=	11000000+10000010+01000010
6	1100	0001	=	11000000+10000001+01000001
7	1011	0000	=	10100000+10010000+00110000
8	1010	1000	=	10100000+10001000+00101000
9	1010	0100	=	10100000+10000100+00100100
10	1010	0010	=	10100000+10000010+00100010
11	1010	0001	=	10100000+10000001+00100001
12	1001	1000	=	10010000+10001000+00011000
13	1001	0100	=	10010000+10000100+00010100
14	1001	0010	=	10010000+10000010+00010010
15	1001	0001	=	10010000+10000001+00010001
16	1000	1100	=	10001000+10000100+00001100
17	1000	1010	=	10001000+10000010+00001010
18	1000	1001	=	10001000+10000001+00001001
19	1000	0110	=	10000100+10000010+00000110
20	1000	0101	=	10000100+10000001+00000101
21	1000	0011	=	10000010+10000001+00000011
22	0111	0000	=	01100000+01010000+00110000
23	0110	1000	=	01100000+01001000+00101000
24	0110	0100	=	01100000+01000100+00100100
25	0110	0010	=	01100000+01000010+00100010
26	0110	0001	=	01100000+01000001+00100001
27	0101	1000	=	01010000+01001000+00011000
28	0101	0100	=	01010000+01000100+00010100
29	0101	0010	=	01010000+01000010+00010010
30	0101	0001	=	01010000+01000001+00010001
31	0100	1100	=	01001000+01000100+00001100
32	0100	1010	=	01001000+01000010+00001010
33	0100	1001	=	01001000+01000001+00001001
34	0100	0110	=	01000100+01000010+00000110
35	0100	0101	=	01000100+01000001+00000101
36	0100	0011	=	01000010+01000001+00000011
37	0011	1000	=	00110000+00101000+00011000
38	0011	0100	=	00110000+00100100+00010100
39	0011	0010	=	00110000+00100010+00010010
40	0011	0001	=	00110000+00100001+00010001

(*Continued*)

Table 3.14.   (*Continued*)

41	0010	1100	=	00101000+00100100+00001100
42	0010	1010	=	00101000+00100010+00001010
43	0010	1001	=	00101000+00100001+00001001
44	0010	0110	=	00100100+00100010+00000110
45	0010	0101	=	00100100+00100001+00000101
46	0010	0011	=	00100010+00100001+00000011
47	0001	1100	=	00011000+00010100+00001100
48	0001	1010	=	00011000+00010010+00001010
49	0001	1001	=	00011000+00010001+00001001
50	0001	0110	=	00010100+00010010+00000110
51	0001	0101	=	00010100+00010001+00000101
52	0001	0011	=	00010010+00010001+00000011
53	0000	1110	=	00001100+00001010+00000110
54	0000	1101	=	00001100+00001001+00000101
55	0000	1011	=	00001010+00001001+00000011
56	0000	0111	=	00000110+00000101+00000011

(that could represent a product of Level 1/4). In other words, the question is whether this single 1 is in the same position of one of the two 1s in the other chunk or not. If there is, we get the subsequent 16 IDs represented by a product of a term and a binary sum. If not, we get the first 16 IDs represented by a sum of a binary and a ternary product. Note that in all cases the single 1 is a mirror of one of the other two. Incidentally, this shows that we deal indeed with single and independent chunks: if not, the correlation would be not between, for example, two 1s both in the second positions from the left (columns b and f), but one in the second and the other in the third positions due to the anticorrelation of the columns b and g.

- In the case in which the chunk with two 1s is represented by a 2D equivalence or contravalence, we have the opposite behaviour: if there is mismatch with the 1 in the other chunk, we get the last 8 IDs represented by the product of a term and a binary sum. If, at the opposite, there is match, we get the last 8 IDs represented each by a sum of a binary and a ternary product.

Table 3.15. The 56 classes of Level 3/8. Each ID of the first form (product of a single element and a sum: 32 cases) selects two rows: the first one is CA1, the second CA2. Note that the first 8 IDs are such that the half is the reverse of the other half. The same is true for the other 24 IDs and also of the last 24 IDs represented by sums of binary and ternary products.

ID				
1110	0000	$X'_{1.1}(X'_{1.4} + X'_{2.2})$	$X'_{1.1}(X'_{1.4} + X_{2.3})$	$X'_{1.1}(X'_{2.2} + X_{2.3})$
		$X'_{1.2}(X'_{1.3} + X'_{2.1})$	$X'_{1.2}(X'_{1.3} + X_{2.4})$	$X'_{1.2}(X'_{2.1} + X_{2.4})$
1101	0000	$X'_{1.1}(X'_{1.4} + X'_{2.2})$	$X'_{1.1}(X'_{1.4} + X'_{2.3})$	$X'_{1.1}(X_{2.2} + X'_{2.3})$
		$X'_{1.3}(X'_{1.2} + X_{2.4})$	$X'_{1.3}(X'_{1.2} + X'_{2.1})$	$X'_{1.3}(X_{2.4} + X'_{2.1})$
1011	0000	$X'_{1.1}(X_{1.4} + X'_{2.2})$	$X'_{1.1}(X_{1.4} + X'_{2.3})$	$X'_{1.1}(X'_{2.2} + X'_{2.3})$
		$X'_{2.1}(X_{2.4} + X'_{1.2})$	$X'_{2.1}(X_{2.4} + X'_{1.3})$	$X'_{2.1}(X'_{1.2} + X'_{1.3})$
0111	0000	$X'_{1.1}(X_{1.4} + X_{2.2})$	$X'_{1.1}(X_{1.4} + X_{2.3})$	$X'_{1.1}(X_{2.2} + X_{2.3})$
		$X_{2.4}(X'_{1.3} + X'_{1.2})$	$X_{2.4}(X'_{2.1} + X'_{1.2})$	$X_{2.4}(X'_{2.1} + X'_{1.3})$
0000	1110	$X_{1.1}(X'_{1.4} + X'_{2.2})$	$X_{1.1}(X'_{1.4} + X'_{2.3})$	$X_{1.1}(X'_{2.2} + X'_{2.3})$
		$X'_{2.4}(X_{1.3} + X_{1.2})$	$X'_{2.4}(X_{2.1} + X_{1.2})$	$X'_{2.4}(X_{2.1} + X_{1.3})$
0000	1101	$X_{1.1}(X'_{1.4} + X_{2.2})$	$X_{1.1}(X'_{1.4} + X_{2.3})$	$X_{1.1}(X_{2.2} + X_{2.3})$
		$X_{2.1}(X'_{2.4} + X_{1.2})$	$X_{2.1}(X'_{2.4} + X_{1.3})$	$X_{2.1}(X_{1.2} + X_{1.3})$
0000	1011	$X_{1.1}(X_{1.4} + X'_{2.2})$	$X_{1.1}(X_{1.4} + X_{2.3})$	$X_{1.1}(X'_{2.2} + X_{2.3})$
		$X_{1.3}(X_{1.2} + X'_{2.4})$	$X_{1.3}(X_{1.2} + X_{2.1})$	$X_{1.3}(X'_{2.4} + X_{2.1})$
0000	0111	$X_{1.1}(X_{1.4} + X_{2.2})$	$X_{1.1}(X_{1.4} + X'_{2.3})$	$X_{1.1}(X_{2.2} + X'_{2.3})$
		$X_{1.2}(X_{1.3} + X_{2.1})$	$X_{1.2}(X_{1.3} + X'_{2.4})$	$X_{1.2}(X_{2.1} + X'_{2.4})$
1100	1000	$X'_{1.4}(X'_{1.1} + X'_{2.2})$	$X'_{1.4}(X'_{1.1} + X_{2.3})$	$X'_{1.4}(X_{2.3} + X'_{2.2})$
		$X'_{1.2}(X'_{1.3} + X'_{2.4})$	$X'_{1.2}(X'_{1.3} + X_{2.1})$	$X'_{1.2}(X_{2.1} + X'_{2.4})$
1100	0100	$X'_{1.4}(X'_{1.1} + X_{2.2})$	$X'_{1.4}(X'_{1.1} + X'_{2.3})$	$X'_{1.4}(X'_{2.3} + X_{2.2})$
		$X'_{1.3}(X'_{1.2} + X_{2.1})$	$X'_{1.3}(X'_{1.2} + X'_{2.4})$	$X'_{1.3}(X'_{2.4} + X_{2.1})$
1000	1100	$X'_{1.4}(X_{1.1} + X'_{2.2})$	$X'_{1.4}(X_{1.1} + X'_{2.3})$	$X'_{1.4}(X'_{2.3} + X'_{2.2})$
		$X'_{2.4}(X_{2.1} + X'_{1.2})$	$X'_{2.4}(X_{2.1} + X'_{1.3})$	$X'_{2.4}(X'_{1.3} + X'_{1.2})$
0100	1100	$X'_{1.4}(X_{1.1} + X_{2.2})$	$X'_{1.4}(X_{1.1} + X_{2.3})$	$X'_{1.4}(X_{2.3} + X_{2.2})$
		$X_{2.1}(X'_{1.2} + X'_{1.3})$	$X_{2.1}(X'_{2.4} + X'_{1.2})$	$X_{2.1}(X'_{2.4} + X'_{1.3})$
0011	0010	$X_{1.4}(X'_{1.1} + X'_{2.2})$	$X_{1.4}(X'_{1.1} + X'_{2.3})$	$X_{1.4}(X'_{2.3} + X'_{2.2})$
		$X'_{2.1}(X_{1.2} + X_{1.3})$	$X'_{2.1}(X_{2.4} + X_{1.2})$	$X'_{2.1}(X_{2.4} + X_{1.3})$
0011	0001	$X_{1.4}(X'_{1.1} + X_{2.2})$	$X_{1.4}(X'_{1.1} + X_{2.3})$	$X_{1.4}(X_{2.3} + X_{2.2})$
		$X_{2.4}(X'_{2.1} + X_{1.2})$	$X_{2.4}(X'_{2.1} + X_{1.3})$	$X_{2.4}(X_{1.3} + X_{1.2})$
0010	0011	$X_{1.4}(X_{1.1} + X'_{2.2})$	$X_{1.4}(X_{1.1} + X_{2.3})$	$X_{1.4}(X_{2.3} + X'_{2.2})$
		$X_{1.3}(X_{1.2} + X'_{2.1})$	$X_{1.3}(X_{1.2} + X_{2.4})$	$X_{1.3}(X_{2.4} + X'_{2.1})$
0001	0011	$X_{1.4}(X_{1.1} + X_{2.2})$	$X_{1.4}(X_{1.1} + X'_{2.3})$	$X_{1.4}(X'_{2.3} + X_{2.2})$
		$X_{1.2}(X_{1.3} + X_{2.4})$	$X_{1.2}(X_{1.3} + X'_{2.1})$	$X_{1.2}(X'_{2.1} + X_{2.4})$

(Continued)

Table 3.15. (*Continued*)

1010	1000	$X'_{2.2}(X'_{1.1} + X'_{1.4})$ $X'_{1.2}(X'_{2.1} + X'_{2.4})$	$X'_{2.2}(X_{2.3} + X'_{1.1})$ $X'_{1.2}(X_{1.3} + X'_{2.1})$	$X'_{2.2}(X_{2.3} + X'_{1.4})$ $X'_{1.2}(X_{1.3} + X'_{2.4})$
1010	0010	$X'_{2.2}(X'_{1.1} + X_{1.4})$ $X'_{2.1}(X'_{1.2} + X_{1.3})$	$X'_{2.2}(X'_{2.3} + X'_{1.1})$ $X'_{2.1}(X'_{2.4} + X'_{1.2})$	$X'_{2.2}(X'_{2.3} + X_{1.4})$ $X'_{2.1}(X'_{2.4} + X_{1.3})$
1000	1010	$X'_{2.2}(X_{1.1} + X'_{1.4})$ $X'_{2.4}(X_{1.3} + X'_{1.2})$	$X'_{2.2}(X'_{2.3} + X_{1.1})$ $X'_{2.4}(X'_{2.1} + X_{1.3})$	$X'_{2.2}(X'_{2.3} + X'_{1.4})$ $X'_{2.4}(X'_{2.1} + X'_{1.2})$
0010	1010	$X'_{2.2}(X_{1.1} + X_{1.4})$ $X_{1.3}(X'_{1.2} + X'_{2.1})$	$X'_{2.2}(X_{2.3} + X_{1.1})$ $X_{1.3}(X'_{1.2} + X'_{2.4})$	$X'_{2.2}(X_{2.3} + X_{1.4})$ $X_{1.3}(X'_{2.4} + X'_{2.1})$
0101	0100	$X_{2.2}(X'_{1.1} + X'_{1.4})$ $X'_{1.3}(X_{1.2} + X_{2.1})$	$X_{2.2}(X'_{2.3} + X'_{1.1})$ $X'_{1.3}(X_{1.2} + X_{2.4})$	$X_{2.2}(X'_{2.3} + X'_{1.4})$ $X'_{1.3}(X_{2.4} + X_{2.1})$
0101	0001	$X_{2.2}(X'_{1.1} + X_{1.4})$ $X_{2.4}(X'_{1.3} + X_{1.2})$	$X_{2.2}(X_{2.3} + X'_{1.1})$ $X_{2.4}(X_{2.1} + X'_{1.3})$	$X_{2.2}(X_{2.3} + X_{1.4})$ $X_{2.4}(X_{2.1} + X_{1.2})$
0100	0101	$X_{2.2}(X_{1.1} + X'_{1.4})$ $X_{2.1}(X_{1.2} + X'_{1.3})$	$X_{2.2}(X_{2.3} + X_{1.1})$ $X_{2.1}(X_{2.4} + X_{1.2})$	$X_{2.2}(X_{2.3} + X'_{1.4})$ $X_{2.1}(X_{2.4} + X'_{1.3})$
0001	0101	$X_{2.2}(X_{1.1} + X_{1.4})$ $X_{1.2}(X_{2.1} + X_{2.4})$	$X_{2.2}(X'_{2.3} + X_{1.1})$ $X_{1.2}(X'_{1.3} + X_{2.1})$	$X_{2.2}(X'_{2.3} + X_{1.4})$ $X_{1.2}(X'_{1.3} + X_{2.4})$
1001	0100	$X'_{2.3}(X'_{1.4} + X'_{1.1})$ $X'_{1.3}(X'_{2.4} + X'_{2.1})$	$X'_{2.3}(X_{2.2} + X'_{1.1})$ $X'_{1.3}(X_{1.2} + X'_{2.1})$	$X'_{2.3}(X_{2.2} + X'_{1.4})$ $X'_{1.3}(X_{1.2} + X'_{2.4})$
1000	0110	$X'_{2.3}(X'_{1.4} + X_{1.1})$ $X'_{2.4}(X'_{1.3} + X_{1.2})$	$X'_{2.3}(X'_{2.2} + X_{1.1})$ $X'_{2.4}(X'_{2.1} + X_{1.2})$	$X'_{2.3}(X'_{2.2} + X'_{1.4})$ $X'_{2.4}(X'_{2.1} + X'_{1.3})$
1001	0010	$X'_{2.3}(X_{1.4} + X'_{1.1})$ $X'_{2.1}(X_{1.2} + X'_{1.3})$	$X'_{2.3}(X'_{2.2} + X'_{1.1})$ $X'_{2.1}(X'_{2.4} + X'_{1.3})$	$X'_{2.3}(X'_{2.2} + X_{1.4})$ $X'_{2.1}(X'_{2.4} + X_{1.2})$
0001	0110	$X'_{2.3}(X_{1.4} + X_{1.1})$ $X_{1.2}(X'_{1.3} + X'_{2.1})$	$X'_{2.3}(X_{2.2} + X_{1.1})$ $X_{1.2}(X'_{1.3} + X'_{2.4})$	$X'_{2.3}(X_{2.2} + X_{1.4})$ $X_{1.2}(X'_{2.1} + X'_{2.4})$
0110	1000	$X_{2.3}(X'_{1.4} + X'_{1.1})$ $X'_{1.2}(X_{1.3} + X_{2.1})$	$X_{2.3}(X'_{2.2} + X'_{1.1})$ $X'_{1.2}(X_{1.3} + X_{2.4})$	$X_{2.3}(X'_{2.2} + X'_{1.4})$ $X'_{1.2}(X_{2.1} + X_{2.4})$
0100	1001	$X_{2.3}(X'_{1.4} + X_{1.1})$ $X_{2.1}(X'_{1.2} + X_{1.3})$	$X_{2.3}(X_{2.2} + X_{1.1})$ $X_{2.1}(X_{2.4} + X_{1.3})$	$X_{2.3}(X_{2.2} + X'_{1.4})$ $X_{2.1}(X_{2.4} + X'_{1.2})$
0110	0001	$X_{2.3}(X_{1.4} + X'_{1.1})$ $X_{2.4}(X_{1.3} + X'_{1.2})$	$X_{2.3}(X_{2.2} + X'_{1.1})$ $X_{2.4}(X_{2.1} + X'_{1.2})$	$X_{2.3}(X_{2.2} + X_{1.4})$ $X_{2.4}(X_{2.1} + X_{1.3})$
0010	1001	$X_{2.3}(X_{1.4} + X_{1.1})$ $X_{1.3}(X_{2.4} + X_{2.1})$	$X_{2.3}(X'_{2.2} + X_{1.1})$ $X_{1.3}(X'_{1.2} + X_{2.1})$	$X_{2.3}(X'_{2.2} + X_{1.4})$ $X_{1.3}(X'_{1.2} + X_{2.4})$

Table 3.15. (*Continued*)

1100	0001	$X'_{1.1}X'_{1.4} + X_{1.1}X_{1.4}X_{2.2}\|X'_{1.1}X'_{1.4} + X_{1.4}X_{1.1}X_{2.3}$ $X_{2.2}X_{2.3} + X'_{2.2}X'_{2.3}X'_{1.1}\|X_{2.2}X_{2.3} + X'_{2.3}X'_{2.2}X'_{1.4}$
1000	0011	$X_{1.1}X_{1.4} + X'_{1.1}X'_{1.4}X'_{2.2}\|X_{1.1}X_{1.4} + X'_{1.4}X'_{1.1}X'_{2.3}$ $X'_{2.2}X'_{2.3} + X_{2.2}X_{2.3}X_{1.1}\|X'_{2.2}X'_{2.3} + X_{2.3}X_{2.2}X_{1.4}$
1010	0001	$X'_{1.1}X'_{2.2} + X_{1.1}X_{1.4}X_{2.2}\|X'_{1.1}X'_{2.2} + X_{1.4}X_{1.1}X_{2.3}$ $X_{1.4}X_{2.3} + X'_{1.4}X'_{1.1}X'_{2.3}\|X_{1.4}X_{2.3} + X'_{2.3}X'_{2.2}X'_{1.4}$
1000	0101	$X_{1.1}X_{2.2} + X'_{1.1}X'_{1.4}X'_{2.2}\|X_{1.1}X_{2.2} + X'_{2.2}X'_{2.3}X'_{1.1}$ $X'_{1.4}X'_{2.3} + X_{1.4}X_{1.1}X_{2.3}\|X'_{1.4}X'_{2.3} + X_{2.3}X_{2.2}X_{1.4}$
1100	0010	$X'_{1.1}X'_{1.4} + X_{1.1}X_{1.4}X'_{2.2}\|X'_{1.1}X'_{1.4} + X_{1.4}X_{1.1}X'_{2.3}$ $X'_{2.2}X'_{2.3} + X_{2.2}X_{2.3}X'_{1.1}\|X'_{2.2}X'_{2.3} + X_{2.3}X_{2.2}X'_{1.4}$
0100	0011	$X_{1.1}X_{1.4} + X'_{1.1}X'_{1.4}X_{2.2}\|X_{1.1}X_{1.4} + X'_{1.4}X'_{1.1}X_{2.3}$ $X_{2.2}X_{2.3} + X'_{2.2}X'_{2.3}X_{1.1}\|X_{2.2}X_{2.3} + X'_{2.3}X'_{2.2}X_{1.4}$
0101	0010	$X'_{1.1}X_{2.2} + X_{1.1}X_{1.4}X'_{2.2}\|X'_{1.1}X_{2.2} + X'_{2.2}X_{1.1}X_{2.3}$ $X'_{2.3}X_{1.4} + X'_{1.4}X'_{1.1}X_{2.3}\|X'_{2.3}X_{1.4} + X_{2.3}X_{2.2}X_{1.4}$ $X'_{1.3}X_{2.4} + X_{1.3}X_{1.2}X'_{2.4}\|X'_{1.3}X_{2.4} + X'_{2.4}X'_{2.1}X_{1.3}$ $X'_{2.1}X_{1.2} + X'_{1.2}X'_{1.3}X_{2.1}\|X'_{2.1}X_{1.2} + X_{2.1}X_{2.4}X'_{1.2}$
0100	1010	$X_{1.1}X'_{2.2} + X'_{1.1}X'_{1.4}X_{2.2}\|X_{1.1}X'_{2.2} + X_{2.2}X_{2.3}X'_{1.1}$ $X_{2.3}X'_{1.4} + X_{1.4}X_{1.1}X'_{2.3}\|X_{2.3}X'_{1.4} + X'_{2.3}X'_{2.2}X_{1.4}$ $X_{1.3}X'_{2.4} + X'_{1.3}X'_{1.2}X_{2.4}\|X_{1.3}X'_{2.4} + X_{2.4}X_{2.1}X'_{1.3}$ $X_{2.1}X'_{1.2} + X_{1.2}X_{1.3}X'_{2.1}\|X_{2.1}X'_{1.2} + X'_{2.1}X'_{2.4}X_{1.2}$
0011	0100	$X'_{1.1}X_{1.4} + X_{1.1}X'_{1.4}X_{2.2}\|X'_{1.1}X_{1.4} + X'_{1.4}X_{1.1}X'_{2.3}$ $X'_{2.3}X_{2.2} + X'_{2.2}X_{2.3}X'_{1.1}\|X'_{2.3}X_{2.2} + X_{2.3}X'_{2.2}X_{1.4}$ $X'_{2.1}X_{2.4} + X_{2.1}X'_{2.4}X_{1.2}\|X'_{2.1}X_{2.4} + X'_{2.4}X_{2.1}X'_{1.3}$ $X'_{1.3}X_{1.2} + X'_{1.2}X_{1.3}X'_{2.1}\|X'_{1.3}X_{1.2} + X_{1.3}X'_{1.2}X_{2.4}$
0010	1100	$X_{1.1}X'_{1.4} + X'_{1.1}X_{1.4}X'_{2.2}\|X_{1.1}X'_{1.4} + X_{1.4}X'_{1.1}X_{2.3}$ $X_{2.3}X'_{2.2} + X_{2.2}X'_{2.3}X_{1.1}\|X_{2.3}X'_{2.2} + X'_{2.3}X_{2.2}X'_{1.4}$ $X_{2.1}X'_{2.4} + X'_{2.1}X_{2.4}X'_{1.2}\|X_{2.1}X'_{2.4} + X_{2.4}X'_{2.1}X_{1.3}$ $X_{1.3}X'_{1.2} + X_{1.2}X'_{1.3}X_{2.1}\|X_{1.3}X'_{1.2} + X'_{1.3}X_{1.2}X'_{2.4}$
1010	0100	$X'_{1.1}X'_{2.2} + X_{1.1}X'_{1.4}X_{2.2}\|X'_{1.1}X'_{2.2} + X_{2.2}X'_{2.3}X_{1.1}$ $X'_{1.4}X'_{2.3} + X_{1.4}X'_{1.1}X_{2.3}\|X'_{1.4}X'_{2.3} + X_{2.3}X'_{2.2}X_{1.4}$
0010	0101	$X_{1.1}X_{2.2} + X'_{1.1}X_{1.4}X'_{2.2}\|X_{1.1}X_{2.2} + X'_{2.2}X_{2.3}X'_{1.1}$ $X_{1.4}X_{2.3} + X'_{1.4}X_{1.1}X'_{2.3}\|X_{1.4}X_{2.3} + X'_{2.3}X_{2.2}X'_{1.4}$

(*Continued*)

Table 3.15.   (*Continued*)

0011	1000	$X'_{1.1}X_{1.4} + X_{1.1}X'_{1.4}X'_{2.2}\vert X'_{1.1}X_{1.4} + X'_{1.4}X_{1.1}X_{2.3}$ $X'_{2.2}X_{2.3} + X_{2.2}X'_{2.3}X'_{1.1}\vert X'_{2.2}X_{2.3} + X'_{2.3}X_{2.2}X_{1.4}$
0001	1100	$X_{1.1}X'_{1.4} + X'_{1.1}X_{1.4}X_{2.2}\vert X_{1.1}X'_{1.4} + X_{1.4}X'_{1.1}X'_{2.3}$ $X_{2.2}X'_{2.3} + X'_{2.2}X_{2.3}X_{1.1}\vert X_{2.2}X'_{2.3} + X_{2.3}X'_{2.2}X'_{1.4}$
0101	1000	$X'_{1.1}X_{2.2} + X_{1.1}X'_{1.4}X'_{2.2}\vert X'_{1.1}X_{2.2} + X'_{2.2}X_{2.3}X_{1.1}$ $X'_{1.4}X_{2.3} + X_{1.4}X'_{1.1}X'_{2.3}\vert X'_{1.4}X_{2.3} + X'_{2.3}X_{2.2}X_{1.4}$
0001	1010	$X_{1.1}X'_{2.2} + X'_{1.1}X_{1.4}X_{2.2}\vert X_{1.1}X'_{2.2} + X_{2.2}X'_{2.3}X'_{1.1}$ $X_{1.4}X'_{2.3} + X'_{1.4}X_{1.1}X_{2.3}\vert X_{1.4}X'_{2.3} + X_{2.3}X'_{2.2}X'_{1.4}$
1001	0001	$X'_{1.1}X'_{2.3} + X_{1.4}X_{1.1}X_{2.3}\vert X'_{1.1}X'_{2.3} + X_{2.2}X_{2.3}X_{1.1}$ $X_{1.4}X_{2.2} + X'_{1.1}X'_{1.4}X'_{2.2}\vert X_{1.4}X_{2.2} + X'_{2.3}X'_{2.2}X'_{1.4}$ $X'_{1.3}X'_{2.1} + X_{1.2}X_{1.3}X_{2.1}\vert X'_{1.3}X'_{2.1} + X_{2.4}X_{2.1}X_{1.3}$ $X_{1.2}X_{2.4} + X'_{1.3}X'_{1.2}X_{2.4}\vert X_{1.2}X_{2.4} + X'_{2.1}X'_{2.4}X'_{1.2}$
0001	1001	$X_{1.1}X_{2.3} + X_{1.4}X'_{1.1}X'_{2.3}\vert X_{1.1}X_{2.3} + X_{2.2}X'_{2.3}X'_{1.1}$ $X_{1.4}X_{2.2} + X_{1.1}X'_{1.4}X'_{2.2}\vert X_{1.4}X_{2.2} + X_{2.3}X'_{2.2}X'_{1.4}$
1001	1000	$X'_{1.1}X'_{2.3} + X'_{1.4}X_{1.1}X_{2.3}\vert X'_{1.1}X'_{2.3} + X'_{2.2}X_{2.3}X_{1.1}$ $X'_{1.4}X'_{2.2} + X'_{1.1}X_{1.4}X_{2.2}\vert X'_{1.4}X'_{2.2} + X'_{2.3}X_{2.2}X_{1.4}$
1000	1001	$X_{1.1}X_{2.3} + X'_{1.4}X'_{1.1}X'_{2.3}\vert X_{1.1}X_{2.3} + X'_{2.2}X'_{2.3}X'_{1.1}$ $X'_{1.4}X'_{2.2} + X_{1.1}X_{1.4}X_{2.2}\vert X'_{1.4}X'_{2.2} + X_{2.3}X_{2.2}X_{1.4}$ $X_{1.3}X_{2.1} + X'_{1.2}X'_{1.3}X'_{2.1}\vert X_{1.3}X_{2.1} + X'_{2.4}X'_{2.1}X'_{1.3}$ $X'_{1.2}X'_{2.4} + X_{1.3}X_{1.2}X_{2.4}\vert X'_{1.2}X'_{2.4} + X_{2.1}X_{2.4}X_{1.2}$
0110	0100	$X'_{1.1}X_{2.3} + X'_{1.4}X_{1.1}X'_{2.3}\vert X'_{1.1}X_{2.3} + X_{2.2}X'_{2.3}X_{1.1}$ $X'_{1.4}X_{2.2} + X'_{1.1}X_{1.4}X'_{2.2}\vert X'_{1.4}X_{2.2} + X_{2.3}X'_{2.2}X_{1.4}$
0100	0110	$X'_{1.4}X_{2.2} + X_{1.1}X_{1.4}X'_{2.2}\vert X'_{1.4}X_{2.2} + X'_{2.3}X'_{2.2}X_{1.4}$ $X_{1.1}X'_{2.3} + X'_{1.4}X'_{1.1}X_{2.3}\vert X_{1.1}X'_{2.3} + X_{2.2}X_{2.3}X'_{1.1}$
0110	0010	$X_{1.4}X'_{2.2} + X'_{1.1}X'_{1.4}X_{2.2}\vert X_{1.4}X'_{2.2} + X_{2.3}X_{2.2}X'_{1.4}$ $X'_{1.1}X_{2.3} + X_{1.4}X_{1.1}X'_{2.3}\vert X'_{1.1}X_{2.3} + X'_{2.2}X'_{2.3}X_{1.1}$
0010	0110	$X_{1.1}X'_{2.3} + X_{1.4}X'_{1.1}X_{2.3}\vert X_{1.1}X'_{2.3} + X'_{2.2}X_{2.3}X'_{1.1}$ $X_{1.4}X'_{2.2} + X_{1.1}X'_{1.4}X_{2.2}\vert X_{1.4}X'_{2.2} + X'_{2.3}X_{2.2}X'_{1.4}$

Note that in all of the last 24 expressions listed in Table 3.15, the terms that appear in the first (binary) product are negated in the second (ternary) product. This negation is a necessity. In fact, if none were negated, such sum would reduce to the binary (first) product. Moreover, suppose that only one term be complemented in the ternary (second) product. In such a case we have something with the form $XY + XY'Z$. However, $\forall X, Y, Z$ this expression can be reduced as follows

$$XY + XY'Z = X(Y + Y'Z) = X(Y + Y')(Y + Z)$$
$$= X(Y + Z), \tag{3.15}$$

where I have used the distributive property of the logical sum. This is a general property of sums of $n \geq 2$ binary products. For instance, $\forall X, Y, Z, W$, the ternary sum of products $XY + XY'W + XY'Z$, can be reduced as follows

$$XY + XY'W + XY'Z = X(Y + Y'W) + XY'Z$$
$$= XY + XW + XY'Z$$
$$= XW + X(Y + Y'Z)$$
$$= XY + XW + XZ$$
$$= X(Y + W + Z). \tag{3.16}$$

Note that each of the 32 (first) products of Table 3.15 is obtained by summing three appropriate components of Level 2/8 (pertaining to different SCAs): $X_{1.1}X_{1.4}, X_{1.1}X_{2.2}, X_{1.4}X'_{2.3}$ contribute to the node

$$X_{1.1}(X_{1.4} + X_{2.2}) = X_{1.1}(X_{1.4} + X'_{2.3}) = X_{1.1}(X_{2.2} + X'_{2.3}). \tag{3.17}$$

It is remarkable that when we have expressions of the kind $X'_{1.1}(X_{1.4} + X_{2.1})$ and $X_{1.1}(X'_{1.4} + X'_{2.1})$ there is inversion of complementation of the relative expressions of CA2 (in fact, an exchange of lines). On the other hand, the 24 last sums are given by summing three appropriate components of Level 2/8 (again pertaining to different SCAs): $X'_{1.1} \sim X_{1.4} \sim X_{2.2} = X'_{2.1} \sim X_{2.4} \sim X_{1.2}$ plus $X_{1.4}X_{2.2} = X_{1.2}X_{2.4}$ and $X_{1.1}X_{2.3} = X_{1.3}X_{2.1}$ gives

$$X_{1.4}X_{2.2} + X_{1.1}X'_{1.4}X'_{2.2} = X_{1.2}X_{2.4} + X'_{1.2}X_{2.1}X'_{2.4}. \tag{3.18}$$

Since each product of two sets can pertain to two different SCAs, both the 2 expression on the left and that on the right are doubled. When considering a single SCA, all expressions of Level 3/8 are obtained through a sum of two expressions of Level 2/8:

$$X'_{1.1}X_{1.4} + X_{1.1}X'_{1.4}X_{2.2} = X'_{1.1}X_{1.4} + (X_{1.1} \sim X'_{1.4} \sim X_{2.2}),$$

$$X_{1.1}(X_{1.4} + X'_{2.2}) = X_{1.1}X_{1.4} + X_{1.1}X'_{1.4}X'_{2.2}$$

$$= X_{1.1}X_{1.4} + (X'_{1.1} \sim X_{1.4} \sim X_{2.2}),$$

where we also (obviously) have $X_{1.1}(X_{1.4} + X'_{2.2}) = X_{1.1}X_{1.4} + X_{1.1}X'_{2.2}$. Note that of the 24 latter classes of Table 3.15, the first 6 are all combinations of some products of Table 3.11 and the first material equivalence there; the second six classes are all combinations of some products of Table 3.11 and the second material equivalence there, and so on. Note also that we have 4 rows for a single ID when there is mismatch between the two main binary products in the same main CA.

Reciprocally, the expressions of Level 2/8 can be considered as products of two expressions of Level 3/8. For instance,

$$X_{1.1}X_{1.4} = (X_{1.1}X_{1.4} + X_{1.1}X'_{1.4}X_{2.2})(X_{1.1}X_{1.4} + X'_{1.1}X_{1.4}X_{2.2}),$$

$$X_{1.1} \sim X_{1.4} \sim X_{2.2} = (X_{1.1}X_{1.4} + X'_{1.1}X'_{1.4}X_{2.2})(X_{1.1}X_{2.2} + X'_{1.1}X'_{1.4}X'_{2.2}).$$

Of course, other derivations are possible for the same two Level 2/8 expressions.

Thus, Level 3/8 introduces two new logical expressions that are basically a sum (the last 24 expressions) and a product (the first 32 expressions). Such a product can be expressed as $4/8 \times 6/8 = 3/8$, as expected. It is important to stress that such a product needs to be written in such a form and not as a sum of two binary products. In fact, the latter pertains to Level 4/8. The reason is that this would represent a disallowed generalisation: $X(Y + Z)$ is in fact only a particular case of sums of binary products. This shows that for each level there are forms that need to be written exactly in one specific way.

We can consider both logical forms (sums of two binary products with a term in common and sums of a binary and a ternary product) as variants of the forms of Level 3/4, or sums of two simple terms. Of course, intermediate levels that are not present in less dimensional algebras present new kinds of relations. They represent an expansion out of previous forms. Level 1/4 presents binary products. However, they are intermediate between SD variables of Level 2/4, which can be understood as products of the variable with itself on the one hand, and **0** that is the product of each variable and its complement, on the other, and similarly for the sums of Level 3/4. Level 1/8 presents ternary products that are intermediate between **0** and binary products. Similar considerations are true for Levels 5/8, 6/8, and 7/8.

The *basic logical form* of Level 3/8 is represented by a sum of triadic product (Level 1/8) and of a binary product (Level 2/8). Note indeed that $1/8 + 2/8 = 3/8$. This sum is evident for the second kind of expressions, but, as we have seen, also the first one (a product of a SD variable and a sum) has this basic form. Later on, when we have accomplished the analysis of the different levels of $\mathscr{B}_3$, a formal definition of this term will be introduced. For now, *basic logical form* is understood as a form that is similar to common forms but present only in odd levels, while the former is typical of even levels.

3.2.3. From Level 3/8 to Level 4/8

All IDs of Level 4/8 are displayed in Table 3.16. From a combinatorial point of view, we have three cases:

- Four 1s in a single chunk: 2 cases.
- Three 1s in a chunk and one in another: 32 cases.
- Two 1s in both chunks: 36 cases.

In other words, we can analyse the combinatorial pattern as

$$\binom{8}{4} = 2\binom{4}{4} + 2\binom{4}{3}\binom{4}{1} + \binom{4}{2}\binom{4}{2}. \qquad (3.19)$$

In a first approximation we need to consider that, of the 70 IDs, 35 are the negations of the other 35. Another way to consider the problem is that all of these expressions are either the neg–reversal of themselves or the neg–reversal of other ones. All the IDs sum two chunks of two different SD variables (or of Level 2/4), apart from

Table 3.16.   The $\binom{8}{4} = 70$ nodes of Level 4/8. Note that the latter 35 IDs are the negation of the former 35 (in inverse order).

1	1111	0000	=	11100000 + 11010000 + 10110000 + 01110000
2	1110	1000	=	11100000 + 11001000 + 10101000 + 01101000
3	1110	0100	=	11100000 + 11000100 + 10100100 + 01100100
4	1110	0010	=	11100000 + 11000010 + 10100010 + 01100010
5	1110	0001	=	11100000 + 11000001 + 10100001 + 01100001
6	1101	1000	=	11010000 + 11001000 + 10011000 + 01011000
7	1101	0100	=	11010000 + 11000100 + 10010100 + 01010100
8	1101	0010	=	11010000 + 11000010 + 10010010 + 01010010
9	1101	0001	=	11010000 + 11000001 + 10010001 + 01010001
10	1100	1100	=	11001000 + 11000100 + 10001100 + 01001100
11	1100	1010	=	11001000 + 11000010 + 10001010 + 01001010
12	1100	1001	=	11001000 + 11000001 + 10001001 + 01001001
13	1100	0110	=	11000100 + 11000010 + 10000110 + 01000110
14	1100	0101	=	11000100 + 11000001 + 10000101 + 01000101
15	1100	0011	=	11000010 + 11000001 + 10000011 + 01000011
16	1011	1000	=	10110000 + 10101000 + 10011000 + 00111000
17	1011	0100	=	10110000 + 10100100 + 10010100 + 00110100
18	1011	0010	=	10110000 + 10100010 + 10010010 + 00110010
19	1011	0001	=	10110000 + 10100001 + 10010001 + 00110001
20	1010	1100	=	10101000 + 10100100 + 10001100 + 00101100
21	1010	1010	=	10101000 + 10100010 + 10001010 + 00101010
22	1010	1001	=	10101000 + 10100001 + 10001001 + 00101001
23	1010	0110	=	10100100 + 10100010 + 10000110 + 00100110
24	1010	0101	=	10100100 + 10100001 + 10000101 + 00100101
25	1010	0011	=	10100010 + 10100001 + 10000011 + 00100011
26	1001	1100	=	10011000 + 10010100 + 10001100 + 00011100
27	1001	1010	=	10011000 + 10010010 + 10001010 + 00011010
28	1001	1001	=	10011000 + 10010001 + 10001001 + 00011001
29	1001	0110	=	10010100 + 10010010 + 10000110 + 00010110
30	1001	0101	=	10010100 + 10010001 + 10000101 + 00010101
31	1001	0011	=	10010010 + 10010001 + 10000011 + 00010011
32	1000	1110	=	10001100 + 10001010 + 10000110 + 00001110
33	1000	1101	=	10001100 + 10001001 + 10000101 + 00001101
34	1000	1011	=	10001010 + 10001001 + 10000011 + 00001011
35	1000	0111	=	10000110 + 10000101 + 10000011 + 00000111
36	0111	1000	=	01110000 + 01101000 + 01011000 + 00111000
37	0111	0100	=	01110000 + 01100100 + 01010100 + 00110100
38	0111	0010	=	01110000 + 01100010 + 01010010 + 00110010
39	0111	0001	=	01110000 + 01100001 + 01010001 + 00110001

Table 3.16. (*Continued*)

40	0110	1100	=	01101000 + 01100100 + 01001100 + 00101100
41	0110	1010	=	01101000 + 01100010 + 01001010 + 00101010
42	0110	1001	=	01101000 + 01100001 + 01001001 + 00101001
43	0110	0110	=	01100100 + 01100010 + 01000110 + 00100110
44	0110	0101	=	01100100 + 01100001 + 01000101 + 00100101
45	0110	0011	=	01100010 + 01100001 + 01000011 + 00100011
46	0101	1100	=	01011000 + 01010100 + 01001100 + 00011100
47	0101	1010	=	01011000 + 01010010 + 01001010 + 00011010
48	0101	1001	=	01011000 + 01010001 + 01001001 + 00011001
49	0101	0110	=	01010100 + 01010010 + 01000110 + 00010110
50	0101	0101	=	01010100 + 01010001 + 01000101 + 00010101
51	0101	0011	=	01010010 + 01010001 + 01000011 + 00010011
52	0100	1110	=	01001100 + 01001010 + 01000110 + 00001110
53	0100	1101	=	01001100 + 01001001 + 01000101 + 00001101
54	0100	1011	=	01001010 + 01001001 + 01000011 + 00001011
55	0100	0111	=	01000110 + 01000101 + 01000011 + 00000111
56	0011	1100	=	00111000 + 00110100 + 00101100 + 00011100
57	0011	1010	=	00111000 + 00110010 + 00101010 + 00011010
58	0011	1001	=	00111000 + 00110001 + 00101001 + 00011001
59	0011	0110	=	00110100 + 00110010 + 00100110 + 00010110
60	0011	0101	=	00110100 + 00110001 + 00100101 + 00010101
61	0011	0011	=	00110010 + 00110001 + 00100011 + 00010011
62	0010	1110	=	00101100 + 00101010 + 00100110 + 00001110
63	0010	1101	=	00101100 + 00101001 + 00100101 + 00001101
64	0010	1011	=	00101010 + 00101001 + 00100011 + 00001011
65	0010	0111	=	00100110 + 00100101 + 00100011 + 00000111
66	0001	1110	=	00011100 + 00011010 + 00010110 + 00001110
67	0001	1101	=	00011100 + 00011001 + 00010101 + 00001101
68	0001	1011	=	00011010 + 00011001 + 00010011 + 00001011
69	0001	0111	=	00010110 + 00010101 + 00010011 + 00000111
70	0000	1111	=	00001110 + 00001101 + 00001011 + 00000111

the case of the variables themselves, where these two halves are those that constitute the variable under consideration. In the case of binary equivalences, these two halves are the chunks of $X_{1.4}$, $X_{2.2}$, $X_{2.3}$, and complements (this shows that CA1 is somehow more fundamental). In Table 3.5, we see that the remaining 48 IDs show a combination of chunks pertaining to the *same* CA (either 1 or 2), while all combinations of Level 3/8 are between chunks pertaining each to a different CA (and the same, as we see, for Level 5/8). This is evident by considering that four 1s can be generated by coupling either two chunks with two 1s each (both pertaining to CA1) or one chunk with a single 1 and another with three 1s (both pertaining to CA2): either even with even or odd with odd. When we take all 70 IDs, we thus remark that we have $8 + 6 + 24 = 38$ combinations of chunks pertaining to CA1 and $8 + 24 = 32$ combinations of chunks pertaining to CA2.

Of course, all coupling of two chunks can be in two forms, depending on permutation (exchange of left and right chunks). Apart from the empty and universal sets, there are only 6 IDs (out of 256 IDs) that are invariant under permutation of chunks: (i) 0011 0011, 0101 0101, 1100   1100, 1010   1010 (represented by the SD variables $X_{1.4}$ and $X_{2.2}$ and their complements) and (ii) the equivalences 1001 1001, 0110 0110. There are other 4 couplings (represented by other binary equivalences) that transform in their complements: 1100 0011, 1010   0101 into 0011   1100, 0101   1010, respectively, and vice versa.

This gives the 70 classes displayed in Table 3.17. Thus, we have three kinds of classes:

- Basic sets (8 as a whole plus complements): they are neg–reversal invariant;
- Binary material equivalences (3 as whole plus their complements): they are reversal invariant;
- Sums of two binary products (24 as a whole plus their complements): the half of each of the two groups is the neg–reversal of the other half of the same group (e.g. $X'_{1.1}X'_{2.2} + X'_{1.4}X_{2.2}$ is the neg–reversal of $X'_{1.1}X_{2.2} + X'_{1.4}X'_{2.2}$ and vice versa). Note that such sums of binary products (or products of binary sums) are a generalisation of the sums of binary products (or products of binary sums) of Level 2/4.

Again, the combinatorial structure and the logical expressions do not coincide completely: the IDs of the SD variables are such that we have 1 case with four 1s in a chunk (plus the complement) and three cases (plus their complements) with two 1s in each chunk. These four cases

Table 3.17. The 70 classes of Level 4/8. When an ID selects two rows, the first is CA1, the second CA2. When there is a single line, the two CAs are fully parallel. For commodity, I have indicated the relative SCA for the last 48 nodes.

0000	1111		$X_{1.1}$		
0011	0011		$X_{1.4}$		
0101	0101		$X_{2.2}$		
0110	1001		$X_{2.3}$		
0001	0111		$X_{1.2}$		
0010	1011		$X_{1.3}$		
0100	1101		$X_{2.1}$		
0111	0001		$X_{2.4}$		
1111	0000		$X'_{1.1}$		
1100	1100		$X'_{1.4}$		
1010	1010		$X'_{2.2}$		
1001	0110		$X'_{2.3}$		
1110	1000		$X'_{1.2}$		
1101	0100		$X'_{1.3}$		
1011	0010		$X'_{2.1}$		
1000	1110		$X'_{2.4}$		
1100	0011		$X_{1.1} \sim X_{1.4}$	$X_{2.2} \sim X_{2.3}$	
1010	0101		$X_{1.1} \sim X_{2.2}$	$X_{1.4} \sim X_{2.3}$	
1001	1001		$X_{1.1} \sim X_{2.3}$	$X_{1.4} \sim X_{2.2}$	
0011	1100		$X_{1.1} \nsim X_{1.4}$	$X_{2.2} \nsim X_{2.3}$	
0101	1010		$X_{1.1} \nsim X_{2.2}$	$X_{1.4} \nsim X_{2.3}$	
0110	0110		$X_{1.1} \nsim X_{2.3}$	$X_{1.4} \nsim X_{2.2}$	
1110	0100	1.a	$X'_{1.1}X'_{2.2} + X'_{1.4}X_{2.2}$	$X'_{1.4}X'_{2.3} + X'_{1.1}X_{2.3}$	1.b
1100	1010	1.a	$X'_{1.1}X'_{1.4} + X_{1.1}X'_{2.2}$	$X'_{2.3}X'_{2.2} + X_{2.3}X'_{1.4}$	1.d
		2.b	$X'_{1.3}X'_{1.2} + X_{1.3}X'_{2.4}$	$X'_{2.1}X'_{2.4} + X_{2.1}X'_{1.2}$	2.c
1010	1100	1.a	$X_{1.1}X'_{1.4} + X'_{1.1}X'_{2.2}$	$X_{2.3}X'_{2.2} + X'_{2.3}X'_{1.4}$	1.d
		2.b	$X_{1.3}X'_{1.2} + X'_{1.3}X'_{2.4}$	$X_{2.1}X'_{2.4} + X'_{2.1}X'_{1.2}$	2.c
1101	0001	1.a	$X'_{1.1}X'_{1.4} + X_{1.4}X_{2.2}$	$X_{2.2}X_{2.3} + X'_{2.3}X'_{1.1}$	1.c
		2.b	$X'_{1.3}X'_{1.2} + X_{1.2}X_{2.4}$	$X_{2.4}X_{2.1} + X'_{2.1}X'_{1.3}$	2.d
1011	0001	1.a	$X'_{1.1}X'_{2.2} + X_{1.4}X_{2.2}$	$X_{1.4}X_{2.3} + X'_{1.1}X'_{2.3}$	1.b
		2.c	$X'_{2.1}X'_{1.2} + X_{2.4}X_{1.2}$	$X_{2.4}X_{1.3} + X'_{2.1}X'_{1.3}$	2.d
1010	0011	1.a	$X_{1.1}X_{1.4} + X'_{1.1}X'_{2.2}$	$X'_{2.3}X'_{2.2} + X_{2.3}X_{1.4}$	1.d
1100	0110	1.b	$X'_{1.4}X'_{1.1} + X_{1.1}X'_{2.3}$	$X'_{2.3}X'_{2.2} + X_{2.2}X'_{1.4}$	1.d
1101	1000	1.a	$X'_{1.1}X_{2.2} + X'_{1.4}X'_{2.2}$	$X'_{1.4}X_{2.3} + X'_{1.1}X'_{2.3}$	1.b
1011	0100	1.b	$X_{1.4}X'_{1.1} + X'_{1.4}X'_{2.3}$	$X_{2.2}X'_{2.3} + X'_{2.2}X'_{1.1}$	1.c
		2.a	$X_{1.2}X'_{1.3} + X'_{1.2}X'_{2.1}$	$X_{2.4}X'_{2.1} + X'_{2.4}X'_{1.3}$	2.d

(Continued)

Table 3.17. (*Continued*)

1001	0011	1.b	$X_{1.4}X_{1.1} + X'_{1.1}X'_{2.3}$	$X'_{2.3}X'_{2.2} + X_{2.2}X_{1.4}$	1.d
		2.a	$X_{1.2}X_{1.3} + X'_{1.3}X'_{2.1}$	$X'_{2.1}X'_{2.4} + X_{2.4}X_{1.2}$	2.c
1000	1101	1.a	$X_{1.1}X_{2.2} + X'_{1.4}X'_{2.2}$	$X'_{1.4}X'_{2.3} + X_{1.1}X_{2.3}$	1.b
		2.c	$X_{2.1}X_{1.2} + X'_{2.4}X'_{1.2}$	$X'_{2.4}X'_{1.3} + X_{2.1}X_{1.3}$	2.d
1110	0001	1.b	$X'_{1.4}X'_{1.1} + X_{1.4}X_{2.3}$	$X_{2.2}X_{2.3} + X'_{2.2}X'_{1.1}$	1.c
1110	0010	1.a	$X'_{1.1}X'_{1.4} + X_{1.4}X'_{2.2}$	$X'_{2.2}X'_{2.3} + X_{2.3}X'_{1.1}$	1.c
1101	0010	1.b	$X'_{1.4}X'_{1.1} + X_{1.4}X'_{2.3}$	$X'_{2.2}X'_{2.3} + X_{2.2}X'_{1.1}$	1.c
		2.a	$X'_{1.2}X'_{1.3} + X_{1.2}X'_{2.1}$	$X'_{2.4}X'_{2.1} + X_{2.4}X'_{1.3}$	2.d
1011	1000	1.a	$X'_{1.1}X_{1.4} + X'_{1.4}X'_{2.2}$	$X'_{2.2}X_{2.3} + X'_{2.3}X'_{1.1}$	1.c
1000	1011	1.a	$X_{1.1}X_{1.4} + X'_{1.4}X'_{2.2}$	$X'_{2.2}X'_{2.3} + X_{2.3}X_{1.1}$	1.c
		2.b	$X_{1.3}X_{1.2} + X'_{1.2}X'_{2.4}$	$X'_{2.4}X'_{2.1} + X_{2.1}X_{1.3}$	2.d
1001	0101	1.c	$X_{2.2}X_{1.1} + X'_{2.3}X'_{1.1}$	$X'_{2.3}X'_{1.4} + X_{2.2}X_{1.4}$	1.d
		2.a	$X_{1.2}X_{2.1} + X'_{1.3}X'_{2.1}$	$X'_{1.3}X'_{2.4} + X_{1.2}X_{2.4}$	2.b
1000	0111	1.b	$X_{1.4}X_{1.1} + X'_{1.4}X'_{2.3}$	$X'_{2.2}X'_{2.3} + X_{2.2}X_{1.1}$	1.c
1010	0110	1.c	$X'_{2.2}X'_{1.1} + X'_{2.3}X_{1.1}$	$X'_{2.3}X'_{1.4} + X'_{2.2}X_{1.4}$	1.d
1001	1010	1.c	$X'_{2.2}X_{1.1} + X'_{2.3}X'_{1.1}$	$X'_{2.3}X_{1.4} + X'_{2.2}X'_{1.4}$	1.d
1001	1100	1.b	$X'_{1.4}X_{1.1} + X'_{1.1}X'_{2.3}$	$X'_{2.3}X_{2.2} + X'_{2.2}X'_{1.4}$	1.d
1100	1001	1.b	$X'_{1.4}X'_{1.1} + X_{1.1}X_{2.3}$	$X_{2.3}X_{2.2} + X'_{2.2}X'_{1.4}$	1.d
		2.a	$X'_{1.2}X'_{1.3} + X_{1.3}X_{2.1}$	$X_{2.1}X_{2.4} + X'_{2.4}X'_{1.2}$	2.c
1010	1001	1.c	$X'_{2.2}X'_{1.1} + X_{2.3}X_{1.1}$	$X_{2.3}X_{1.4} + X'_{2.2}X'_{1.4}$	1.d
		2.a	$X'_{1.2}X'_{2.1} + X_{1.3}X_{2.1}$	$X_{1.3}X_{2.4} + X'_{1.2}X'_{2.4}$	2.b
1100	0101	1.a	$X'_{1.1}X'_{1.4} + X_{1.1}X_{2.2}$	$X_{2.3}X_{2.2} + X'_{2.3}X'_{1.4}$	1.d
0001	1011	1.a	$(X_{1.1} + X_{2.2})(X_{1.4} + X'_{2.2})$	$(X_{1.4} + X_{2.3})(X_{1.1} + X'_{2.3})$	1.b
0011	0101	1.a	$(X'_{1.1} + X_{2.2})(X_{1.1} + X_{1.4})$	$(X'_{2.3} + X_{1.4})(X_{2.3} + X_{2.2})$	1.d
		2.b	$(X'_{1.3} + X_{2.4})(X_{1.3} + X_{1.2})$	$(X'_{2.1} + X_{1.2})(X_{2.1} + X_{2.4})$	2.c
0101	0011	1.a	$(X_{1.1} + X_{2.2})(X'_{1.1} + X_{1.4})$	$(X_{2.3} + X_{1.4})(X'_{2.3} + X_{2.2})$	1.d
		2.b	$(X_{1.3} + X_{2.4})(X'_{1.3} + X_{1.2})$	$(X_{2.1} + X_{1.2})(X'_{2.1} + X_{2.4})$	2.c
0010	1110	1.a	$(X'_{1.1} + X_{1.4})(X'_{1.4} + X'_{2.2})$	$(X'_{2.2} + X'_{2.3})(X_{2.3} + X_{1.1})$	1.c
		2.b	$(X_{1.3} + X_{1.2})(X'_{1.2} + X'_{2.4})$	$(X'_{2.4} + X'_{2.1})(X_{2.1} + X_{1.3})$	2.d
0100	1110	1.a	$(X_{1.1} + X_{2.2})(X'_{1.4} + X'_{2.2})$	$(X'_{1.4} + X'_{2.3})(X_{1.1} + X_{2.3})$	1.b
		2.c	$(X_{2.1} + X_{1.2})(X'_{2.4} + X'_{1.2})$	$(X'_{2.4} + X'_{1.3})(X_{2.1} + X_{1.3})$	2.d
0101	1100	1.a	$(X'_{1.1} + X'_{1.4})(X_{1.1} + X_{2.2})$	$(X_{2.3} + X_{2.2})(X'_{2.3} + X'_{1.4})$	1.d
0011	1001	1.b	$(X_{1.4} + X_{1.1})(X'_{1.1} + X_{2.3})$	$(X_{2.3} + X_{2.2})(X'_{2.2} + X_{1.4})$	1.d
0010	0111	1.a	$(X_{1.1} + X'_{2.2})(X_{1.4} + X_{2.2})$	$(X_{1.4} + X'_{2.3})(X_{1.1} + X_{2.3})$	1.b
0100	1011	1.b	$(X'_{1.4} + X_{1.1})(X_{1.4} + X_{2.3})$	$(X'_{2.2} + X_{2.3})(X_{2.2} + X_{1.1})$	1.c
		2.a	$(X'_{1.2} + X_{1.3})(X_{1.2} + X_{2.1})$	$(X'_{2.4} + X_{2.1})(X_{2.4} + X_{1.3})$	2.d
0110	1100	1.b	$(X'_{1.4} + X'_{1.1})(X_{1.1} + X_{2.3})$	$(X_{2.3} + X_{2.2})(X'_{2.2} + X'_{1.4})$	1.d
		2.a	$(X'_{1.2} + X'_{1.3})(X_{1.3} + X_{2.1})$	$(X_{2.1} + X_{2.4})(X'_{2.4} + X'_{1.2})$	2.c

Table 3.17. (*Continued*)

0111	0010	1.a	$(X'_{1.1} + X_{2.2})(X_{1.4} + X_{2.2})$	$(X_{1.4} + X_{2.3})(X'_{1.1} + X'_{2.3})$	1.b
		2.c	$(X'_{2.1} + X'_{1.2})(X_{2.4} + X_{1.2})$	$(X_{2.4} + X_{1.3})(X'_{2.1} + X'_{1.3})$	2.d
0001	1110	1.b	$(X_{1.4} + X_{1.1})(X'_{1.4} + X'_{2.3})$	$(X'_{2.2} + X'_{2.3})(X_{2.2} + X_{1.1})$	1.c
0001	1101	1.a	$(X_{1.1} + X_{1.4})(X'_{1.4} + X'_{2.2})$	$(X_{2.2} + X_{2.3})(X'_{2.3} + X_{1.1})$	1.c
0010	1101	1.b	$(X_{1.4} + X_{1.1})(X'_{1.4} + X_{2.3})$	$(X_{2.2} + X_{2.3})(X'_{2.2} + X_{1.1})$	1.c
		2.a	$(X_{1.2} + X_{1.3})(X'_{1.2} + X_{2.1})$	$(X_{2.4} + X_{2.1})(X'_{2.4} + X_{1.3})$	2.d
0100	0111	1.a	$(X_{1.1} + X'_{1.4})(X_{1.4} + X_{2.2})$	$(X_{2.2} + X'_{2.3})(X_{2.3} + X_{1.1})$	1.c
0111	0100	1.a	$(X'_{1.1} + X'_{1.4})(X_{1.4} + X_{2.2})$	$(X_{2.2} + X_{2.3})(X'_{2.3} + X'_{1.1})$	1.c
		2.b	$(X'_{1.3} + X'_{1.2})(X_{1.2} + X_{2.4})$	$(X_{2.4} + X_{2.1})(X'_{2.1} + X'_{1.3})$	2.d
0110	1010	1.c	$(X'_{2.2} + X'_{1.1})(X_{2.3} + X_{1.1})$	$(X_{2.3} + X_{1.4})(X'_{2.2} + X'_{1.4})$	1.d
		2.a	$(X'_{1.2} + X'_{2.1})(X_{1.3} + X_{2.1})$	$(X_{1.3} + X_{2.4})(X'_{1.2} + X'_{2.4})$	2.b
0111	1000	1.b	$(X'_{1.4} + X'_{1.1})(X_{1.4} + X_{2.3})$	$(X_{2.2} + X_{2.3})(X'_{2.2} + X'_{1.1})$	1.c
0101	1001	1.c	$(X_{2.2} + X_{1.1})(X_{2.3} + X'_{1.1})$	$(X_{2.3} + X_{1.4})(X_{2.2} + X'_{1.4})$	1.d
0110	0101	1.c	$(X_{2.2} + X'_{1.1})(X_{2.3} + X_{1.1})$	$(X_{2.3} + X'_{1.4})(X_{2.2} + X_{1.4})$	1.d
0110	0011	1.b	$(X_{1.4} + X'_{1.1})(X_{1.1} + X_{2.3})$	$(X_{2.3} + X'_{2.2})(X_{2.2} + X_{1.4})$	1.d
0011	0110	1.b	$(X_{1.4} + X_{1.1})(X'_{1.1} + X'_{2.3})$	$(X'_{2.3} + X'_{2.2})(X_{2.2} + X_{1.4})$	1.d
		2.a	$(X_{1.2} + X_{1.3})(X'_{1.3} + X'_{2.1})$	$(X'_{2.1} + X'_{2.4})(X_{2.4} + X_{1.2})$	2.c
0101	0110	1.c	$(X_{2.2} + X_{1.1})(X'_{2.3} + X'_{1.1})$	$(X'_{2.3} + X'_{1.4})(X_{2.2} + X_{1.4})$	1.d
		2.a	$(X_{1.2} + X_{2.1})(X'_{1.3} + X'_{2.1})$	$(X'_{1.3} + X'_{2.4})(X_{1.2} + X_{2.4})$	2.b
0011	1010	1.a	$(X_{1.1} + X_{1.4})(X'_{1.1} + X'_{2.2})$	$(X'_{2.3} + X'_{2.2})(X_{2.3} + X_{1.4})$	1.d

constitute CA1. Then, we have other four cases with three 1s in a chunk and one in the other: they correspond to CA2. The 6 equivalences show again two 1s in each chunk, but with one half of the ID being the reverse (and not the neg–reverse) of the other part. The other 48 expressions show both three 1s in a chunk and another in the other (24 nodes) and two 1s in each chunk (again 24 nodes), according to what was said about the IDs, but always without any correlation between the two halves. In other words, correlations between chunks play a major role in the logical expressions. The 6 combinations that are expressed by equivalence can be logically formulated in any of the two CAs, but the chunks pertain only to CA1. In fact, material equivalences are logically invariant across CAs 1 and 2.

All classes are generated by summing 4 appropriate expressions of Level 3/8. For the first kind of classes we have the combination of 4 (out of the 32) products listed in Table 3.15. For instance, $X_{1.1}$ is generated by summing

$$X_{1.1}(X'_{1.4} + X'_{2.2}), \quad X_{1.1}(X'_{1.4} + X_{2.2}),$$
$$X_{1.1}(X_{1.4} + X'_{2.2}), \quad X_{1.1}(X_{1.4} + X_{2.2}). \tag{3.20}$$

Material equivalences are generated by summing 4 (out of 24) sums of duplets and triplets. For instance, $X_{1.1} \sim X_{1.4}$ is generated by combining

$$X'_{1.1}X'_{1.4} + X_{1.1}X_{1.4}X'_{2.2}, \quad X'_{1.1}X'_{1.4} + X_{1.1}X_{1.4}X_{2.2}, \tag{3.21}$$
$$X_{1.1}X_{1.4} + X'_{1.1}X'_{1.4}X'_{2.2}, \quad X_{1.1}X_{1.4} + X'_{1.1}X'_{1.4}X_{2.2}. \tag{3.22}$$

However, sums of two expressions of Level 3/8 do suffice. Indeed,

$$
\begin{aligned}
X_{1.1} &= X_{1.1}(X_{1.4} + X_{2.2}) + X_{1.1}(X'_{1.4} + X'_{2.2}) \\
&= X_{1.1}(X_{1.4} + X'_{2.2}) + X_{1.1}(X'_{1.4} + X_{2.2}), \\
X_{1.1} \sim X_{1.4} &= (X_{1.1}X_{1.4} + X'_{1.1}X'_{1.4}X'_{2.2}) \\
&\quad + (X'_{1.1}X'_{1.4} + X_{1.1}X_{1.4}X_{2.2}) \\
&= (X_{1.1}X_{1.4} + X'_{1.1}X'_{1.4}X_{2.2}) \\
&\quad + (X'_{1.1}X'_{1.4} + X_{1.1}X_{1.4}X'_{2.2}). \tag{3.23}
\end{aligned}
$$

Of course, as it happens for Level 2/4, SD variables can be considered both as sums of two binary products and as products of two binary sums. For instance,

$$X_{1.1} = X_{1.1}X_{1.4} + X_{1.1}X'_{1.4} = (X_{1.1} + X_{1.4})(X_{1.1} + X'_{1.4}).$$

The same is true for binary equivalences. Of course, expressions of Level 3/8 can be generated through products of two expressions of Level 4/8, e.g.:

$$
\begin{aligned}
&X'_{1.1}X_{1.4} + X_{1.1}X'_{1.4}X_{2.2} \\
&\quad = (X_{1.1} + X_{1.4})(X'_{1.1} + X'_{1.4})(X'_{1.1} + X_{2.2})(X_{1.4} + X_{2.2}), \\
&X_{1.1}(X_{1.4} + X_{2.2}) \\
&\quad = (X_{1.1} + X_{1.4})(X_{1.1} + X'_{1.4})(X_{1.1} + X_{2.2})(X_{1.4} + X_{2.2}).
\end{aligned}
$$

Sums of two binary products and products of two binary sums are the common logical forms of Level 4/8. Note that $2/8 + 2/8 = 4/8$.

Another way to consider the problem is to generate the logical expressions of Level 4/8 by summing the products of Level 2/8 (look at Table 3.11 in order to parallelise what happens for Level 2/4). Of course, all SD variables are generated by sums of the kind $X = XY + XY'$. In particular, there are three of such sums for any SD variable:

$$X_{1.1} = X_{1.1}X_{1.4} + X_{1.1}X'_{1.4} = X_{1.1}X_{2.2} + X_{1.1}X'_{2.2}$$

$$= X_{1.1}X_{2.3} + X_{1.1}X'_{2.3}. \tag{3.24}$$

We can easily obtain a similar result for binary equivalences by summing either appropriate products or appropriate ternary equivalences.

The other 48 logical expressions can be equivalently formulated as sums of two binary products or as products of two binary sums. I have chosen to represent 24 of them in the first form and the other 24 in the other form simply because they turn out to be the negation of each other (however, in the next chapter I show that the most general way to consider expressions of Levels 1/2 is to apply the neg–reversal operation: $X'_{1.1}X'_{2.2} + X'_{1.4}X_{2.2}$ (1110 0100) is the neg–reversal of $X'_{1.1}X_{2.2} + X'_{1.4}X_{2.2}$ (1101 1000)). But this form is purely conventional. In the first form's general structure is $X'Y + XZ$, that is, it has three variables involved: one of them occurs in the first product in its primed form while not in the other; the other two variables can occur both in their unprimed form and as complements. It may be helpful to understanding all of these expressions to recall Equation (2.6), that is, $XY + X'Z + YZ = XY + X'Z$ as well as $(X' + Y)(X + Z)(Y + Z) = (X' + Y)(X + Z)$. Moreover, $(X' + Y)(X + Z) = XY + X'Z$.

Now, it is easy to see that, considering the primed forms alone (since the unprimed one is automatically included in all sums), we have 8 sums for each of such primed form (here $X'_{1.1}$) in conjunction with another variable $(X_{1.4})$ and its complement $(X'_{1.4})$, e.g.

$$X'_{1.1}X'_{1.4} + X_{1.1}X'_{2.2}, \qquad X'_{1.1}X_{1.4} + X_{1.1}X'_{2.2},$$
$$+ X_{1.1}X_{2.2}, \qquad\qquad + X_{1.1}X_{2.2},$$
$$\text{and}$$
$$+ X_{1.1}X'_{2.3}, \qquad\qquad + X_{1.1}X'_{2.3}, \qquad (3.25)$$
$$+ X_{1.1}X_{2.3}, \qquad\qquad + X_{1.1}X_{2.3}.$$

Since we have 6 different primed cases, this makes $6 \times 8 = 48$ cases as a whole, as expected.

The emergence of new logical forms may turn out to be surprising when going from less dimensional algebra to an algebra with more dimensions. Things are not exactly in this way. The middle level of $\mathscr{B}_1$ is constituted by X_0 and X'_0 only, while in the middle level of $\mathscr{B}_2$ we find not only SD variables but also material equivalences. However, expressions like X_0 or X_1 can be understood as self–relations, while equivalences can be considered as the same kind of relation but with the second term shifted. In other words, material equivalences are a weaker form of self–relation: the stronger form is logical identity (e.g. $X_0 = X_0$), or logical equivalence, while the weaker form is material equivalence. On the other hand, in the middle level of $\mathscr{B}_3$ we find not only SD variables and (binary) equivalences but also irreducible expressions like the latter 48 in Table 3.17. Now, it turns out that these two different kinds of sums share some important properties: (i) only binary products are involved and (ii) at least one term in at least one of the products must be negative. The only difference is that we have on one side two products either with the same terms or with their complements and on the other two products with three different terms (otherwise the sum would be reducible to less complex expressions). Indeed, let us consider $X_{1.1}X_{1.4} + X'_{1.1}X'_{2.2}$ and replace $X_{1.4}$ by $X'_{2.2}$; then it reduces to $X'_{2.2}$. Now, it is impossible to have an irreducible sum of two products with the same terms (or their complements) in both products that are not a material equivalence (or contravalence, what is the same). Indeed, $XY + XY'$ reduces to X, which also shows another connection between SD variables and material equivalence.

The same is true for the ternary equivalences met in Level 2/8. First, they are the sum of two ternary products as binary products are (the latter are reduced forms). Moreover, again, when a (irreducible) sum of two ternary products involves only the terms in one product and their complements in the other, automatically a material equivalence follows. In fact, consider the following ternary equivalences

$$XYZ + X'Y'Z', \ XYZ' + X'Y'Z, \tag{3.26}$$

$$XY'Z + X'YZ', \ X'YZ + XY'Z'. \tag{3.27}$$

It can be easily seen that by trying to get a similar sum that did not involve an equivalence is impossible: $X'YZ' + X'Y'Z = X'(Y \nsim Z)$. Similar considerations are true for a number $n > 3$ terms. Thus, we have three kinds of sums of two n-ary products (of course, similar considerations are true for products of sums):

- Material equivalences: in one product some terms, in the other product their complements.
- Other irreducible sums of products: all terms occurring in one product need to be complemented in the other one, but at least one term is different for the two products.
- Reducible forms: in such a case there is at least one term occurring in the same form in both products. This case is evident for SD variables (but also for sums of binary products of Level 3/8).

Thus, for any level of any $\mathscr{B}_n$ algebra, the difference between irreducible and reducible forms is essentially due to whether terms occurring in the different products are complemented or not. We also find here the nexus with the mirror 1s. For Level 4/8 we have binary equivalences display two mirror 1s, irreducible sums of two binary products one mirror 1, SD variables show no mirror 1.

The standard definition of material equivalence is that it is a reflexive, symmetric, and transitive relation. The reflexive property (a weaker form of $X = X$) is easily proved since $\forall X$

$$X \sim X = XX + X'X' = X + X' = 1. \tag{3.28}$$

The symmetric property is obvious, since both sum and product are commutative. The transitive property can be proved in this way: assume that both $XY + X'Y'$ and $YZ + Y'Z'$ are true, then, $\forall X, Y, Z$, we have

$$(XY + X'Y')(YZ + Y'Z') = XYZ + X'Y'Z', \tag{3.29}$$

what implies $XZ + X'Z'$. However, things stand in a different way for contravalence. It is indeed still symmetric, due again to commutativity, but it is obviously not reflexive

$$X \nsim X = XX' + X'X = 0, \tag{3.30}$$

and is neither transitive, since $\forall X, Y, Z$

$$(X \nsim Y)(Y \nsim Z) = (XY' + X'Y)(YZ' + Y'Z) = XY'Z + X'YZ', \tag{3.31}$$

which logically implies that $X \sim Z$. Of course, if we reiterate this procedure with a fourth term (say W), such a term will be contravalent to both X and Z and equivalent to Y. This property could be called "alternate" transitivity. Thus, the only property common to equivalence and contravalence is symmetry. What is the reason? The property introduced in the first bullet above (that we have in one product some terms, in the other product their complements) is true for both equivalence and contravalence (in fact the strict connection between this property and symmetry is due to the mirror-like disposition of truth values), but is even more general since it is what allows us to distinguish material equivalence (or contravalence) from *any* other sum (product) of n–ary products (sums).

3.2.4.   *From Level 4/8 to Level 5/8*

From a combinatorial point of view, we have (i) 8 cases with four 1s in a chunk and a 1 in the other, and (ii) 48 cases with three 1s in a chunk

Table 3.18. The $\binom{8}{5} = 56$ nodes of Level 5/8.

1	1111	1000	=	11110000 + 11101000 + 11011000 + 10111000 + 01111000
2	1111	0100	=	11110000 + 11100100 + 11010100 + 10110100 + 01110100
3	1111	0010	=	11110000 + 11100010 + 11010010 + 10110010 + 01110010
4	1111	0001	=	11110000 + 11100001 + 11010001 + 10110001 + 01110001
5	1110	1100	=	11101000 + 11100100 + 11001100 + 10101100 + 01101100
6	1110	1010	=	11101000 + 11100010 + 11001010 + 10101010 + 01101010
7	1110	1001	=	11101000 + 11100001 + 11001001 + 10101001 + 01101001
8	1110	0110	=	11100100 + 11100010 + 11000110 + 10100110 + 01100110
9	1110	0101	=	11100100 + 11100001 + 11000101 + 10100101 + 01100101
10	1110	0011	=	11100010 + 11100001 + 11000011 + 10100011 + 01100011
11	1101	1100	=	11011000 + 11010100 + 11001100 + 10011100 + 01011100
12	1101	1010	=	11011000 + 11010010 + 11001010 + 10011010 + 01011010
13	1101	1001	=	11011000 + 11010001 + 11001001 + 10011001 + 01011001
14	1101	0110	=	11010100 + 11010010 + 11000110 + 10010110 + 01010110
15	1101	0101	=	11010100 + 11010001 + 11000101 + 10010101 + 01010101
16	1101	0011	=	11010010 + 11010001 + 11000011 + 10010011 + 01010011
17	1100	1110	=	11001100 + 11001010 + 11000110 + 10001110 + 01001110
18	1100	1101	=	11001100 + 11001001 + 11000101 + 10001101 + 01001101
19	1100	1011	=	11001010 + 11001001 + 11000011 + 10001011 + 01001011
20	1100	0111	=	11000110 + 11000101 + 11000011 + 10000111 + 01000111
21	1011	1100	=	10111000 + 10110100 + 10101100 + 10011100 + 00111100
22	1011	1010	=	10111000 + 10110010 + 10101010 + 10011010 + 00111010
23	1011	1001	=	10111000 + 10110001 + 10101001 + 10011001 + 00111001
24	1011	0110	=	10110100 + 10110010 + 10100110 + 10010110 + 00110110
25	1011	0101	=	10110100 + 10110001 + 10100101 + 10010101 + 00110101
26	1011	0011	=	10110010 + 10110001 + 10100011 + 10010011 + 00110011
27	1010	1110	=	10101100 + 10101010 + 10100110 + 10001110 + 00101110
28	1010	1101	=	10101100 + 10101001 + 10100101 + 10001101 + 00101101
29	1010	1011	=	10101010 + 10101001 + 10100011 + 10001011 + 00101011
30	1010	0111	=	10100110 + 10100101 + 10100011 + 10000111 + 00100111
31	1001	1110	=	10011100 + 10011010 + 10010110 + 10001110 + 00011110
32	1001	1101	=	10011100 + 10011001 + 10010101 + 10001101 + 00011101
33	1001	1011	=	10011010 + 10011001 + 10010011 + 10001011 + 00011011
34	1001	0111	=	10010110 + 10010101 + 10010011 + 10000111 + 00010111
35	1000	1111	=	10001110 + 10001101 + 10001011 + 10000111 + 00001111
36	0111	1100	=	01111000 + 01110100 + 01101100 + 01011100 + 00111100
37	0111	1010	=	01111000 + 01110010 + 01101010 + 01011010 + 00111010
38	0111	1001	=	01111000 + 01110001 + 01101001 + 01011001 + 00111001
39	0111	0110	=	01110100 + 01110010 + 01100110 + 01010110 + 00110110
40	0111	0101	=	01110100 + 01110001 + 01100101 + 01010101 + 00110101
41	0111	0011	=	01110010 + 01110001 + 01100011 + 01010011 + 00110011
42	0110	1110	=	01101100 + 01101010 + 01100110 + 01001110 + 00101110
43	0110	1101	=	01101100 + 01101001 + 01100101 + 01001101 + 00101101
44	0110	1011	=	01101010 + 01101001 + 01100011 + 01001011 + 00101011
45	0110	0111	=	01100110 + 01100101 + 01100011 + 01000111 + 00100111

(*Continued*)

Table 3.18.   (*Continued*)

46	0101	1110	=	01011100 + 01011010 + 01010110 + 01001110 + 00011110
47	0101	1101	=	01011100 + 01011001 + 01010101 + 01001101 + 00011101
48	0101	1011	=	01011010 + 01011001 + 01010011 + 01001011 + 00011011
49	0101	0111	=	01010110 + 01010101 + 01010011 + 01000111 + 00010111
50	0100	1111	=	01001110 + 01001101 + 01001011 + 01000111 + 00001111
51	0011	1110	=	00111100 + 00111010 + 00110110 + 00101110 + 00011110
52	0011	1101	=	00111100 + 00111001 + 00110101 + 00101101 + 00011101
53	0011	1011	=	00111010 + 00111001 + 00110011 + 00101011 + 00011011
54	0011	0111	=	00110110 + 00110101 + 00110011 + 00100111 + 00010111
55	0010	1111	=	00101110 + 00101101 + 00101011 + 00100111 + 00001111
56	0001	1111	=	00011110 + 00011101 + 00011011 + 00010111 + 00001111

and two in the other. Of course, the same considerations developed for Level 3/8 about the combination of halves of SD variables are true here (but in a complementary way):

$$\binom{8}{5} = 2\binom{4}{4}\binom{4}{1} + 2\binom{4}{3}\binom{4}{2}. \tag{3.32}$$

While Level 3/8 summed two chunks of 3D SD variables (one pertaining to Level 2/4, the other to Level 1/4) that gives three 1s and Level 4/8 summed two chunks of Level 2/4 (of 3D SD variables) that sums to four 1s, here we again sum two chunks of 3D SD variables but pertaining to Levels 2/4 and 3/4, and whose sum gives five 1s (Table 3.18). Note that all IDs of these levels are complementary to all the IDs of Level 3/8 and, as already mentioned, similarly for Levels 6/8 and 2/8 as well as for Levels 7/8 and 1/8. As said, the first 8 cases show all 4 combinations of a 1 in each chunk alternatively, while the other 48 cases are again a sum of two chunks representing each the half of a 3D SD variable. For instance, 0101 1101 is given by the sum of the first chunk of the ID of $X_{2.2}$ and the second chunk of the ID of $X_{2.1}$, while 1110 0110 is given by the sum of the first chunk of the ID of $X'_{1.2}$ and the second chunk of the ID of $X'_{2.3}$. All the combinations between two chunks of Level 3/8 are 56 and also all similar combinations of Level 5/8 are 56, what makes 112. Now, it turns out that if we take the 14 chunks of the 3D SD variables that can give rise to three and five 1s, we can subdivide them in 7 elements of CA1 and 7 elements of CA2. Taking CA1 as fixed, we can

Table 3.19. The 7 chunks of CA1 giving rise to IDs of Levels 3/8 and 5/8.

3/8	0000	0011	0101	0110	1100	1010	1001
5/8	1111	0011	0101	0110	1100	1010	1001

get $4 \times 2 = 8$ combinations for each element (depending on whether the chunk is left or right), what makes $7 \times 8 = 56$ combinations for level, as displayed by Table 3.19.

Note that, apart from the first 8 rows of Table 3.20, the other 24 sums of a variable and a binary product show a behaviour that is in agreement with what is observed for Level 3/8. For instance, 0011 (X_1) goes together with either 0111 ($X_1 + X_2$) or 1011 ($X_1 + X_2'$), where it should be noted that the 1s need to coincide for two of the three 1s. As expected, for the other 24 expressions we have the opposite behaviour. Of course, the last 24 expressions show complementations of variables occurring in the two sums: the two terms in the first summed couple are negated in the ternary sum, always in agreement with what is said for Level 3/8 and 4/8, while the first 32 expressions do not. The difference between first and second kinds of logical forms is reflected in whether they share a position of a 0 or not (from the middle level onwards it is suitable to consider the shared 0s).

Since the 3D Level 5/8 is somehow intermediate between the 2D Levels 2/4 and 3/4, it displays a sum (the dominant operation) of a single element and a binary product. They are the counterpart of the product of a term and a binary sum found at Level 3/8. As already remarked for the latter expression in Subsection 3.2.2, such a sum cannot be written as a product of two binary sums since such expressions pertain to Level 4/8. On the other hand, expressions like $(X_{1.1} + X_{1.4})(X_{1.1}' + X_{1.4}' + X_{2.2}')$ are the product of a binary sum (pertaining to Level 6/8) and a ternary sum (pertaining to Level 7/8), in reversed way relative to the complementary expressions of Level 3/8 that sum a binary product (out of Level 2/8) and a ternary product (out of Level 1/8). Thus, as it occurs for Level 3/8, these two new logical expressions are basically a sum and a product.

Table 3.20. The 56 classes of Level 5/8. Each ID of the first form (a sum of a single element and a binary product) selects two rows: the first one is CA1, the second CA2. Note that, being each the negation of a class of Level 3/8, they are disposed in the same order of Table 3.15. Moreover, the half of the first 8 IDs is the reversal of the other half, and the same for the other 24 sums of SD variables and a product as well as for the last 24 IDs.

0001	1111	$X_{1.1} + X_{1.4}X_{2.2}$	$X_{1.1} + X_{1.4}X'_{2.3}$	$X_{1.1} + X_{2.2}X'_{2.3}$
		$X_{1.2} + X_{1.3}X_{2.1}$	$X_{1.2} + X_{1.3}X'_{2.4}$	$X_{1.2} + X_{2.1}X'_{2.4}$
0010	1111	$X_{1.1} + X_{1.4}X'_{2.2}$	$X_{1.1} + X_{1.4}X_{2.3}$	$X_{1.1} + X'_{2.2}X_{2.3}$
		$X_{1.3} + X_{1.2}X'_{2.4}$	$X_{1.3} + X_{1.2}X_{2.1}$	$X_{1.3} + X'_{2.4}X_{2.1}$
0100	1111	$X_{1.1} + X'_{1.4}X_{2.2}$	$X_{1.1} + X'_{1.4}X_{2.3}$	$X_{1.1} + X_{2.2}X_{2.3}$
		$X_{2.1} + X'_{2.4}X_{1.2}$	$X_{2.1} + X'_{2.4}X_{1.3}$	$X_{2.1} + X_{1.2}X_{1.3}$
1000	1111	$X_{1.1} + X'_{1.4}X'_{2.2}$	$X_{1.1} + X'_{1.4}X'_{2.3}$	$X_{1.1} + X'_{2.2}X'_{2.3}$
		$X'_{2.4} + X_{1.3}X_{1.2}$	$X_{2.4} + X_{2.1}X_{1.2}$	$X'_{2.4} + X_{2.1}X_{1.3}$
1111	0001	$X'_{1.1} + X_{1.4}X_{2.2}$	$X'_{1.1} + X_{1.4}X_{2.3}$	$X'_{1.1} + X_{2.2}X_{2.3}$
		$X_{2.4} + X'_{1.3}X'_{1.2}$	$X_{2.4} + X'_{2.1}X'_{1.2}$	$X_{2.4} + X'_{2.1}X'_{1.3}$
1111	0010	$X'_{1.1} + X_{1.4}X'_{2.2}$	$X'_{1.1} + X_{1.4}X'_{2.3}$	$X'_{1.1} + X'_{2.2}X'_{2.3}$
		$X'_{2.1} + X_{2.4}X'_{1.2}$	$X'_{2.1} + X_{2.4}X'_{1.3}$	$X'_{2.1} + X'_{1.2}X'_{1.3}$
1111	0100	$X'_{1.1} + X'_{1.4}X_{2.2}$	$X'_{1.1} + X'_{1.4}X'_{2.3}$	$X'_{1.1} + X_{2.2}X'_{2.3}$
		$X'_{1.3} + X'_{1.2}X_{2.4}$	$X'_{1.3} + X'_{1.2}X'_{2.1}$	$X'_{1.3} + X_{2.4}X'_{2.1}$
1111	1000	$X'_{1.1} + X'_{1.4}X'_{2.2}$	$X'_{1.1}X'_{1.4}X_{2.3}$	$X'_{1.1} + X'_{2.2}X_{2.3}$
		$X'_{1.2} + X'_{1.3}X'_{2.1}$	$X'_{1.2} + X'_{1.3}X_{2.4}$	$X'_{1.2} + X'_{2.1}X_{2.4}$
0011	0111	$X_{1.4} + X_{1.1}X_{2.2}$	$X_{1.4} + X_{1.1}X'_{2.3}$	$X_{1.4} + X'_{2.3}X_{2.2}$
		$X_{1.2} + X_{1.3}X_{2.4}$	$X_{1.2} + X_{1.3}X'_{2.1}$	$X_{1.2} + X'_{2.1}X_{2.4}$
0011	1011	$X_{1.4} + X_{1.1}X'_{2.2}$	$X_{1.4} + X_{1.1}X_{2.3}$	$X_{1.4} + X_{2.3}X'_{2.2}$
		$X_{1.3} + X_{1.2}X'_{2.1}$	$X_{1.3} + X_{1.2}X_{2.4}$	$X_{1.3} + X_{2.4}X'_{2.1}$
0111	0011	$X_{1.4} + X'_{1.1}X_{2.2}$	$X_{1.4} + X'_{1.1}X_{2.3}$	$X_{1.4} + X_{2.3}X_{2.2}$
		$X_{2.4} + X'_{2.1}X_{1.2}$	$X_{2.4} + X'_{2.1}X_{1.3}$	$X_{2.4} + X_{1.3}X_{1.2}$
1011	0011	$X_{1.4} + X'_{1.1}X'_{2.2}$	$X_{1.4} + X'_{1.1}X'_{2.3}$	$X_{1.4} + X'_{2.3}X'_{2.2}$
		$X'_{2.1} + X_{1.2}X_{1.3}$	$X'_{2.1} + X_{2.4}X_{1.2}$	$X'_{2.1} + X_{2.4}X_{1.3}$
1100	1101	$X'_{1.4} + X_{1.1}X_{2.2}$	$X'_{1.4} + X_{1.1}X_{2.3}$	$X'_{1.4} + X_{2.3}X_{2.2}$
		$X_{2.1} + X'_{1.2}X'_{1.3}$	$X_{2.1} + X'_{2.4}X'_{1.2}$	$X_{2.1} + X'_{2.4}X'_{1.3}$
1100	1110	$X'_{1.4} + X_{1.1}X'_{2.2}$	$X'_{1.4} + X_{1.1}X'_{2.3}$	$X'_{1.4} + X_{2.3}X'_{2.2}$
		$X'_{2.4} + X_{2.1}X'_{1.2}$	$X'_{2.4} + X_{2.1}X'_{1.3}$	$X'_{2.4} + X'_{1.3}X'_{1.2}$
1101	1100	$X'_{1.4} + X'_{1.1}X_{2.2}$	$X'_{1.4} + X'_{1.1}X'_{2.3}$	$X'_{1.4} + X'_{2.3}X_{2.2}$
		$X'_{1.3} + X'_{1.2}X_{2.1}$	$X'_{1.3} + X'_{1.2}X'_{2.4}$	$X'_{1.3} + X'_{2.4}X_{2.1}$
1110	1100	$X'_{1.4} + X'_{1.1}X'_{2.2}$	$X'_{1.4} + X'_{1.1}X_{2.3}$	$X'_{1.4} + X_{2.3}X'_{2.2}$
		$X'_{1.2} + X'_{1.3}X'_{2.4}$	$X'_{1.2} + X'_{1.3}X_{2.1}$	$X_{1.2} + X_{2.1}X'_{2.4}$

Table 3.20. (*Continued*)

0101	0111	$X_{2.2} + X_{1.1}X_{1.4}$ $X_{1.2} + X_{2.1}X_{2.4}$	$X_{2.2} + X'_{2.3}X_{1.1}$ $X_{1.2} + X'_{1.3}X_{2.1}$	$X_{2.2} + X'_{2.3}X_{1.4}$ $X_{1.2} + X'_{1.3}X_{2.4}$
0101	1101	$X_{2.2} + X'_{1.1}X'_{1.4}$ $X_{2.1} + X_{1.2}X'_{1.3}$	$X_{2.2} + X_{2.3}X_{1.1}$ $X_{2.1} + X_{2.4}X_{1.2}$	$X_{2.2} + X_{2.3}X'_{1.4}$ $X_{2.1} + X_{2.4}X'_{1.3}$
0111	0101	$X_{2.2} + X'_{1.1}X_{1.4}$ $X_{2.4} + X'_{1.3}X_{1.2}$	$X_{2.2} + X_{2.3}X'_{1.1}$ $X_{2.4} + X_{2.1}X'_{1.3}$	$X_{2.2} + X_{2.3}X_{1.4}$ $X_{2.4} + X_{2.1}X_{1.2}$
1101	0101	$X_{2.2} + X'_{1.1}X'_{1.4}$ $X'_{1.3} + X_{1.2}X_{2.1}$	$X_{2.2} + X'_{2.3}X'_{1.1}$ $X'_{1.3} + X_{1.2}X_{2.4}$	$X_{2.2} + X'_{2.3}X'_{1.4}$ $X'_{1.3} + X_{2.4}X_{2.1}$
1010	1011	$X'_{2.2} + X_{1.1}X_{1.4}$ $X_{1.3} + X'_{1.2}X'_{2.1}$	$X'_{2.2} + X_{2.3}X_{1.1}$ $X_{1.3} + X'_{1.2}X'_{2.4}$	$X'_{2.2} + X_{2.3}X_{1.4}$ $X_{1.3} + X'_{2.4}X'_{2.1}$
1010	1110	$X'_{2.2} + X_{1.1}X'_{1.4}$ $X'_{2.4} + X_{1.3}X'_{1.2}$	$X'_{2.2} + X'_{2.3}X_{1.1}$ $X'_{2.4} + X'_{2.1}X_{1.3}$	$X'_{2.2} + X'_{2.3}X'_{1.4}$ $X'_{2.4} + X'_{2.1}X'_{1.2}$
1011	1010	$X'_{2.2} + X'_{1.1}X_{1.4}$ $X'_{2.1} + X'_{1.2}X_{1.3}$	$X'_{2.2} + X'_{2.3}X'_{1.1}$ $X'_{2.1} + X'_{2.4}X'_{1.2}$	$X'_{2.2} + X'_{2.3}X_{1.4}$ $X'_{2.1} + X'_{2.4}X_{1.3}$
1110	1010	$X'_{2.2} + X'_{1.1}X'_{1.4}$ $X'_{1.2} + X'_{2.1}X'_{2.4}$	$X'_{2.2} + X_{2.3}X'_{1.1}$ $X'_{1.2} + X_{1.3}X'_{2.1}$	$X'_{2.2} + X_{2.3}X'_{1.4}$ $X'_{1.2} + X_{1.3}X'_{2.4}$
0110	1011	$X_{2.3} + X_{1.4}X_{1.1}$ $X_{1.3} + X_{2.4}X_{2.1}$	$X_{2.3} + X'_{2.2}X_{1.1}$ $X_{1.3} + X'_{1.2}X_{2.1}$	$X_{2.3} + X'_{2.2}X_{1.4}$ $X_{1.3} + X'_{1.2}X_{2.4}$
0111	1001	$X_{2.3} + X_{1.4}X'_{1.1}$ $X_{2.4} + X_{1.3}X'_{1.2}$	$X_{2.3} + X_{2.2}X'_{1.1}$ $X_{2.4} + X_{2.1}X'_{1.2}$	$X_{2.3} + X_{2.2}X_{1.4}$ $X_{2.4} + X_{2.1}X_{1.3}$
0110	1101	$X_{2.3} + X'_{1.4}X_{1.1}$ $X_{2.1} + X'_{1.2}X_{1.3}$	$X_{2.3} + X_{2.2}X_{1.1}$ $X_{2.1} + X_{2.4}X_{1.3}$	$X_{2.3} + X_{2.2}X'_{1.4}$ $X_{2.1} + X_{2.4}X'_{1.2}$
1110	1001	$X_{2.3} + X'_{1.4}X'_{1.1}$ $X'_{1.2} + X_{1.3}X_{2.1}$	$X_{2.3} + X'_{2.2}X'_{1.1}$ $X'_{1.2} + X_{1.3}X_{2.4}$	$X_{2.3} + X'_{2.2}X'_{1.4}$ $X'_{1.2} + X_{2.1}X_{2.4}$
1001	0111	$X'_{2.3} + X_{1.4}X_{1.1}$ $X_{1.2} + X'_{1.3}X'_{2.1}$	$X'_{2.3} + X_{2.2}X_{1.1}$ $X_{1.2} + X'_{1.3}X'_{2.4}$	$X'_{2.3} + X_{2.2}X_{1.4}$ $X_{1.2} + X'_{2.1}X'_{2.4}$
1011	0110	$X'_{2.3} + X_{1.4}X'_{1.1}$ $X'_{2.1} + X_{1.2}X'_{1.3}$	$X'_{2.3} + X_{2.2}X'_{1.1}$ $X'_{2.1} + X'_{2.4}X'_{1.3}$	$X'_{2.3} + X'_{2.2}X_{1.4}$ $X'_{2.1} + X'_{2.4}X_{1.2}$
1001	1110	$X'_{2.3} + X'_{1.4}X_{1.1}$ $X'_{2.4} + X'_{1.3}X_{1.2}$	$X'_{2.3} + X'_{2.2}X_{1.1}$ $X'_{2.4} + X'_{2.1}X_{1.2}$	$X'_{2.3} + X'_{2.2}X'_{1.4}$ $X'_{2.4} + X'_{2.1}X'_{1.3}$
1101	0110	$X'_{2.3} + X'_{1.4}X'_{1.1}$ $X'_{1.3} + X'_{2.4}X'_{2.1}$	$X'_{2.3} + X_{2.2}X'_{1.1}$ $X'_{1.3} + X_{1.2}X'_{2.1}$	$X'_{2.3} + X_{2.2}X'_{1.4}$ $X'_{1.3} + X_{1.2}X'_{2.4}$
0011	1110	$(X_{1.1} + X_{1.4})(X'_{1.1} + X'_{1.4} + X'_{2.2}) \mid (X_{1.1} + X_{1.4})(X'_{1.4} + X'_{1.1} + X'_{2.3})$ $(X'_{2.2} + X'_{2.3})(X_{2.2} + X_{2.3} + X_{1.1}) \mid (X'_{2.2} + X'_{2.3})(X_{2.3} + X_{2.2} + X_{1.4})$			
0111	1100	$(X'_{1.1} + X'_{1.4})(X_{1.1} + X_{1.4} + X_{2.2}) \mid (X'_{1.1} + X'_{1.4})(X_{1.4} + X_{1.1} + X_{2.3})$ $(X_{2.2} + X_{2.3})(X'_{2.2} + X'_{2.3} + X'_{1.1}) \mid (X_{2.2} + X_{2.3})(X'_{2.3} + X'_{2.2} + X'_{1.4})$			

(*Continued*)

Table 3.20.   (*Continued*)

0101 1110	$(X_{1.1} + X_{2.2})(X'_{1.1} + X'_{1.4} + X'_{2.2}) \mid (X_{1.1} + X_{2.2})(X'_{1.4} + X'_{1.1} + X'_{2.3})$ $(X'_{1.4} + X'_{2.3})(X_{1.4} + X_{1.1} + X_{2.3}) \mid (X'_{1.4} + X'_{2.3})(X_{2.3} + X_{2.2} + X_{1.4})$
0111 1010	$(X'_{1.1} + X'_{2.2})(X_{1.1} + X_{1.4} + X_{2.2}) \mid (X'_{1.1} + X'_{2.2})(X_{2.2} + X_{2.3} + X_{1.1})$ $(X_{1.4} + X_{2.3})(X'_{1.4} + X'_{1.1} + X'_{2.3}) \mid (X_{1.4} + X_{2.3})(X'_{2.3} + X'_{2.2} + X'_{1.4})$
0011 1101	$(X_{1.1} + X_{1.4})(X'_{1.1} + X'_{1.4} + X_{2.2}) \mid (X_{1.1} + X_{1.4})(X'_{1.4} + X'_{1.1} + X_{2.3})$ $(X_{2.2} + X_{2.3})(X'_{2.2} + X'_{2.3} + X_{1.1}) \mid (X_{2.2} + X_{2.3})(X'_{2.3} + X'_{2.2} + X_{1.4})$
1011 1100	$(X'_{1.1} + X'_{1.4})(X_{1.1} + X_{1.4} + X'_{2.2}) \mid (X'_{1.1} + X'_{1.4})(X_{1.4} + X_{1.1} + X'_{2.3})$ $(X'_{2.2} + X'_{2.3})(X_{2.2} + X_{2.3} + X'_{1.1}) \mid (X'_{2.2} + X'_{2.3})(X_{2.3} + X_{2.2} + X'_{1.4})$
1010 1101	$(X_{1.1} + X'_{2.2})(X'_{1.1} + X'_{1.4} + X_{2.2}) \mid (X_{1.1} + X'_{2.2})(X_{2.2} + X'_{1.1} + X'_{2.3})$ $(X_{2.3} + X'_{1.4})(X_{1.4} + X_{1.1} + X'_{2.3}) \mid (X_{2.3} + X'_{1.4})(X'_{2.3} + X'_{2.2} + X'_{1.4})$ $(X_{1.3} + X'_{2.4})(X'_{1.3} + X'_{1.2} + X_{2.4}) \mid (X_{1.3} + X'_{2.4})(X_{2.4} + X_{2.1} + X'_{1.3})$ $(X_{2.1} + X'_{1.2})(X_{1.2} + X_{1.3} + X'_{2.1}) \mid (X_{2.1} + X'_{1.2})(X'_{2.1} + X'_{2.4} + X_{1.2})$
1011 0101	$(X'_{1.1} + X_{2.2})(X_{1.1} + X_{1.4} + X'_{2.2}) \mid (X'_{1.1} + X_{2.2})(X'_{2.2} + X'_{2.3} + X_{1.1})$ $(X'_{2.3} + X_{1.4})(X'_{1.4} + X'_{1.1} + X_{2.3}) \mid (X'_{2.3} + X_{1.4})(X_{2.3} + X_{2.2} + X'_{1.4})$ $(X'_{1.3} + X_{2.4})(X_{1.3} + X_{1.2} + X'_{2.4}) \mid (X'_{1.3} + X_{2.4})(X'_{2.4} + X'_{2.1} + X_{1.3})$ $(X'_{2.1} + X_{1.2})(X'_{1.2} + X'_{1.3} + X_{2.1}) \mid (X'_{2.1} + X_{1.2})(X_{2.1} + X_{2.4} + X'_{1.2})$
1100 1011	$(X_{1.1} + X'_{1.4})(X'_{1.1} + X_{1.4} + X'_{2.2}) \mid (X_{1.1} + X'_{1.4})(X_{1.4} + X'_{1.1} + X_{2.3})$ $(X_{2.3} + X'_{2.2})(X_{2.2} + X'_{2.3} + X_{1.1}) \mid (X_{2.3} + X'_{2.2})(X'_{2.3} + X_{2.2} + X'_{1.4})$ $(X_{2.1} + X'_{2.4})(X'_{2.1} + X_{2.4} + X'_{1.2}) \mid (X_{2.1} + X'_{2.4})(X_{2.4} + X'_{2.1} + X_{1.3})$ $(X_{1.3} + X'_{1.2})(X_{1.2} + X'_{1.3} + X_{2.1}) \mid (X_{1.3} + X'_{1.2})(X'_{1.3} + X_{1.2} + X'_{2.4})$
1101 0011	$(X'_{1.1} + X_{1.4})(X_{1.1} + X'_{1.4} + X_{2.2}) \mid (X'_{1.1} + X_{1.4})(X'_{1.4} + X_{1.1} + X'_{2.3})$ $(X'_{2.3} + X_{2.2})(X'_{2.2} + X_{2.3} + X'_{1.1}) \mid (X'_{2.3} + X_{2.2})(X_{2.3} + X'_{2.2} + X_{1.4})$ $(X'_{2.1} + X_{2.4})(X_{2.1} + X'_{2.4} + X_{1.2}) \mid (X'_{2.1} + X_{2.4})(X'_{2.4} + X_{2.1} + X'_{1.3})$ $(X'_{1.3} + X_{1.2})(X'_{1.2} + X_{1.3} + X'_{2.1}) \mid (X'_{1.3} + X_{1.2})(X_{1.3} + X'_{1.2} + X_{2.4})$
0101 1011	$(X_{1.1} + X_{2.2})(X'_{1.1} + X_{1.4} + X'_{2.2}) \mid (X_{1.1} + X_{2.2})(X'_{2.2} + X_{2.3} + X'_{1.1})$ $(X_{1.4} + X_{2.3})(X'_{1.4} + X_{1.1} + X'_{2.3}) \mid (X_{1.4} + X_{2.3})(X'_{2.3} + X_{2.2} + X'_{1.4})$
1101 1010	$(X'_{1.1} + X'_{2.2})(X_{1.1} + X'_{1.4} + X_{2.2}) \mid (X'_{1.1} + X'_{2.2})(X_{2.2} + X'_{2.3} + X_{1.1})$ $(X'_{1.4} + X'_{2.3})(X_{1.4} + X'_{1.1} + X_{2.3}) \mid (X'_{1.4} + X'_{2.3})(X_{2.3} + X'_{2.2} + X_{1.4})$
1100 0111	$(X_{1.1} + X'_{1.4})(X'_{1.1} + X_{1.4} + X_{2.2}) \mid (X_{1.1} + X'_{1.4})(X_{1.4} + X'_{1.1} + X'_{2.3})$ $(X_{2.2} + X'_{2.3})(X'_{2.2} + X_{2.3} + X_{1.1}) \mid (X_{2.2} + X'_{2.3})(X_{2.3} + X'_{2.2} + X'_{1.4})$
1110 0011	$(X'_{1.1} + X_{1.4})(X_{1.1} + X'_{1.4} + X'_{2.2}) \mid (X'_{1.1} + X_{1.4})(X'_{1.4} + X_{1.1} + X_{2.3})$ $(X'_{2.2} + X_{2.3})(X_{2.2} + X'_{2.3} + X'_{1.1}) \mid (X'_{2.2} + X_{2.3})(X'_{2.3} + X_{2.2} + X_{1.4})$
1010 0111	$(X_{1.1} + X'_{2.2})(X'_{1.1} + X_{1.4} + X_{2.2}) = (X_{1.1} + X'_{2.2})(X_{2.2} + X'_{2.3} + X_{1.1})$ $(X_{1.4} + X'_{2.3})(X'_{1.4} + X_{1.1} + X_{2.3}) \mid (X_{1.4} + X'_{2.3})(X_{2.3} + X'_{2.2} + X_{1.4})$
1110 0101	$(X'_{1.1} + X_{2.2})(X_{1.1} + X'_{1.4} + X'_{2.2}) \mid (X'_{1.1} + X_{2.2})(X'_{2.2} + X_{2.3} + X_{1.1})$ $(X'_{1.4} + X_{2.3})(X_{1.4} + X'_{1.1} + X'_{2.3}) \mid (X'_{1.4} + X_{2.3})(X'_{2.3} + X_{2.2} + X_{1.4})$

Table 3.20. (*Continued*)

0110 1110	$(X_{1.1} + X_{2.3})(X'_{1.4} + X'_{1.1} + X'_{2.3})\|(X_{1.1} + X_{2.3})(X'_{2.2} + X'_{2.3} + X_{1.1})$ $(X'_{1.4} + X'_{2.2})(X_{1.1} + X_{1.4} + X_{2.2})\|(X'_{1.4} + X'_{2.2})(X_{2.3} + X_{2.2} + X_{1.4})$ $(X_{1.3} + X_{2.1})(X'_{1.2} + X'_{1.3} + X'_{2.1})\|(X_{1.3} + X_{2.1})(X'_{2.4} + X'_{2.1} + X'_{1.3})$ $(X'_{1.2} + X'_{2.4})(X_{1.3} + X_{1.2} + X_{2.4})\|(X'_{1.2} + X'_{2.4})(X_{2.1} + X_{2.4} + X_{1.2})$
1110 0110	$(X'_{1.1} + X'_{2.3})(X'_{1.4} + X_{1.1} + X_{2.3})\|(X'_{1.1} + X'_{2.3})(X'_{2.2} + X_{2.3} + X_{1.1})$ $(X'_{1.4} + X'_{2.2})(X'_{1.1} + X_{1.4} + X_{2.2})\|(X'_{1.4} + X'_{2.2})(X'_{2.3} + X_{2.2} + X_{1.4})$
0110 0111	$(X_{1.1} + X_{2.3})(X_{1.4} + X'_{1.1} + X'_{2.3})\|(X_{1.1} + X_{2.3})(X_{2.2} + X'_{2.3} + X'_{1.1})$ $(X_{1.4} + X_{2.2})(X_{1.1} + X'_{1.4} + X'_{2.2})\|(X_{1.4} + X_{2.2})(X_{2.3} + X'_{2.2} + X'_{1.4})$
0111 0110	$(X'_{1.1} + X'_{2.3})(X_{1.4} + X_{1.1} + X_{2.3})\|(X'_{1.1} + X'_{2.3})(X_{2.2} + X_{2.3} + X_{1.1})$ $(X_{1.4} + X_{2.2})(X'_{1.1} + X'_{1.4} + X'_{2.2})\|(X_{1.4} + X_{2.2})(X'_{2.3} + X'_{2.2} + X_{1.4})$ $(X'_{1.3} + X'_{2.1})(X_{1.2} + X_{1.3} + X_{2.1})\|(X'_{1.3} + X'_{2.1})(X_{2.4} + X_{2.1} + X_{1.3})$ $(X_{1.2} + X_{2.4})(X'_{1.3} + X'_{1.2} + X'_{2.4})\|(X_{1.2} + X_{2.4})(X'_{2.1} + X'_{2.4} + X_{1.2})$
1001 1011	$(X_{1.1} + X'_{2.3})(X_{1.4} + X'_{1.1} + X_{2.3})\|(X_{1.1} + X'_{2.3})(X'_{2.2} + X_{2.3} + X'_{1.1})$ $(X_{1.4} + X'_{2.2})(X_{1.1} + X'_{1.4} + X_{2.2})\|(X_{1.4} + X'_{2.2})(X'_{2.3} + X_{2.2} + X'_{1.4})$
1011 1001	$(X_{1.4} + X'_{2.2})(X'_{1.1} + X'_{1.4} + X_{2.2})\|(X_{1.4} + X'_{2.2})(X_{2.3} + X_{2.2} + X'_{1.4})$ $(X_{2.3} + X'_{1.1})(X_{1.4} + X_{1.1} + X'_{2.3})\|(X_{2.3} + X'_{1.1})(X'_{2.2} + X'_{2.3} + X_{1.1})$
1001 1101	$(X_{1.1} + X'_{2.3})(X'_{1.4} + X'_{1.1} + X_{2.3})\|(X_{1.1} + X'_{2.3})(X_{2.2} + X_{2.3} + X'_{1.1})$ $(X_{2.2} + X'_{1.4})(X_{1.1} + X_{1.4} + X'_{2.2})\|(X_{2.2} + X'_{1.4})(X'_{2.3} + X'_{2.2} + X_{1.4})$
1101 1001	$(X'_{1.1} + X_{2.3})(X'_{1.4} + X_{1.1} + X'_{2.3})\|(X'_{1.1} + X_{2.3})(X_{2.2} + X'_{2.3}X_{1.1})$ $(X'_{1.4} + X_{2.2})(X'_{1.1} + X_{1.4} + X'_{2.2})\|(X'_{1.4} + X_{2.2})(X_{2.3} + X'_{2.2} + X_{1.4})$

All expressions of Level 5/8 are generated by summing two expressions of Level 4/8. For instance,

$$(X_{1.1} + X_{1.4})(X'_{1.1} + X'_{1.4} + X'_{2.2})$$
$$= X'_{1.1}X_{1.4} + X_{1.1}X'_{1.4} + X_{1.1}X'_{2.2} + X_{1.4}X'_{2.2},$$
$$X_{1.1} + X_{1.4}X_{2.2}$$
$$= (X_{1.1} + X_{1.4})(X_{1.1} + X'_{1.4} + X_{2.2})$$
$$= X_{1.1}X_{1.4} + X_{1.1}X'_{1.4} + X_{1.1}X_{2.2} + X_{1.4}X_{2.2}.$$

Conversely, all expressions of Level 4/8 can be generated through products of two expressions of Level 5/8:

$$X_{1.1} = (X_{1.1} + X_{1.4}X_{2.2})(X'_{1.4} + X_{1.1}X_{1.4}),$$
$$X_{1.1} \sim X_{1.4} = (X_{1.1} + X'_{1.1}X'_{1.4})(X'_{1.4} + X'_{1.1}X'_{1.4}),$$
$$X'_{11}X'_{2.2} + X'_{1.4}X_{2.2} = (X'_{1.1} + X'_{1.4}X_{2.2})(X'_{2.2} + X'_{1.4}X_{2.2}).$$

The basic form is represented here by a product between a binary and a ternary sum.

Note also that the previous expressions can be also logically reformulated:

$$(X_{1.1} + X_{1.4})(X'_{1.1} + X'_{1.4} + X'_{2.2})$$

$$= X_{1.1}X'_{1.4} + X_{1.1}X'_{2.2} + X'_{1.1}X_{1.4} + X_{1.4}X'_{2.2}$$

$$= (X_{1.1} \approx X_{1.4}) + X'_{2.2}(X_{1.1} + X_{1.4})$$

$$= X_{1.1}(X'_{1.4} + X'_{2.2}) + X_{1.4}(X'_{1.1} + X'_{2.2}). \tag{3.33}$$

3.2.5. *From Level 5/8 to Level 6/8*

From a combinatorial point of view, we have

$$\binom{8}{6} = 2\binom{4}{4}\binom{4}{2} + \binom{4}{3}\binom{4}{3}. \tag{3.34}$$

The IDs are of two kinds: either we combine a chunk from Level 2/4 with the universal set 1111 or we combine two chunks of Level 3/4 (Table 3.21). Table 3.22 displays the IDs and the relative logical expressions.

As for the 2D Level 3/4, we find here binary sums. These sums are again regulated by the rules mentioned in Subsection 3.2.1 for products. Additionally, there are also ternary material contravalences. It is understood that material contravalence, for example, $X_{1.1} \approx X_{1.4} \approx X_{2.2}$, is defined as $(X_{1.1} \sim X_{1.4} \sim X_{2.2})'$. Similarly to what said in Subsection 3.2.1 about ternary equivalences, here the relation between a first term and both a second and a third is sufficient to determine the relation between the second and the third term. For instance,

$$(X_{1.1} + X_{1.4} + X_{2.2})(X'_{1.1} + X'_{1.4} + X'_{2.2})$$

$$= (X'_{1.1}X_{1.4} + X_{1.1}X'_{1.4})(X'_{1.1}X_{2.2} + X_{1.1}X'_{2.2}). \tag{3.35}$$

Binary sums display two mirror 1s, while contravalences three mirror 1s.

Table 3.21. The $\binom{8}{6} = 28$ nodes of Level 6/8.

#				sum
1	1111	1100	=	11111000 + 11110100 + 11110010 + 11101100 + 11101010 + 11100110 + 11011100 + 11011010 + 11010110 + 11001110 + 01111100 + 01111010 + 01110110 + 01101110 + 00111110
2	1111	1010	=	11111000 + 11110100 + 11110001 + 11101100 + 11101001 + 11100101 + 11011100 + 11011001 + 11010101 + 11001101 + 01111010 + 01111001 + 01110101 + 01101101 + 00111101
3	1111	1001	=	11111000 + 11110010 + 11110001 + 11101010 + 11101001 + 11100011 + 11011010 + 11011001 + 11010011 + 11001011 + 01111001 + 01110011 + 01101011 + 01100111 + 00111011
4	1111	0110	=	11110100 + 11110010 + 11100110 + 11101100 + 11100110 + 11100101 + 10111100 + 10110110 + 10101110 + 10011110 + 01110110 + 01110101 + 01101110 + 01100111 + 00110111
5	1111	0101	=	11110100 + 11110001 + 11100101 + 11101010 + 11100101 + 11100011 + 10111010 + 10110101 + 10101101 + 10011101 + 01110101 + 01100111 + 01101101 + 01011101 + 00110111
6	1111	0011	=	11110010 + 11110001 + 11100011 + 11101001 + 11100011 + 11100110 + 10110110 + 10110011 + 10101011 + 10011011 + 01110011 + 01100111 + 01101011 + 01011011 + 00110111
7	1110	1110	=	11101100 + 11101010 + 11100110 + 11011100 + 11011010 + 11010110 + 10111100 + 10111010 + 10110110 + 10011110 + 01101110 + 01101101 + 01100111 + 01011110 + 00101111
8	1110	1101	=	11101100 + 11101001 + 11100101 + 11011100 + 11011001 + 11010101 + 10111100 + 10111001 + 10110101 + 10011101 + 01101101 + 01100111 + 01011101 + 01010111 + 00101111
9	1110	1011	=	11101010 + 11101001 + 11100011 + 11011010 + 11011001 + 11010011 + 10111010 + 10111001 + 10110011 + 10011011 + 01101011 + 01100111 + 01011011 + 01010111 + 00101111
10	1110	0111	=	11100110 + 11100101 + 11100011 + 11010110 + 11010101 + 11010011 + 10110110 + 10110101 + 10110011 + 10011110 + 01100111 + 01010111 + 01011011 + 01011101 + 00100111
11	1101	1110	=	11011100 + 11011010 + 11001110 + 10111100 + 10111010 + 10101110 + 11011110 + 10011110 + 11001110 + 10011011 + 01011110 + 01011101 + 01011011 + 01010111 + 00011110
12	1101	1101	=	11011100 + 11011001 + 11001101 + 10111100 + 10111001 + 10101101 + 11011101 + 10011101 + 11001101 + 10011001 + 01011101 + 01010111 + 01011011 + 01001111 + 00011101
13	1101	1011	=	11011010 + 11011001 + 11001011 + 10111010 + 10111001 + 10101011 + 11011011 + 10011011 + 11001011 + 10011011 + 01011011 + 01010111 + 01001111 + 01011101 + 00011011
14	1101	0111	=	11010110 + 11010101 + 11010011 + 10110110 + 10110101 + 10110011 + 11010111 + 10010111 + 11000111 + 10010111 + 01010111 + 01001111 + 01011011 + 01011101 + 00010111
15	1100	1111	=	11001110 + 11001101 + 11001011 + 10101110 + 10101101 + 10101011 + 11000111 + 10011110 + 10011101 + 10011011 + 01001111 + 01011011 + 01011101 + 01011110 + 00001111
16	1011	1110	=	10111100 + 10111010 + 10110110 + 10011110 + 10101110 + 10111110 + 10111101 + 10111011 + 10110111 + 10011111 + 00111110 + 00111101 + 00111011 + 00110111 + 00101111
17	1011	1101	=	10111100 + 10111001 + 10110101 + 10011101 + 10101101 + 10111101 + 10111110 + 10111011 + 10110111 + 10011111 + 00111101 + 00111110 + 00111011 + 00110111 + 00101111
18	1011	1011	=	10111010 + 10111001 + 10110011 + 10011011 + 10101011 + 10111011 + 10111110 + 10111101 + 10110111 + 10011111 + 00111011 + 00111110 + 00111101 + 00110111 + 00101111
19	1011	0111	=	10110110 + 10110101 + 10110011 + 10011110 + 10101110 + 10110111 + 10111110 + 10111101 + 10111011 + 10011111 + 00110111 + 00111110 + 00111101 + 00111011 + 00100111
20	1010	1111	=	10101110 + 10101101 + 10101011 + 10011110 + 10011101 + 10011011 + 10101111 + 10111110 + 10111101 + 10111011 + 00101111 + 00111110 + 00111101 + 00111011 + 00001111
21	1001	1111	=	10011110 + 10011101 + 10011011 + 10010111 + 10001111 + 10011111 + 10101111 + 10111110 + 10111101 + 10111011 + 00011111 + 00111110 + 00111101 + 00111011 + 00001111
22	0111	1110	=	01111100 + 01111010 + 01110110 + 01101110 + 01011110 + 00111110 + 01111110 + 01111101 + 01111011 + 01110111 + 01101111 + 01011111 + 00111110 + 00101111 + 00011111
23	0111	1101	=	01111100 + 01111001 + 01110101 + 01101101 + 01011101 + 00111101 + 01111101 + 01111110 + 01111011 + 01110111 + 01101111 + 01011111 + 00111101 + 00101111 + 00011111
24	0111	1011	=	01111010 + 01111001 + 01110011 + 01101011 + 01011011 + 00111011 + 01111011 + 01111110 + 01111101 + 01110111 + 01101111 + 01011111 + 00111011 + 00101111 + 00011111
25	0111	0111	=	01110110 + 01110101 + 01110011 + 01101110 + 01101101 + 01101011 + 01110111 + 01111110 + 01111101 + 01111011 + 00110111 + 00111110 + 00111101 + 00111011 + 00010111
26	0110	1111	=	01101110 + 01101101 + 01101011 + 01100111 + 01011110 + 00101111 + 01101111 + 01111110 + 01111101 + 01111011 + 00011111 + 00111110 + 00111101 + 00111011 + 00001111
27	0101	1111	=	01011110 + 01011101 + 01011011 + 01010111 + 01001111 + 00011111 + 01011111 + 01111110 + 01111101 + 01111011 + 00101111 + 00111110 + 00111101 + 00111011 + 00001111
28	0011	1111	=	00111110 + 00111101 + 00111011 + 00110111 + 00101111 + 00011111 + 00111111 + 01111110 + 01111101 + 01111011 + 01110111 + 01101111 + 01011111 + 00111111 + 00001111

Table 3.22. The 28 classes of Level 6/8. In each group of four expressions the half is the reversal of the other half, apart from the contravalences that are reversal invariant.

Binary ID		CA1	CA2
0011	1111	$X_{1.1} + X_{1.4}$	$X_{1.2} + X_{1.3}$
1100	1111	$X_{1.1} + X'_{1.4}$	$X_{2.1} + X'_{2.4}$
1111	0011	$X'_{1.1} + X_{1.4}$	$X'_{2.1} + X_{2.4}$
1111	1100	$X'_{1.1} + X'_{1.4}$	$X'_{1.2} + X'_{1.3}$
0101	1111	$X_{1.1} + X_{2.2}$	$X_{1.2} + X_{2.1}$
1010	1111	$X_{1.1} + X'_{2.2}$	$X_{1.3} + X'_{2.4}$
1111	0101	$X'_{1.1} + X_{2.2}$	$X'_{1.3} + X_{2.4}$
1111	1010	$X'_{1.1} + X'_{2.2}$	$X'_{1.2} + X'_{2.1}$
0110	1111	$X_{1.1} + X_{2.3}$	$X_{1.3} + X_{2.1}$
1001	1111	$X_{1.1} + X'_{2.3}$	$X_{1.2} + X'_{2.4}$
1111	1001	$X'_{1.1} + X_{2.3}$	$X'_{1.2} + X_{2.4}$
1111	0110	$X'_{1.1} + X'_{2.3}$	$X'_{1.3} + X'_{2.1}$
0111	0111	$X_{1.4} + X_{2.2}$	$X_{1.2} + X_{2.4}$
1011	1011	$X_{1.4} + X'_{2.2}$	$X_{1.3} + X'_{2.1}$
1101	1101	$X'_{1.4} + X_{2.2}$	$X'_{1.3} + X_{2.1}$
1110	1110	$X'_{1.4} + X'_{2.2}$	$X'_{1.2} + X'_{2.4}$
0111	1011	$X_{1.4} + X_{2.3}$	$X_{1.3} + X_{2.4}$
1011	0111	$X_{1.4} + X'_{2.3}$	$X_{1.2} + X'_{2.1}$
1110	1101	$X'_{1.4} + X_{2.3}$	$X'_{1.2} + X_{2.1}$
1101	1110	$X'_{1.4} + X'_{2.3}$	$X'_{1.3} + X'_{2.4}$
0111	1101	$X_{2.2} + X_{2.3}$	$X_{2.1} + X_{2.4}$
1101	0111	$X_{2.2} + X'_{2.3}$	$X_{1.2} + X'_{1.3}$
1110	1011	$X'_{2.2} + X_{2.3}$	$X'_{1.2} + X_{1.3}$
1011	1110	$X'_{2.2} + X'_{2.3}$	$X'_{2.1} + X'_{2.4}$
0111	1110	$X_{1.1} \approx X_{1.4} \approx X_{2.2}$	
1011	1101	$X_{1.1} \approx X_{1.4} \approx X'_{2.2}$	
1101	1011	$X_{1.1} \approx X'_{1.4} \approx X_{2.2}$	
1110	0111	$X'_{1.1} \approx X_{1.4} \approx X_{2.2}$	

All expressions of Level 6/8 can be derived through sums of two expressions of Level 5/8:

$$X_{1.1} + X_{1.4} = [(X_{1.1} + X_{1.4})(X_{1.1} + X'_{1.4} + X_{2.2})]$$

$$+ [(X_{1.1} + X'_{1.4})(X_{1.1} + X'_{1.4} + X_{2.2})]$$

$$= (X_{1.1} + X_{1.4}X_{2.2}) + (X_{1.1} + X_{1.4}X'_{2.2}),$$

$$X_{1.1} \wr X_{1.4} \wr X_{2.2} = X_{1.1}X'_{1.4} + X'_{1.1}X_{1.4} + X_{1.1}X'_{2.2}$$
$$+ X'_{1.1}X_{2.2} + X_{1.4}X'_{2.2} + X'_{1.4}X_{2.2}$$
$$= [(X_{1.1} + X_{1.4})(X'_{1.1} + X'_{1.4} + X'_{2.2})]$$
$$\times [(X_{1.1} + X_{2.2})(X'_{1.1} + X'_{1.4} + X'_{2.2})].$$

Conversely, all expressions of Level 5/8 can be derived through products of two expressions of Level 6/8:

$$(X_{1.1} + X_{1.4})(X'_{1.1} + X'_{1.4} + X'_{2.2}) = (X_{1.1} + X_{1.4})(X_{1.1} \wr X_{1.4} \wr X_{2.2}),$$
$$X_{1.1} + X_{1.4}X_{2.2} = (X_{1.1} + X_{1.4})(X_{1.1} + X_{2.2}).$$

All of these expressions can be conceived as products of two ternary sums:

$$X_{1.1} + X_{1.4} = (X_{1.1} + X_{1.4} + X_{2.2})(X_{1.1} + X_{1.4} + X'_{2.2}),$$

$$\tag{3.36a}$$

$$X_{1.1} \wr X_{1.4} \wr X'_{2.2} = (X_{1.1} + X_{1.4} + X'_{2.2})(X'_{1.1} + X'_{1.4} + X_{2.2}).$$

$$\tag{3.36b}$$

Therefore, these are the common forms. Note finally that we can reformulate these expressions as

$$(X_{1.1} + X_{1.4})(X'_{1.1} + X'_{1.4} + X'_{2.2}) = (X_{1.1} \wr X_{1.4}) + X'_{2.2}(X_{1.1} + X_{1.4}).$$

$$\tag{3.37}$$

The generation of the IDs of the 3D binary sums from the IDs of 2D ones is displayed in Table 3.23. For the self–sums see Table 3.24.

3.2.6. *From Level 6/8 to Level 7/8*

From a combinatorial point of view, we have

$$\binom{8}{7} = 2\binom{4}{4}\binom{4}{3}. \tag{3.38}$$

The IDs combine a chunk of Level 3/4 with 1111. Tables 3.25, 3.26 and 3.27 display the IDs and relative logical expressions for CA1 and CA2, respectively.

Table 3.23. Comparison between IDs of Level 3/4 and IDs of Level 6/8.

Level 1/4		Level 2/8		
$X_1' + X_2'$	1110	$X_{1.1}' + X_{2.2}'$	1111	1010
		$X_{1.1}' + X_{2.3}'$	1111	0110
		$X_{1.4}' + X_{2.2}'$	1110	1110
		$X_{1.4}' + X_{2.3}'$	1101	1110
$X_1' + X_2$	1101	$X_{1.1}' + X_{2.2}$	1111	0101
		$X_{1.1}' + X_{2.3}$	1111	1001
		$X_{1.4}' + X_{2.2}$	1101	1101
		$X_{1.4}' + X_{2.3}$	1110	1101
$X_1 + X_2'$	1011	$X_{1.1} + X_{2.2}'$	1010	1111
		$X_{1.1} + X_{2.3}'$	1001	1111
		$X_{1.4} + X_{2.2}'$	1011	1011
		$X_{1.4} + X_{2.3}'$	1011	0111
$X_1 + X_2$	0111	$X_{1.1} + X_{2.2}$	0101	1111
		$X_{1.1} + X_{2.3}$	0110	1111
		$X_{1.4} + X_{2.2}$	0111	0111
		$X_{1.4} + X_{2.3}$	0111	1011

Table 3.24. Further comparison between IDs of Level 1/4 and IDs of Level 2/8.

$X_{1.1}' + X_{1.4}'$	1111	1100	1011	1110	$X_{2.2}' X_{2.3}'$
$X_{1.1}' + X_{1.4}$	1111	0011	1110	1011	$X_{2.2}' X_{2.3}$
$X_{1.1} + X_{1.4}'$	1100	1111	1101	0111	$X_{2.2} X_{2.3}'$
$X_{1.1} + X_{1.4}$	0011	1111	0111	1101	$X_{2.2} X_{2.3}$

Of course, we have

$$X_{1.1} + X_{1.4} + X_{2.2} = (X_{1.1} + X_{1.4}) + (X_{1.1} + X_{2.2}) + (X_{1.4} + X_{2.2}).$$
$$(3.39)$$

Conversely,

$$X_{1.1} + X_{1.4} = (X_{1.1} + X_{1.4} + X_{2.2})(X_{1.1} + X_{1.4} + X_{2.2}'). \quad (3.40)$$

Table 3.25. The $\binom{8}{7} = 8$ nodes of Level 7/8.

1	1111	1110	$=$	1111100 + 1111010 + 1110110 + 1101110 + 1011110 + 0111110
2	1111	1101	$=$	1111100 + 1111001 + 1110101 + 1101101 + 1011101 + 0111101
3	1111	1011	$=$	1111010 + 1111001 + 1110011 + 1101011 + 1011011 + 0111011
4	1111	0111	$=$	1110110 + 1110101 + 1110011 + 1100111 + 1010111 + 0110111
5	1110	1111	$=$	1101110 + 1101101 + 1101011 + 1100111 + 1001111 + 0101111
6	1101	1111	$=$	1011110 + 1011101 + 1011011 + 1010111 + 1001111 + 0011111
7	1011	1111	$=$	0111110 + 0111101 + 0111011 + 0110111 + 0101111 + 0011111
8	0111	1111	$=$	0011111 + 0101111 + 0110111 + 0111011 + 0111101 + 0111110

Table 3.26. The 8 classes of Level 7/8 using CA1.

0111	1111	$X_{1,1}+X_{1,4}+X_{2,2}$	$X_{1,4}+X_{1,1}+X_{2,3}$	$X_{2,2}+X_{2,3}+X_{1,1}$	$X_{2,3}+X_{2,2}+X_{1,4}$
1011	1111	$X_{1,1}+X_{1,4}+X'_{2,2}$	$X_{1,4}+X_{1,1}+X'_{2,3}$	$X'_{2,2}+X_{2,3}+X_{1,1}$	$X'_{2,3}+X_{2,2}+X_{1,4}$
1101	1111	$X_{1,1}+X'_{1,4}+X_{2,2}$	$X'_{1,4}+X_{1,1}+X_{2,3}$	$X_{2,2}+X'_{2,3}+X_{1,1}$	$X_{2,3}+X'_{2,2}+X_{1,4}$
1110	1111	$X_{1,1}+X'_{1,4}+X_{2,2}$	$X'_{1,4}+X_{1,1}+X_{2,3}$	$X_{2,2}+X_{2,3}+X_{1,1}$	$X_{2,3}+X_{2,2}+X_{1,4}$
1111	0111	$X'_{1,1}+X_{1,4}+X_{2,2}$	$X_{1,4}+X'_{1,1}+X_{2,3}$	$X_{2,2}+X_{2,3}+X'_{1,1}$	$X_{2,3}+X_{2,2}+X_{1,4}$
1111	1011	$X'_{1,1}+X_{1,4}+X_{2,2}$	$X_{1,4}+X'_{1,1}+X_{2,3}$	$X_{2,2}+X_{2,3}+X'_{1,1}$	$X_{2,3}+X_{2,2}+X'_{1,4}$
1111	1101	$X'_{1,1}+X_{1,4}+X_{2,2}$	$X_{1,4}+X'_{1,1}+X_{2,3}$	$X_{2,2}+X_{2,3}+X'_{1,1}$	$X_{2,3}+X_{2,2}+X'_{1,4}$
1111	1110	$X_{1,1}+X'_{1,4}+X_{2,2}$	$X'_{1,4}+X_{1,1}+X_{2,3}$	$X_{2,2}+X_{2,3}+X_{1,1}$	$X_{2,3}+X_{2,2}+X'_{1,4}$

Table 3.27. The 8 classes of Level 7/8 for CA2.

0111	1111	$X_{1.2} + X_{1.3} + X_{2.1}$	$X_{1.3} + X_{1.2} + X_{2.4}$	$X_{2.1} + X_{2.4} + X_{1.2}$	$X_{2.4} + X_{2.1} + X_{1.3}$
1011	1111	$X_{1.2} + X_{1.3} + X'_{2.1}$	$X_{1.3} + X_{1.2} + X'_{2.4}$	$X'_{2.1} + X'_{2.4} + X_{1.2}$	$X'_{2.4} + X'_{2.1} + X_{1.3}$
1101	1111	$X_{1.2} + X'_{1.3} + X_{2.1}$	$X'_{1.3} + X_{1.2} + X_{2.4}$	$X_{2.1} + X'_{2.4} + X_{1.2}$	$X'_{2.4} + X_{2.1} + X'_{1.3}$
1110	1111	$X'_{1.2} + X_{1.3} + X'_{2.1}$	$X_{1.3} + X'_{1.2} + X'_{2.4}$	$X_{2.1} + X'_{2.4} + X'_{1.2}$	$X'_{2.4} + X_{2.1} + X_{1.3}$
1111	0111	$X'_{1.2} + X_{1.3} + X'_{2.1}$	$X'_{1.3} + X_{1.2} + X_{2.4}$	$X'_{2.1} + X_{2.4} + X_{1.2}$	$X_{2.4} + X'_{2.1} + X'_{1.3}$
1111	1011	$X'_{1.2} + X_{1.3} + X'_{2.1}$	$X_{1.3} + X'_{1.2} + X'_{2.4}$	$X'_{2.1} + X_{2.4} + X'_{1.2}$	$X_{2.4} + X'_{2.1} + X_{1.3}$
1111	1101	$X'_{1.2} + X'_{1.3} + X_{2.1}$	$X'_{1.3} + X_{1.2} + X_{2.4}$	$X_{2.1} + X_{2.4} + X'_{1.2}$	$X_{2.4} + X_{2.1} + X'_{1.3}$
1111	1110	$X'_{1.2} + X'_{1.3} + X_{2.1}$	$X'_{1.3} + X'_{1.2} + X'_{2.4}$	$X'_{2.1} + X'_{2.4} + X'_{1.2}$	$X'_{2.4} + X'_{2.1} + X'_{1.3}$

Resuming (and recalling Table 3.5), the levels of $\mathscr{B}_3$ show a characteristic sequence: the IDs of Level 1/8 are generated by coupling a chunk of CA1 (0000) and another of CA2; Level 2/8 by coupling two chunks either of CA1 or CA2; Level 3/8 by coupling again a chunk of CA1 and another of CA2; Level 4/8 shows again endogenous pairing, while Level 5/8 is again an exogenous one, and similarly Levels 6/8 and 7/8, respectively. Of course, Levels 0/8 and 8/8 also are endogenous. In other words, levels with even numerator are endogenous (even), while levels with odd numerator are exogenous (odd). This character is maintained across all nD algebras (it is also true for the 2D algebra: Levels 1/4 and 3/4 combine an element [two digits] coming from X_1 and another from X_2). Note finally, that all endogenous pairings are 128 (one half of all nodes), 64 for CA1, and 64 for CA2. Of course, 128 nodes are the result of exogenous pairing.

3.2.7. *Summary*

Although every SCA is able to cover all 256 nodes of $\mathscr{B}_3$, in some cases this would happen in ways that are not fully satisfactory from a logical point of view. In fact, each SCA can cover only 182 nodes (following the order of levels: 1, 8, 16, 48, 36, 48, 16, 8, 1) expressing the logical forms that have been reported here. This order is due to the fact that each SCA has distinctive patterns: in order to have the most reduced logical expression of a node, we need in general to consider which columns can be paired in such a way that e.g. one of the three involved variables is cut off, so that we get a binary product from ternary products (a further reduction can be performed for obtaining SD variables). This requires that the two columns share all truth values but a single one that is opposite. Excluding the trivial cases of the expressions of Levels 1/8 and 7/8 (which do not require such reductions), and making reference to Table 3.2, we see that, for each SCA, an "a" column can be paired in this way with three other columns (those with a single 1) as well as an "h" column can be paired with three other columns (those with two 1s). Of course, there are also alternative combinations, so that for each choice of pairing

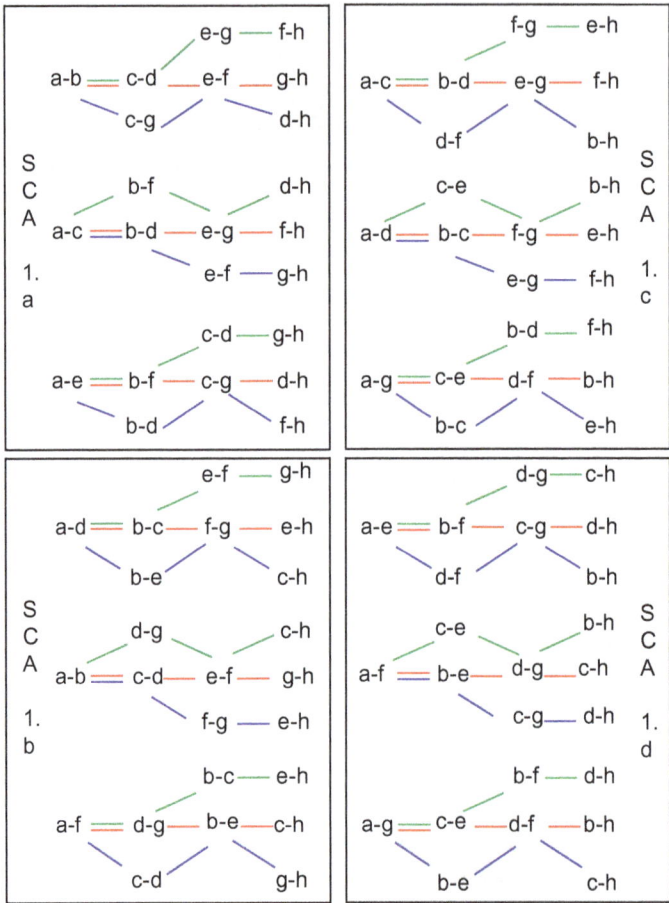

Figure 3.3.   SCA1.a–d's patterns.

of "a" with another column (starting from the left), we have three different but crossing patterns (each covering all 8 columns, without superpositions of columns), as displayed in Figures 3.3–3.4. In such a way, it is easy to understand how we get the logical expressions, for instance, product $X_{1.1}X_{2.2}$ (expressing node 0000 0101): in both SCA 1.a and SCA 1.c there is the pairing f-h. Similarly, the expression $X_{1.1}(X_{1.4}+X_{2.2})$ (node 0000 0111) results from the fact that in SCA 1.a we have both the pairings f-h and g-h (with a superposition of

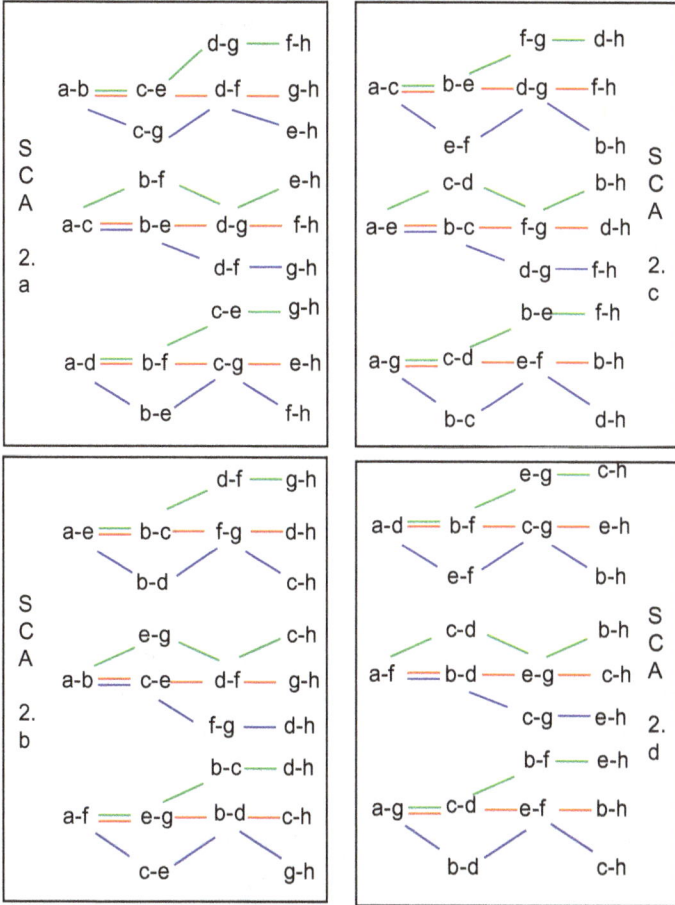

Figure 3.4.   SCA2.a-d's patterns.

columns for each SCA). As last example, consider the expression $X'_{1.1}X'_{2.2} + X'_{1.4}X_{2.2}$ (node 1110 0100): it turns out that in SCA 1.a we have (in the same pattern) both a-c and b-f.

This explains the fact that, when we apply a SCA to nodes for which there are no such pairings, this gives rise to expressions that are not false but logically weird. For instance, if we would express the node 1000 0010 using the SCA 1.a, the relative logical expression was $X'_{2.2}(X_{1.1} \sim X_{1.4})$, which is certainly not so straight as $X'_{2.2}X'_{2.3}$.

However, the two expressions are obviously logically equivalent. In fact, recalling the transformation rules (Subsection 3.1.3), we can write

$$X'_{2.2}X'_{2.3} = X'_{2.2}\left[X_{1.1}(X_{1.4} \sim X_{2.2}) + X'_{1.1}(X_{1.4} \approx X_{2.2})\right]'$$

$$= X'_{2.2}\left[X_{1.1}(X_{1.4} \sim X_{2.2})\right]'\left[X'_{1.1}(X_{1.4} \approx X_{2.2})\right]'$$

$$= X'_{2.2}(X'_{1.1} + X_{1.4}X'_{2.2} + X'_{1.4}X_{2.2})$$

$$\times (X_{1.1} + X_{1.4}X_{2.2} + X'_{1.4}X'_{2.2})$$

$$= X'_{2.2}(X'_{1.1}X_{1.4}X_{2.2} + X'_{1.1}X'_{1.4}X'_{2.2}$$

$$+ X_{1.1}X_{1.4}X'_{2.2} + X_{1.1}X'_{1.4}X_{2.2})$$

$$= X'_{2.2}\left[X_{2.2}(X_{1.1} \approx X_{1.4}) + X'_{2.2}(X_{1.1} \sim X_{1.4})\right]$$

$$= X'_{2.2}(X_{1.1} \sim X_{1.4}). \tag{3.41}$$

Note that, in general, all logical forms that are not straight involve an equivalence as part of the expression. This could be taken as a simple aesthetic consideration; nevertheless, there are two points here:

(i) As I have mentioned, an equivalence is a weaker form of self–relation ($X \sim Y$ instead of $X = X$), so that we can conceive an equivalence as a kind of weaker SD variable and the expression $X'_{2.2}(X_{1.1} \sim X_{1.4})$ as a binary product of two variables. But $X_{1.1} \sim X_{1.4}$ is not a true basic set, failing to be self–dual (its ID having only the property to be the reversal of itself). Thus, it is a kind of inappropriate product. This case is clear when we consider that the equivalence $X_{1.1} \sim X_{1.4} \sim X_{2.2}$ could be considered as the product of the two weak variables $X_{1.1} \sim X_{1.4}$ and $X_{1.1} \sim X_{2.2}$, and thus having the form of the 24 binary products of Level 2/8. The fact that in the unsatisfactory expressions material equivalences occur is due to the transformation rules among SD variables (Subsection 3.1.3) that I have in fact used in the previous derivation.

(ii) The previous form would obscure the symmetry and therefore the general combinatorics of the algebra. This will be better understood when we go to higher dimensions.

I have said that the 1s present in an ID can show superpositions of columns. By now keep in mind that for all odd levels (apart from $1/m$ and $(m-1)/m$) such superpositions are present (here Levels 3/8 and 5/8). However, not every logical expression shows here such a superposition. In fact, sums of binary and ternary products (for Level 3/8) as well as products of binary and ternary sums (for Level 5/8) do not show this behaviour. This situation is why they are the *basic* forms for those levels. At the opposite, we do not have such a situation for even levels, and that is why I have distinguished them from the previous ones and called them *common* forms.

Let us now consider the arrows of the 3D algebra, as displayed in Table 3.28. It may be noted that the distributions shown there follow a precise order:

- The series for each level starts with number 1 and goes through the series of natural numbers up to the limit determined by the level. For example, the arrows from Level 3/8 to Level 4/8 need to be five per node of Level 3/8 (this is the example I consider here). This is the first series (from 1 to 5).
- The second item repeats the first element of the previous series and goes further with series where it was previously truncated. This is the second series (1, 6-9).
- The third item starts with the second element of the first series, takes the first element of the second series, and starts the third series where the latter was truncated (2,6,10-12), and so on. Note that the gaps between elements reduces step by step. The second item had a gap 4 (between 1 and 6); the third item two gaps 3 (between 2 and 6 and between 6 and 10); the fourth item a gap 3 (between 3 and 7) and two gaps 2 (between 7 and 10 as well as between 10 and 13); the fifth item a gap 3 (between 4 and 8), a gap 2 (between 8 and 11), and two gaps 1 (between 11 and 13 as well as between 13 and 15); the sixth item a gap 3 (between 5 and 9),

Table 3.28. Derivation of all logical relations (arrows). The table needs to be read from below and thinking that in each box arrows go from the bottom line to the top. The numeration is extracted from Tables 3.7 or 3.8, 3.10, 3.14, 3.16, 3.18, 3.21, and 3.25. To a certain extent, the present table is an inverse relative to those listed: they tell us how a node is generated from previous nodes, while the present table tells us which nodes a given node generates (see also (2.14)).

Level 8-8		**1**							
Level 7-8		1,2,3,4,5,6,7,8							
Level 7-8	Level 6-8	4,8 / 25	5,8 / 26	6,8 / 27	7,8 / 28	6,7 / 21	1,8 / 22	2,8 / 23	3,8 / 24
Level 7-8	Level 6-8	2,7 / 17	3,7 / 18	4,7 / 19	5,7 / 20	3,6 / 13	4,6 / 14	5,6 / 15	1,7 / 16
Level 7-8	Level 6-8	3,5 / 9	4,5 / 10	1,6 / 11	2,6 / 12	2,4 / 5	3,4 / 6	1,5 / 7	2,5 / 8
Level 7-8	**Level 6-8**	1,2 / 1	1,3 / 2	2,3 / 3	1,4 / 4				
Level 6-8	Level 5-8	14,25,27 / 49	15,26,27 / 50	16,22,28 / 51	17,23,28 / 52	18,24,28 / 53	19,25,28 / 54	20,26,28 / 55	21,27,28 / 56
Level 6-8	Level 5-8	6,24,25 / 41	7,22,26 / 42	8,23,26 / 43	9,24,26 / 44	10,25,26 / 45	11,22,27 / 46	12,23,27 / 47	13,24,27 / 48
Level 6-8	Level 5-8	13,18,21 / 33	14,19,21 / 34	15,20,21 / 35	1,22,23 / 36	2,22,24 / 37	3,23,24 / 38	4,22,25 / 39	5,23,25 / 40
Level 6-8	Level 5-8	5,17,19 / 25	6,18,19 / 26	7,16,20 / 27	8,17,20 / 28	9,18,20 / 29	10,19,20 / 30	11,16,21 / 31	12,17,21 / 32
Level 6-8	Level 5-8	7,11,15 / 17	8,12,15 / 18	9,13,15 / 19	10,14,15 / 20	1,16,17 / 21	2,16,18 / 22	3,17,18 / 23	4,16,19 / 24

Table 3.28. (*Continued*)

Level 6-8 / Level 5-8	5,8,10 / 9	6,9,10 / 10	1,11,12 / 11	2,11,13 / 12	3,12,13 / 13	4,11,14 / 14	5,12,14 / 15	6,13,14 / 16
Level 6-8 / Level 5-8	1,2,3 / 1	1,4,5 / 2	2,4,6 / 3	3,5,6 / 4	4,7,8 / 5	2,7,9 / 6	3,8,10 / 7	4,7,10 / 8
Level 5-8 / Level 4-8	30,45,54,55 / 65	31,46,51,56 / 66	32,47,52,56 / 67	33,48,53,56 / 68	34,49,54,56 / 69	35,50,55,56 / 70		
Level 5-8 / Level 4-8	22,37,51,53 / 57	23,38,52,53 / 58	24,39,51,54 / 59	25,40,52,54 / 60	26,41,53,54 / 61	27,42,51,55 / 62	28,43,52,55 / 63	29,44,53,55 / 64
Level 5-8 / Level 4-8	14,39,46,49 / 49	15,40,47,49 / 50	16,41,48,49 / 51	17,42,46,50 / 52	18,43,47,50 / 53	19,44,48,50 / 54	20,45,49,50 / 55	21,36,51,52 / 56
Level 5-8 / Level 4-8	6,37,42,44 / 41	7,38,43,44 / 42	8,39,42,45 / 43	9,40,43,45 / 44	10,41,44,45 / 45	11,36,46,47 / 46	12,37,46,48 / 47	13,38,47,48 / 48
Level 5-8 / Level 4-8	18,28,32,35 / 33	19,29,33,35 / 34	20,30,34,35 / 35	1,36,37,38 / 36	2,36,39,40 / 37	3,37,39,41 / 38	4,38,40,41 / 39	5,36,42,43 / 40
Level 5-8 / Level 4-8	10,26,29,30 / 25	11,21,31,32 / 26	12,22,31,33 / 27	13,23,32,33 / 28	14,24,31,34 / 29	15,25,32,34 / 30	16,26,33,34 / 31	17,27,31,35 / 32
Level 5-8 / Level 4-8	2,21,24,25 / 17	3,22,24,26 / 18	4,23,25,26 / 19	5,21,27,28 / 20	6,22,27,29 / 21	7,23,28,29 / 22	8,24,27,30 / 23	9,25,28,30 / 24
Level 5-8 / Level 4-8	4,13,15,16 / 9	5,11,17,18 / 10	6,12,17,19 / 11	7,13,18,19 / 12	8,14,17,20 / 13	9,15,18,20 / 14	10,16,19,20 / 15	1,21,22,23 / 16
Level 5-8 / Level 4-8	1,2,3,4 / 1	1,5,6,7 / 2	2,5,8,9 / 3	3,6,8,10 / 4	4,7,9,10 / 5	5,11,12,13 / 6	6,11,14,15 / 7	7,12,14,16 / 8
Level 4-8 / Level 3-8	32,52,62,66,70 / 53		33,53,63,67,70 / 54		34,54,64,68,70 / 55		35,55,65,69,70 / 56	
Level 4-8 / Level 3-8	28,48,58,67,68 / 49		29,49,59,66,69 / 50		30,50,60,67,69 / 51		31,51,61,68,69 / 52	

(*Continued*)

Table 3.28.   (*Continued*)

Level 4-8	24,44,60,63,65	25,45,61,64,65	26,46,56,66,67	27,47,57,66,68
Level 3-8	45	46	47	48
Level 4-8	20,40,56,62,63	21,41,57,62,64	22,42,58,63,64	23,43,59,62,65
Level 3-8	41	42	43	44
Level 4-8	16,36,56,57,58	17,37,56,59,60	18,38,57,59,61	19,39,58,60,61
Level 3-8	37	38	39	40
Level 4-8	12,42,48,53,54	13,43,49,52,55	14,44,50,53,55	15,45,51,54,55
Level 3-8	33	34	35	36
Level 4-8	8,38,47,49,51	9,39,48,50,51	10,40,49,52,53	11,41,47,52,54
Level 3-8	29	30	31	32
Level 4-8	4,38,41,43,45	5,39,42,44,45	6,36,40,47,48	7,37,46,49,50
Level 3-8	25	26	27	28
Level 4-8	15,25,31,34,35	1,36,37,38,39	2,36,40,41,42	3,37,40,43,44
Level 3-8	21	22	23	24
Level 4-8	11,21,27,32,34	12,22,28,33,34	13,23,29,33,35	14,24,30,33,35
Level 3-8	17	18	19	20
Level 4-8	7,17,26,29,30	8,18,27,29,31	9,19,28,30,31	10,20,26,32,33
Level 3-8	13	14	15	16
Level 4-8	3,17,20,23,24	4,18,21,23,25	5,19,22,24,25	6,16,26,27,28
Level 3-8	9	10	11	12
Level 4-8	4,8,11,13,15	5,9,12,14,15	1,16,17,18,19	2,16,20,21,22
Level 3-8	5	6	7	8
Level 4-8	1,2,3,4,5	1,6,7,8,9	2,6,10,11,12	3,7,10,13,14
Level 3-8	1	2	3	4

Table 3.28. (Continued)

Level 3-8	18,33,43,49,54,55	19,34,44,50,53,56	20,35,45,51,54,56	21,36,46,52,55,56
Level 2-8	25	26	27	28
Level 3-8	14,29,39,48,50,52	15,30,40,49,51,52	16,31,41,47,53,54	17,32,42,48,53,55
Level 2-8	21	22	23	24
Level 3-8	10,25,39,42,44,46	11,26,40,43,45,46	12,27,37,47,48,49	13,28,38,47,50,51
Level 2-8	17	18	19	20
Level 3-8	6,26,30,33,35,36	7,22,37,38,39,40	8,23,37,41,42,43	9,24,38,41,44,45
Level 2-8	13	14	15	16
Level 3-8	2,22,27,28,29,30	3,23,27,31,32,33	4,24,28,31,34,35	5,25,29,32,34,36
Level 2-8	9	10	11	12
Level 3-8	4,9,13,16,19,20	5,10,14,17,19,21	6,11,15,18,20,21	1,22,23,24,25,26
Level 2-8	5	6	7	8
Level 3-8	1,2,3,4,5,6	1,7,8,9,10,11	2,7,12,13,14,15	3,8,12,16,17,18
Level 2-8	1	2	3	4
Level 2-8	4,10,15,19,23,24,25	5,11,16,20,23,26,27	6,12,17,21,24,26,28	7,13,18,22,25,27,28
Level 1-8	5	6	7	8
Level 2-8	1,2,3,4,5,6,7	1,8,9,10,11,12,13	2,8,14,15,16,17,18	3,9,14,19,20,21,22
Level 1-8	1	2	3	4
Level 1-8	1,2,3,4,5,6,7,8			
Level 0-8	1			

a gap 2 (between 9 and 12), a gap 1 (between 12 and 14), and a gap 0 (14-15).

- When one completes all the previous series and elements of the previous series comes to form a new series, like 5,9,12,14-15 (where 5 is the last element of the first series, 9 the last element of the second series, 12 the last element of the third series, 14 the last element of the fourth series, and 15 the first element of the fifth series), the subsequent item starts again from the first item that is still available and after that gives rise to a full new series (1,16-19). The numbers (like 1 here) for each level need to be repeated as many times as the arrow that converge on that node (here four).

3.3. Distribution

The two basic rules of distribution can be cast as follows:

$$X_{1.1} + X_{1.4}X_{2.2} = (X_{1.1} + X_{1.4})(X_{1.1} + X_{2.2}), \qquad (3.42a)$$

$$X_{1.1}X_{1.4} + X_{1.1}X_{2.2} = X_{1.1}(X_{1.4} + X_{2.2}). \qquad (3.42b)$$

As said, we need $\mathscr{B}_3$ for showing its validity, since three variables are involved here. However, a subnetwork of the whole algebra does suffice, as displayed in Figure 3.5. The first rule can be shown by considering following diagram:

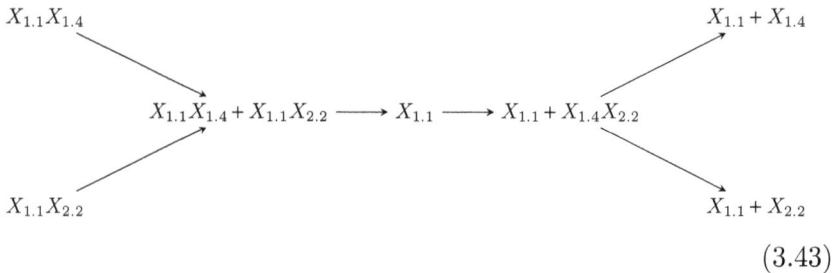

$X_{1.1}X_{1.4}$ $X_{1.1} + X_{1.4}$

$X_{1.1}X_{1.4} + X_{1.1}X_{2.2} \longrightarrow X_{1.1} \longrightarrow X_{1.1} + X_{1.4}X_{2.2}$

$X_{1.1}X_{2.2}$ $X_{1.1} + X_{2.2}$

$$(3.43)$$

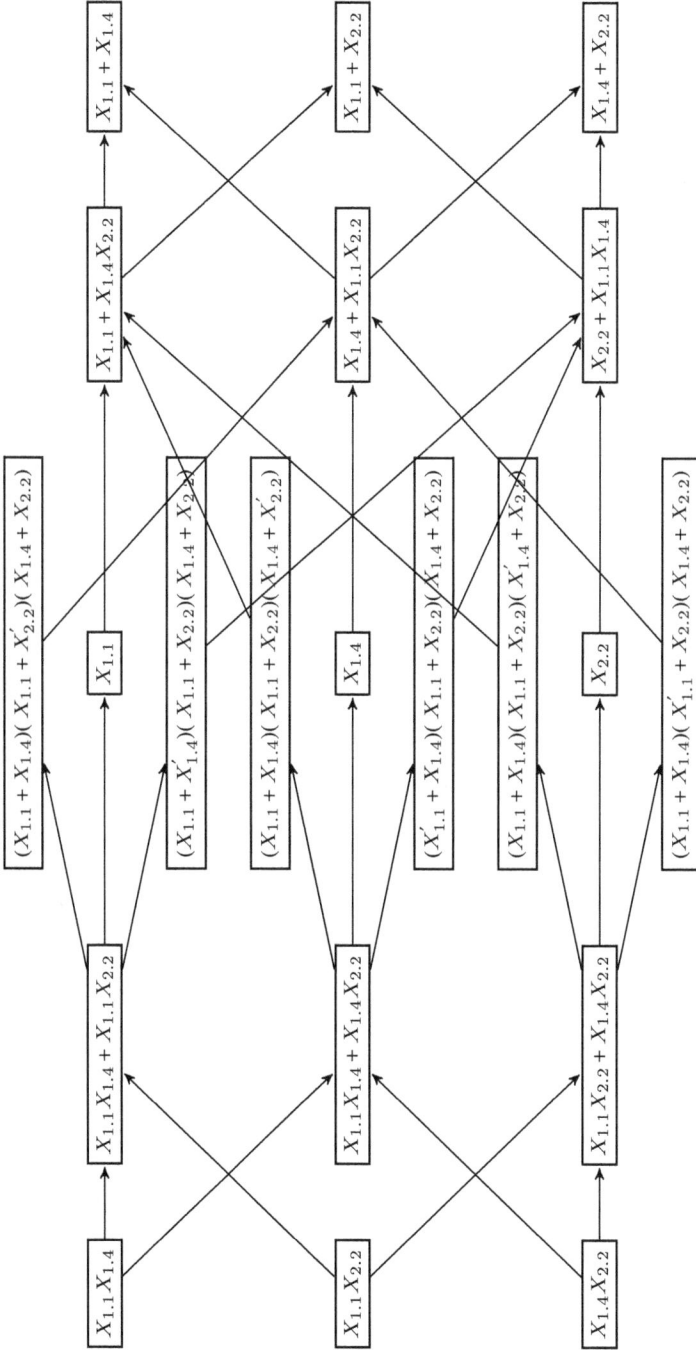

Figure 3.5. A selection out of the 5 central levels is shown with 21 nodes as a whole.

Since in the central row of the diagram we have a sequence of logical implications, due to transitivity we can shrink the sequence to the last term.

$$X_{1.1}X_{1.4} \qquad\qquad\qquad X_{1.1}+X_{1.4}$$

$$X_{1.1}+X_{1.4}X_{2.2}$$

$$X_{1.1}X_{2.2} \qquad\qquad\qquad X_{1.1}+X_{2.2} \qquad (3.44)$$

Thus, we have shown that $X_{1.1} + X_{1.4}X_{2.2} = (X_{1.1} + X_{1.4})(X_{1.1} + X_{2.2})$. To display the second distribution rule is more complicated. The crucial point is to get a final product that is equivalent to an initial sum. Since we have three different branches from each initial sum, we need to perform a product between a single node and the sum of the other two (which go to a single node, establishing in this way a closed circuit, although a single path be sufficient):

$$X_{1.1}$$

$$X_{1.1}X_{1.4} + X_{1.1}X_{2.2} \longrightarrow (X_{1.1} + X_{1.4})(X_{1.1} + X'_{2.2})(X_{1.4} + X_{2.2}) \longrightarrow X_{1.4} + X_{1.1}X_{2.2}$$

$$X_{1.4} + X_{2.2}$$

$$(X_{1.1} + X'_{1.4})(X_{1.1} + X_{2.2})(X_{1.4} + X_{2.2}) \longrightarrow X_{2.2} + X_{1.1}X_{1.4}$$

$$(3.45)$$

This allows us to reduce the whole to

$$X_{1.1}$$

$$X_{1.1}X_{1.4} + X_{1.1}X_{2.2}$$

$$X_{1.4} + X_{2.2} \qquad (3.46)$$

where it is important that there be no path from the first term through the second term to the last term. This diagram shows that $X_{1.1}X_{1.4} + X_{1.1}X_{2.2} = X_{1.1}(X_{1.4} + X_{2.2})$. Similarly, we have $X_{1.1}X_{1.4} + X_{1.4}X_{2.2} = X_{1.4}(X_{1.1} + X_{2.2})$ and $X_{1.1}X_{2.2} + X_{1.4}X_{2.2} = X_{2.2}(X_{1.1} + X_{1.4})$.

3.4. Other Cases

3.4.1. *Reduction of other algebras to complete algebras*

The reader may wonder why I have not considered algebras with IDs of say 3 digits or 5, so that it seems that the algebras considered in this book are a particular case. Of course, it is always possible to build algebras with the previous (as well as other) digits. But they cover only a part of the possible combinations, and in fact can be easily reduced to the cases considered here, as I can show by some examples. Let us consider 3 digits. In this case, we have the series

$$\sum_{x=0}^{x=3} \binom{3}{x} = 1 + 3 + 3 + 1. \tag{3.47}$$

This case can be led to $\mathcal{B}_2$ as displayed in Table 3.29. Note that the elements of $\mathcal{B}_2$ simply add the last digit (of course, I could have chosen the opposite convention by adding the first digit). Although the elements with three digits cover only a part of the IDs of $\mathcal{B}_2$, such an algebra is closed under complementation, reversal, and neg–reversal (one half of the IDs is the negative of the other part, while for reversal and neg–reversal some are obviously invariant under the one or the other operation). We can consider the relative logical expressions of $\mathcal{B}_2$ as valid also for the relative 3-digit ID. Of course, we cannot speak here of a set of SD variables whose combination determines all nodes. This is evident for this case (and all the odd ones) since the IDs have a odd number of digits and *a fortiori* there are no self–dual IDs.

Table 3.29. Maps of 3-digit
IDs into $\mathscr{B}_2$'s IDs.

$000 \mapsto 0000$	$111 \mapsto 1111$
$100 \mapsto 1000$	$011 \mapsto 0111$
$010 \mapsto 0100$	$101 \mapsto 1011$
$001 \mapsto 0010$	$110 \mapsto 1101$

Table 3.30. Maps of 5-digit IDs into $\mathscr{B}_3$'s IDs.

$00000 \mapsto 00000000$	$11111 \mapsto 11111111$
$10000 \mapsto 10000000$	$01111 \mapsto 01111111$
$01000 \mapsto 01000000$	$10111 \mapsto 10111111$
$00100 \mapsto 00100000$	$11011 \mapsto 11011111$
$00010 \mapsto 00010000$	$11101 \mapsto 11101111$
$00001 \mapsto 00001000$	$11110 \mapsto 11110111$
$11000 \mapsto 11000000$	$00111 \mapsto 00111111$
$10100 \mapsto 10100000$	$01011 \mapsto 01011111$
$10010 \mapsto 10010000$	$01101 \mapsto 01101111$
$10001 \mapsto 10001000$	$01110 \mapsto 01110111$
$01100 \mapsto 01100000$	$10011 \mapsto 10011111$
$01010 \mapsto 01010000$	$10101 \mapsto 10101111$
$01001 \mapsto 01001000$	$10110 \mapsto 10110111$
$00110 \mapsto 00110000$	$11001 \mapsto 11001111$
$00101 \mapsto 00101000$	$11010 \mapsto 11010111$
$00011 \mapsto 00011000$	$11100 \mapsto 11100111$

Similarly, we can reduce 5-digit IDs to $\mathscr{B}_3$. In such a case, we add 3 digits on the right. The series is given by

$$\sum_{x=0}^{x=5} \binom{5}{x} = 1 + 5 + 10 + 10 + 5 + 1. \tag{3.48}$$

Since we have here 6 levels (as before, the central one fails), we could map the central 10+10 IDs either to Levels 2/8 and 6/8 or to Levels 3/8 and 5/8. However, the first choice is convenient, as displayed in Table 3.30.

The last example is with 6-digit IDs. We have the following series

$$\sum_{x=0}^{x=6} \binom{6}{x} = 1 + 6 + 15 + 20 + 15 + 6 + 1, \tag{3.49}$$

Table 3.31. Maps of 6-digit IDs into $\mathscr{B}_3$'s IDs.

000000 $\mapsto$ 00000000	111111 $\mapsto$ 11111111
100000 $\mapsto$ 10000000	011111 $\mapsto$ 01111111
010000 $\mapsto$ 01000000	101111 $\mapsto$ 10111111
001000 $\mapsto$ 00100000	110111 $\mapsto$ 11011111
000100 $\mapsto$ 00010000	111011 $\mapsto$ 11101111
000010 $\mapsto$ 00001000	111101 $\mapsto$ 11110111
000001 $\mapsto$ 00000100	111110 $\mapsto$ 11111011
110000 $\mapsto$ 11000000	001111 $\mapsto$ 00111111
101000 $\mapsto$ 10100000	010111 $\mapsto$ 01011111
100100 $\mapsto$ 10010000	011011 $\mapsto$ 01101111
100010 $\mapsto$ 10001000	011101 $\mapsto$ 01110111
100001 $\mapsto$ 10000100	011110 $\mapsto$ 01111011
011000 $\mapsto$ 01100000	100111 $\mapsto$ 10011111
010100 $\mapsto$ 01010000	101011 $\mapsto$ 10101111
010010 $\mapsto$ 01001000	101101 $\mapsto$ 10110111
010001 $\mapsto$ 01000100	101110 $\mapsto$ 10111011
001100 $\mapsto$ 00110000	110011 $\mapsto$ 11001111
001010 $\mapsto$ 00101000	110101 $\mapsto$ 11010111
001001 $\mapsto$ 00100100	110110 $\mapsto$ 11011011
000110 $\mapsto$ 00011000	111001 $\mapsto$ 11100111
000101 $\mapsto$ 00010100	111010 $\mapsto$ 11101011
000011 $\mapsto$ 00001100	111100 $\mapsto$ 11110011
111000 $\mapsto$ 11100010	000111 $\mapsto$ 00011101
110100 $\mapsto$ 11010010	001011 $\mapsto$ 00101101
110010 $\mapsto$ 11001010	001101 $\mapsto$ 00110101
110001 $\mapsto$ 11000110	001110 $\mapsto$ 00111001
101100 $\mapsto$ 10110010	010011 $\mapsto$ 01001101
101010 $\mapsto$ 10101010	010101 $\mapsto$ 01010101
101001 $\mapsto$ 10100110	010110 $\mapsto$ 01011001
100110 $\mapsto$ 10011010	011001 $\mapsto$ 01100101
100101 $\mapsto$ 10010110	011010 $\mapsto$ 01101001
100011 $\mapsto$ 10001110	011100 $\mapsto$ 01110001

where it is noted that we have here a central distribution. Apparently, it is convenient to follow the same strategy across all cases and so to add the last 2 digits, as displayed in Table 3.31.

Note that the first, second, fifth, and sixth rows in the right column of the last group (mappings into Level 4/8) display on the right the SD variables $X_{2.4}$, $X_{2.3}$, $X_{2.2}$, $X_{2.1}$. However, a rapid calculation shows that we can never obtain the other 4 SD variables and the same occurs if we add the two supplementary digits on the left

Table 3.32.   Maps of 6-digit IDs into Level 4/8.

$111000 \mapsto 01110001$	$111000 \mapsto 11110000$
$110100 \mapsto 01101001$	$110100 \mapsto 11101000$
$110010 \mapsto 01100101$	$110010 \mapsto 11100100$
$110001 \mapsto 01100011$	$110001 \mapsto 11100010$
$101100 \mapsto 01011001$	$101100 \mapsto 11011000$
$101010 \mapsto 01010101$	$101010 \mapsto 11010100$
$101001 \mapsto 01010011$	$101001 \mapsto 11010010$
$100110 \mapsto 01001101$	$100110 \mapsto 11001100$
$100101 \mapsto 01001011$	$100101 \mapsto 11001010$
$100011 \mapsto 01000111$	$100011 \mapsto 11000110$
$000111 \mapsto 10001110$	$000111 \mapsto 00001111$
$001011 \mapsto 10010110$	$001011 \mapsto 00010111$
$001101 \mapsto 10011010$	$001101 \mapsto 00011011$
$001110 \mapsto 10011100$	$001110 \mapsto 00011101$
$010011 \mapsto 10100110$	$010011 \mapsto 00100111$
$010101 \mapsto 10101010$	$010101 \mapsto 00101011$
$010110 \mapsto 10101100$	$010110 \mapsto 00101101$
$011001 \mapsto 10110010$	$011001 \mapsto 00110011$
$011010 \mapsto 10110100$	$011010 \mapsto 00110101$
$011100 \mapsto 10111000$	$011100 \mapsto 00111001$

instead of on the right. Then, due to the symmetry of the even number of digits, we could choose to add the first and last digit. Since this means to have a 0 on the left and 1 on the right and vice versa, but with two different orders depending on which of the 10 IDs we apply to the first option (and on which 10 we apply the second one), let us explore these two possibilities, as displayed in Table 3.32 (where I only considered the mappings into Level 4/8). In the left column we have an option, in the right one the other. Note that on the first, second, sixth, and eighth rows on the left column in the first group we have $X_{2.4}, X_{2.3}, X_{2.2}, X_{2.1}$, while on the first, second, sixth, and eighth rows on the right column in the second group we have $X_{1.1}, X_{1.2}, X_{1.3}, X_{1.4}$. Of course, neither of these two sets of variables can constitute a CA for the reasons explained in Subsection 3.1.2.

3.4.2.   *A closer look to 6 digits*

It may be interesting to consider in more detail the case with 6 digits. As mentioned, this case constitutes a network of 64 nodes, and a quick look at the two previous tables allows us to individuate the SD variables of this network: 000111, 010101, 001011, 011001. For the sake of simplicity only, label these SD variables as X_1, X_2, X_3, X_4. The first two are clearly extensions of X_1, X_2 pertaining to $\mathcal{B}_2$, respectively, while the last two are more hybrid. Now, if we ask about their logical combinations, we get the results shown in Table 3.33.

For triplets I have used only some of the possibilities, but in the case of binary products it is evident that some simply do not exist. For instance, the product $X_1'X_2$ would produce the wrong ID 010000. The same occurs for other expressions. Note in particular the quaternary equivalences (that pertain to $\mathcal{B}_4$) and the sums of three quaternary products.

What we learn here is something important: all the algebras considered in this book (with nodes given by 2^m [with $m = 2^n$]) are closed under logical sum and product, complementation, reversal, and neg–reversal. Instead the algebras characterised by $m = 1, 2, 3, \ldots$ and that do not fall under the previous case (and can be therefore conceived of as extrapolations of the previous group of algebras) are closed only under complementation, reversal, and neg–reversal (although only two of these operations are sufficient). This shows that (1) the latter three operations are more fundamental, and (2) the algebras considered in this book are complete in a strong sense (what also allows the closeness under both logical operations and logical laws).

3.4.3. *Anomalous sets and subsets*

It seems that, with 6–digit ID, the SD variables are at the same time too few and too much: too few for building two distinct CAs and too much because 3 variables would largely suffice to cover the 64 nodes. As mentioned, the situation is worse with odd IDs, like the 3–digit or the 5–digit case, since, here, by definition we do not have SD variables at all (in fact, the middle level does not exist). Such a situation determines a fundamental ambiguity in expressing the nodes in logical terms: we not only do not know for certain how to deal with such

Table 3.33. The 64 nodes with 6 digits. Each row displays complementary nodes, while (when the ID is not symmetric or SD) each row displays the relative reverse and neg–reverse in the row that is complementary to the previous one at any level of the algebra.

000000	**0**	111111	**1**
000001	$X_1 X_2 X_3$	111110	$X_1' + X_2' + X_3'$
000010	$X_1 X_2' X_4'$	111101	$X_1' + X_2 + X_4$
000100	$X_1 X_3' X_4'$	111011	$X_1' + X_3 + X_4$
001000	$X_1' X_3 X_4$	110111	$X_1 + X_3' + X_4'$
010000	$X_1' X_2 X_4$	101111	$X_1 + X_2' + X_4'$
100000	$X_1' X_2' X_3'$	011111	$X_1 + X_2 + X_3$
000011	$X_1 X_3$	111100	$X_1' + X_3'$
000101	$X_1 X_2$	111010	$X_1' + X_2'$
000110	$X_1 X_4'$	111001	$X_1' + X_4$
001001	$X_3 X_4$	110110	$X_3' + X_4'$
001010	$X_2' X_3$	110101	$X_2 + X_3'$
001100	$X_1 \sim X_2 \sim X_3' \sim X_4'$	110011	$(X_1 \sim X_2 \sim X_3' \sim X_4')'$
010001	$X_2 X_4$	101110	$X_2' + X_4'$
010010	$X_1 \sim X_2' \sim X_3 \sim X_4'$	101101	$(X_1 \sim X_2' \sim X_3 \sim X_4')'$
100010	$X_2' X_4'$	011101	$X_2 + X_4$
100001	$X_1 \sim X_2 \sim X_3 \sim X_4$	011110	$(X_1 \sim X_2 \sim X_3 \sim X_4)'$
010100	$X_2 X_3'$	101011	$X_2' + X_3$
100100	$X_3' X_4'$	011011	$X_3 + X_4$
011000	$X_1' X_4$	100111	$X_1 + X_4'$
101000	$X_1' X_2'$	010111	$X_1 + X_2$
110000	$X_1' X_3'$	001111	$X_1 + X_3$
000111	X_1	111000	X_1'
001011	X_3	110100	X_3'
001101	$X_1' X_2' X_3 X_4 + X_1 X_2 X_3' X_4'$ $+ X_1 X_2 X_3 X_4$	110010	$X_1' X_2' X_3' X_4'$ $+ X_1' X_2 X_3' X_4 + X_1 X_2' X_3 X_4'$
001110	$X_1' X_2' X_3 X_4 + X_1 X_2 X_3' X_4'$ $+ X_1 X_2' X_3 X_4'$	110001	$X_1' X_2' X_3' X_4'$ $+ X_1' X_2 X_3' X_4 + X_1 X_2 X_3 X_4$
010110	$X_1' X_2 X_3' X_4 + X_1 X_2 X_3' X_4'$ $+ X_1 X_2' X_3 X_4'$	101001	$X_1' X_2' X_3' X_4'$ $+ X_1' X_2' X_3 X_4 + X_1 X_2 X_3 X_4$
011010	$X_1' X_2 X_3' X_4 + X_1' X_2' X_3 X_4$ $+ X_1 X_2' X_3 X_4'$	100101	$X_1' X_2' X_3' X_4'$ $+ X_1 X_2 X_3' X_4' + X_1 X_2 X_3 X_4$
011100	$X_1' X_2 X_3' X_4 + X_1' X_2' X_3 X_4$ $+ X_1 X_2 X_3' X_4'$	100011	$X_1' X_2' X_3' X_4'$ $+ X_1 X_2' X_3 X_4' + X_1 X_2 X_3 X_4$
101100	$X_1' X_2' X_3' X_4' + X_1' X_2' X_3 X_4$ $+ X_1 X_2 X_3' X_4'$	010011	$X_1' X_2 X_3' X_4$ $+ X_1 X_2' X_3 X_4' + X_1 X_2 X_3' X_4'$
010101	X_2	101010	X_2'
011001	X_4	100110	X_4'

a situation but, if we tried, we would get logical expressions with the same general form (e.g. binary products) but expressed with IDs that should pertain to different levels of the algebra. In such a case we would cut the connection of nodes–logical expressions.

What we can do is to make use of the IDs in order to establish some subsets. The reason is simple. Suppose that we like to build an algebra $\mathscr{B}(S)$. Even if we know that the "set" S (that should be considered rather as a class) has the form $X + Y$, whatever X and Y are, without knowing the ID we certainly cannot compute the subsets. For instance, in $\mathscr{B}_2$ we have 7 subsets (e.g. 0000, 0001, 0010, 0100, 0011, 0101, 0110), but in $\mathscr{B}_3$ they are obviously much more, and so on. Of course, we can say that for $n \geq 3$ we are involving other variables that were not considered in the previous definition. However, this case seems *ad hoc* unless we specify the dimension of the algebra.

Thus, let us make use of the ID and consider, for instance, the 5-digit case (Table 3.30). Then, we can build a sort of algebra $\mathscr{B}(S)$ by considering, for instance, the ID 01111. Of course, the subsets of this "set" S are 15: 00000, 00001, 00010, 00100, 01000, 00011, 00101, 01001, 00110, 01010, 01100, 00111, 01011, 01101, 01110. Now, the analysis of the previous algebras tells us that the first 4 IDs should be represented by some products (and the last 4 IDs by some sum). However, which kind of product (sum), binary or ternary? Moreover, until we determine at least the number (if not the definition) of the involved variables (or basic sets), we cannot know how many products and sums we have. Still worse is the situation with the expressions with two 1s: what should they represent? Impossible to make a guess. It is therefore suitable to consider networks characterised by odd IDs not as independent algebras but as subnetworks of other networks following the procedure introduced here. Of course, we could always choose to bypass the problem of the IDs and introduce arbitrary variables $X, Y, Z, \ldots$, and in this way build an algebra. This is the traditional view of logic as dealing with arbitrary "propositions". In the introductory chapter I have explained that the approach of the present book is different: either you rely on arbitrariness or you need to follow an approach in which IDs play a central role. Of course, similar considerations are valid also for other kinds of particular algebras, like those lacking the *supremum* or *infimum*, and so on. Nevertheless, such extrapolations are very helpful when we deal with problems of classification of data, as we shall see later.

Chapter 4

$\mathscr{B}_4$

4.1. Self–dual Variables

The complete analysis of $\mathscr{B}_3$ in Chapter 3 provides the basis for understanding the fundamental characters of Boolean algebra. Thus, we do not need to reiterate such a complete analysis for dimensions >3. Such an analysis would largely exceed the scope not only of a chapter but even of a whole book! Indeed, $\mathscr{B}_4$ is characterised by the following numbers: $m(4) = 16, s'(4) = 128, l(4) = 17$,

$$k(4) = 1 + 16 + 120 + 560 + 1820 + 4368 + 8008 + 11440 + 12870$$
$$+ 11440 + 8008 + 4368 + 1820 + 560 + 120 + 16 + 1$$
$$= 65,536, \tag{4.1}$$

and

$$r(4) = 2(16 + 240 + 1680 + 7280 + 21840 + 48048 + 80080 + 102960)$$
$$= 524,288 = 2^{19} = 2^{m(4)+3}. \tag{4.2}$$

The number of paths is exceedingly high:

$$p(4) = 16! = 20,922,789,888,000. \tag{4.3}$$

Let us first summarise the 16 areas of $\mathscr{B}_4$ as in Table 4.1. In Table 4.2 it is shown how the variables $X_{1.1.x}$'s $(1 \leq x \leq 16)$ can be built. Note

Table 4.1.   The 16 4D areas.

a	b	c	d	e	f	g	h	i	j	k	l	m	n	o	p
0	0	0	0	0	0	0	0	1	1	1	1	1	1	1	1
0	0	0	0	1	1	1	1	0	0	0	0	1	1	1	1
0	0	1	1	0	0	1	1	0	0	1	1	0	0	1	1
0	1	0	1	0	1	0	1	0	1	0	1	0	1	0	1

Table 4.2.   Building of $\mathscr{B}_4$'s variables. Note that the second column displays all the 16 nodes of $\mathscr{B}_2$ according to the series of non–negative integers and in agreement with Table 4.1 (while the 3rd column shows the same elements but in almost reversed order).

$X_{1.1.1}$	0000	0000	1111	1111
$X_{1.1.2}$	0000	0001	0111	1111
$X_{1.1.3}$	0000	0010	1011	1111
$X_{1.1.4}$	0000	0011	0011	1111
$X_{1.1.5}$	0000	0100	1101	1111
$X_{1.1.6}$	0000	0101	0101	1111
$X_{1.1.7}$	0000	0110	1001	1111
$X_{1.1.8}$	0000	0111	0001	1111
$X_{1.1.9}$	0000	1000	1110	1111
$X_{1.1.10}$	0000	1001	0110	1111
$X_{1.1.11}$	0000	1010	1010	1111
$X_{1.1.12}$	0000	1011	0010	1111
$X_{1.1.13}$	0000	1100	1100	1111
$X_{1.1.14}$	0000	1101	0100	1111
$X_{1.1.15}$	0000	1110	1000	1111
$X_{1.1.16}$	0000	1111	0000	1111

that the internal part of the ID (the central 8 numbers) reproduces all the 2D SD variables and their complements in the sequence 0–first (the natural number sequence). On the same outline we build all of the 128 SD variables.

First of all, let us cast the 128 SD variables (without considering the 128 complements) as in Table 4.3. As expected, the sum of the last (third index) for each doublet makes $17 = 16 + 1$ (Table 4.4). Note that we have in fact two series for each CA, A and B, that

Table 4.3. The sixteen main CAs for $\mathscr{B}_4$ (the left column is an expansion of CA1 of $\mathscr{B}_3$ and the right column is an expansion of CA2 of $\mathscr{B}_3$). Note that all ID columns in any set have four 1s and four 0s apart from the first and the last. The 16 CAs can be easily generated by focussing on the last eight values and first considering the last four values of each row. All the possible combinations for four truth–values are 16: 1 for four 1s, 1 for four 0s, 4 for three 1s and one 1 and vice versa, 6 for two 1s and two 0s. Thus, columns a and p have all 0s and 1s, respectively; columns b and o have the four 1s below and above the dashed line (in accordance with $\mathscr{B}_3$); columns c–d and m–n have two 1s above and below the central dashed line, respectively (here the elements come in duplets).

CA	abcd	efgh	ijkl	mnop		CA	abcd	efgh	ijkl	mnop	
	0000	0000	1111	1111	$X_{1.1.1}$		0001	0000	1111	0111	$X_{1.2.1}$
	0000	1111	0000	1111	$X_{1.1.16}$		0001	1111	0000	0111	$X_{1.2.16}$
	0011	0011	0011	0011	$X_{1.4.4}$		0010	0011	0011	1011	$X_{1.3.4}$
1.1	0011	1100	1100	0011	$X_{1.4.13}$	2.1	0010	1100	1100	1011	$X_{1.3.13}$
	0101	0101	0101	0101	$X_{2.2.6}$		0100	0101	0101	1101	$X_{2.1.6}$
	0101	1010	1010	0101	$X_{2.2.11}$		0100	1010	1010	1101	$X_{2.1.11}$
	0110	0110	1001	1001	$X_{2.3.7}$		0111	0110	1001	0001	$X_{2.4.7}$
	0110	1001	0110	1001	$X_{2.3.10}$		0111	1001	0110	0001	$X_{2.4.10}$
	0000	0001	0111	1111	$X_{1.1.2}$		0001	0001	0111	0111	$X_{1.2.2}$
	0000	1110	1000	1111	$X_{1.1.15}$		0001	1110	1000	0111	$X_{1.2.15}$
	0011	0010	1011	0011	$X_{1.4.3}$		0010	0010	1011	1011	$X_{1.3.3}$
1.2	0011	1101	0100	0011	$X_{1.4.14}$	2.2	0010	1101	0100	1011	$X_{1.3.14}$
	0101	0100	1101	0101	$X_{2.2.5}$		0100	0100	1101	1101	$X_{2.1.5}$
	0101	1011	0010	0101	$X_{2.2.12}$		0100	1011	0010	1101	$X_{2.1.12}$
	0110	0111	0001	1001	$X_{2.3.8}$		0111	0111	0001	0001	$X_{2.4.8}$
	0110	1000	1110	1001	$X_{2.3.9}$		0111	1000	1110	0001	$X_{2.4.9}$
	0000	0010	1011	1111	$X_{1.1.3}$		0001	0010	1011	0111	$X_{1.2.3}$
	0000	1101	0100	1111	$X_{1.1.14}$		0001	1101	0100	0111	$X_{1.2.14}$
	0011	0001	0111	0011	$X_{1.4.2}$		0010	0001	0111	1011	$X_{1.3.2}$
1.3	0011	1110	1000	0011	$X_{1.4.15}$	2.3	0010	1110	1000	1011	$X_{1.3.15}$
	0101	0111	0001	0101	$X_{2.2.8}$		0100	0111	0001	1101	$X_{2.1.8}$
	0101	1000	1110	0101	$X_{2.2.9}$		0100	1000	1110	1101	$X_{2.1.9}$
	0110	0100	1101	1001	$X_{2.3.5}$		0111	0100	1101	0001	$X_{2.4.5}$
	0110	1011	0010	1001	$X_{2.3.12}$		0111	1011	0010	0001	$X_{2.4.12}$
	0000	0011	0011	1111	$X_{1.1.4}$		0001	0011	0011	0111	$X_{1.2.4}$
	0000	1100	1100	1111	$X_{1.1.13}$		0001	1100	1100	0111	$X_{1.2.13}$
	0011	0000	1111	0011	$X_{1.4.1}$		0010	0000	1111	1011	$X_{1.3.1}$
1.4	0011	1111	0000	0011	$X_{1.4.16}$	2.4	0010	1111	0000	1011	$X_{1.3.16}$
	0101	0110	1001	0101	$X_{2.2.7}$		0100	0110	1001	1101	$X_{2.1.7}$
	0101	1001	0110	0101	$X_{2.2.10}$		0100	1001	0110	1101	$X_{2.1.10}$
	0110	0101	0101	1001	$X_{2.3.6}$		0111	0101	0101	0001	$X_{2.4.6}$
	0110	1010	1010	1001	$X_{2.3.11}$		0111	1010	1010	0001	$X_{2.4.11}$

(*Continued*)

Table 4.3.   (*Continued*)

	0000	0100	1101	1111	$X_{1.1.5}$		0001	0100	1101	0111	$X_{1.2.5}$
	0000	1011	0010	1111	$X_{1.1.12}$		0001	1011	0010	0111	$X_{1.2.12}$
	0011	0111	0001	0011	$X_{1.4.8}$		0010	0111	0001	1011	$X_{1.3.8}$
1.5	0011	1000	1110	0011	$X_{1.4.9}$	2.5	0010	1000	1110	1011	$X_{1.3.9}$
	0101	0001	0111	0101	$X_{2.2.2}$		0100	0001	0111	1101	$X_{2.1.2}$
	0101	1110	1000	0101	$X_{2.2.15}$		0100	1110	1000	1101	$X_{2.1.15}$
	0110	0010	1011	1001	$X_{2.3.3}$		0111	0010	1011	0001	$X_{2.4.3}$
	0110	1101	0100	1001	$X_{2.3.14}$		0111	1101	0100	0001	$X_{2.4.14}$
	0000	0101	0101	1111	$X_{1.1.6}$		0001	0101	0101	0111	$X_{1.2.6}$
	0000	1010	1010	1111	$X_{1.1.11}$		0001	1010	1010	0111	$X_{1.2.11}$
	0011	0110	1001	0011	$X_{1.4.7}$		0010	0110	1001	1011	$X_{1.3.7}$
1.6	0011	1001	0110	0011	$X_{1.4.10}$	2.6	0010	1001	0110	1011	$X_{1.3.10}$
	0101	0000	1111	0101	$X_{2.2.1}$		0100	0000	1111	1101	$X_{2.1.1}$
	0101	1111	0000	0101	$X_{2.2.16}$		0100	1111	0000	1101	$X_{2.1.16}$
	0110	0011	0011	1001	$X_{2.3.4}$		0111	0011	0011	0001	$X_{2.4.4}$
	0110	1100	1100	1001	$X_{2.3.13}$		0111	1100	1100	0001	$X_{2.4.13}$
	0000	0110	1001	1111	$X_{1.1.7}$		0001	0110	1001	0111	$X_{1.2.7}$
	0000	1001	0110	1111	$X_{1.1.10}$		0001	1001	0110	0111	$X_{1.2.10}$
	0011	0101	0101	0011	$X_{1.4.6}$		0010	0101	0101	1011	$X_{1.3.6}$
1.7	0011	1010	1010	0011	$X_{1.4.11}$	2.7	0010	1010	1010	1011	$X_{1.3.11}$
	0101	0011	0011	0101	$X_{2.2.4}$		0100	0011	0011	1101	$X_{2.1.4}$
	0101	1100	1100	0101	$X_{2.2.13}$		0100	1100	1100	1101	$X_{2.1.13}$
	0110	0000	1111	1001	$X_{2.3.1}$		0111	0000	1111	0001	$X_{2.4.1}$
	0110	1111	0000	1001	$X_{2.3.16}$		0111	1111	0000	0001	$X_{2.4.16}$
	0000	0111	0001	1111	$X_{1.1.8}$		0001	0111	0001	0111	$X_{1.2.8}$
	0000	1000	1110	1111	$X_{1.1.9}$		0001	1000	1110	0111	$X_{1.2.9}$
	0011	0100	1101	0011	$X_{1.4.5}$		0010	0100	1101	1011	$X_{1.3.5}$
1.8	0011	1011	0010	0011	$X_{1.4.12}$	2.8	0010	1011	0010	1011	$X_{1.3.12}$
	0101	0010	1011	0101	$X_{2.2.3}$		0100	0010	1011	1101	$X_{2.1.3}$
	0101	1101	0100	0101	$X_{2.2.14}$		0100	1101	0100	1101	$X_{2.1.14}$
	0110	0001	0111	1001	$X_{2.3.2}$		0111	0001	0111	0001	$X_{2.4.2}$
	0110	1110	1000	1001	$X_{2.3.15}$		0111	1110	1000	0001	$X_{2.4.15}$

alternate across the CAs according to *ABBA BAAB*:

$$A : 1 - 4 - 6 - 7 - 10 - 11 - 13 - 16,$$

$$B : 2 - 3 - 5 - 8 - 9 - 12 - 14 - 15, \tag{4.4}$$

where the 8 basic couples are also represented. Note that 1–4 and 2–3 constitute the two groups of last indexes in Table 3.3.

Table 4.4. Each row represents a CA. Note that every column has either ascending or descending order (with alternation of raising and lowering indices), with one interruption in the second column of couples and three interruptions in the last two. Dividing the 16 third indices in 4 equal segments, we have that when the couple $X_{1.1.x}$–$X_{1.4.y}$ occupies the first segment ($1 \leq x, y \leq 4$), the couple $X_{2.2.w}$–$X_{2.3.z}$ occupies the second segment ($5 \leq w, z \leq 8$), and vice versa (analogically for the complementary, i.e. third and fourth segments). Note that when there is a break, we have inversion of the third index in the first or second couple. Moreover, the sum of the third indices in both the first and second couple remain constant between breaks: for instance, $X_{1.1.x}$ and $X_{1.4.y}$ is 5 for the first 4 rows and $5 + 8 = 13$ in the last ones. Also the second couple is either 5 or 13.

$X_{1.1.1} - X_{1.1.16}$	$X_{1.4.4} - X_{1.4.13}$	$X_{2.2.6} - X_{2.2.11}$	$X_{2.3.7} - X_{2.3.10}$
$X_{1.1.2} - X_{1.1.15}$	$X_{1.4.3} - X_{1.4.14}$	$X_{2.2.5} - X_{2.2.12}$	$X_{2.3.8} - X_{2.3.9}$
$X_{1.1.3} - X_{1.1.14}$	$X_{1.4.2} - X_{1.4.15}$	$X_{2.2.8} - X_{2.2.9}$	$X_{2.3.5} - X_{2.3.12}$
$X_{1.1.4} - X_{1.1.13}$	$X_{1.4.1} - X_{1.4.16}$	$X_{2.2.7} - X_{2.2.10}$	$X_{2.3.6} - X_{2.3.11}$
$X_{1.1.5} - X_{1.1.12}$	$X_{1.4.8} - X_{1.4.9}$	$X_{2.2.2} - X_{2.2.15}$	$X_{2.3.3} - X_{2.3.14}$
$X_{1.1.6} - X_{1.1.11}$	$X_{1.4.7} - X_{1.4.10}$	$X_{2.2.1} - X_{2.2.16}$	$X_{2.3.4} - X_{2.3.13}$
$X_{1.1.7} - X_{1.1.10}$	$X_{1.4.6} - X_{1.4.11}$	$X_{2.2.4} - X_{2.2.13}$	$X_{2.3.1} - X_{2.3.16}$
$X_{1.1.8} - X_{1.1.9}$	$X_{1.4.5} - X_{1.4.12}$	$X_{2.2.3} - X_{2.2.14}$	$X_{2.3.2} - X_{2.3.15}$

On this basis, we can say that the rule that governs the generation of main sets is the following

$$f(n) = f(n-1)s'(n-1), \qquad (4.5)$$

where $f(n)$ gives the number of such sets. The first of such functions is, of course, $f(2) = 1$, so that we have $f(3) = 1 \times 2 = 2, f(4) = 2 \times 8 = 16, f(5) = 16 \times 128 = 2048$, and so on. Let us come back to Table 4.3. Note that there is a characteristic order in the way in which the main CAs are built. Let us consider all CAs having CA1 as mother code alphabet (exactly the same considerations are true for the other 8). Let us in particular focus on the internal part of the ID, that is on the 8 central numbers. It is evident that in CA1.1 this internal part corresponds to the external one: for example, $X_{1.1.1}$ has an internal part constituted by $X_{1.1}$ and the same for the two external halves (while $X_{1.1.16}$ shows its complement as internal part), so for $X_{1.4.4}$ with $X_{1.4}$, and so on. Instead CA1.2 shows internal parts all constituted by elements of CA2. This behaviour is true for all main CAs, with the sole difference of permutations. In fact, CA1.3, relative to CA1.2, shows exchange of (the internal part of) rows for each doublet, and so for CA1.4 relative to CA1.1. CA1.5, relative to CA1.2, shows exchange of the two halves (four internal parts with four internal parts), and the same for CA1.6 relative to CA1.1. CA1.7 shows both kinds of exchange relative to CA1.1 and so for CA1.8 relative to CA1.2. For this reason, the whole of the 16 expansion of a $\mathscr{B}_3$ variable performs a cycle across the eight CAs 1 and 2. These pathways are segmented in two-element sequences.

Such rotation for CA1 can be also expressed as exchange of columns:

$$1.1 \xmapsto{h \leftrightarrow i} 1.2 \xmapsto{g \leftrightarrow j, h \leftrightarrow i} 1.3 \xmapsto{h \leftrightarrow i} 1.4 \xmapsto{f \leftrightarrow k, g \leftrightarrow j, h \leftrightarrow i} 1.5$$

$$\xmapsto{h \leftrightarrow i} 1.6 \xmapsto{g \leftrightarrow j, h \leftrightarrow i} 1.7 \xmapsto{h \leftrightarrow i} 1.8.$$

Of course, each repetition of the exchange will bring back to the previous situation, so that at the end CA1.8 differs from CA1.1 for 3 exchanges: $f \leftrightarrow k, g \leftrightarrow j, h \leftrightarrow i$. The same scheme is followed by CA2.

4.2.   SCAs

Let us now consider alternative codifications. For each of the 16 main CAs displayed here there are 70 theoretical choices given by the binomial coefficient

$$\binom{8}{4} = 70. \tag{4.6}$$

However, not all combinations will work. In fact, we need to follow the fundamental combinatorial series determining the number of columns for each number of 1s per column (0, 1, 2, 3, 4, respectively)

$$\binom{4}{0}, \binom{4}{1}, \binom{4}{2}, \binom{4}{3}, \binom{4}{4}. \tag{4.7}$$

Recall that in Table 3.3 I subdivided the two main CAs into two subsets through a central horizontal line. In particular, any SCA in the framework of either CA1 or CA2 was generated by picking an element from one of these subsets and two from the other subset. This framework may seem forced, as we need to choose three elements out of four. However, it responds to a general way in which CAs are formed through combinatorics. To understand how this works, we need first to find a general way to denote the eight elements of each of the 16 main CAs. Since the 8 SD variables come in doublets, we can denote them by 1a, 2a, 3a, 4a, and 1b, 2b, 3b, 4b (where the numbers denote second index), according to whether we are picking out the first or the second element of each doublet. Then, we distribute the eight elements of each of 16 main CAs into two as well as four boxes, as displayed in Tables 4.5–4.6. Now, the resulting combinations that fulfil the combinatorial requirements are 56:

- We pick a single element of the left column of Table 4.5 and 3 elements of the right column (not located on the same row of the left one) and vice versa. Using a semicolon for separating the first 4 SD variables from the second ones, we have either a,b; b,b or a,b; a,a and the two resulting from exchange of left and right side. This gives $2 \times 4 = 8$ SCAs.
- We pick a whole column of Table 4.6 and combine it with any two other elements not in the same column: (i) a-b,a; a and a-b; a,a with exchange of left and right: $2 \times (4 + 2) = 12$ combinations, and similarly for (ii) a-b,b; b and a-b; b,b. Instead of, (iii) a-b,a; b

Table 4.5.   First kind of combinations for $\mathscr{B}_4$.

1a	1b
2a	2b
3a	3b
4a	4b

Table 4.6.   Second kind of combinations for $\mathscr{B}_4$.

1a	2a	3a	4a
1b	2b	3b	4b

and a-b,b; a give $4 \times 4 = 16$ combinations, while (iv) a-b; a,b give $2 \times 4 = 8$ combinations. That is 48 SCAs as a whole.

At the opposite, the following 14 combinations do not work:

- When we have either all as or all bs: 2 cases.
- When we have all elements in the first or second set of 4 SD variables: 2 cases.
- When we have two doublets, i.e. two columns of Table 4.6 not in the same half: 2 cases.
- When we have two as and two bs that are not both doublets and do not occur in the same half: 6 cases.

Thus, as said, we get 56 SCAs as whole for each of the 16 main CAs (896 as a whole).[1] Note that for the 3D algebra there is no restriction on the possible combinations within a main CA. In fact, the above restrictions tell us that we cannot have all as or all bs and that we cannot have all elements in the first half (here 4 rows) or in the second half (again 4 rows) of a CA (it is evident that in the last two cases each column is doubled), and it is evident that in the 3D case these requirements are automatically satisfied for whatever choice. Of course, with growing dimensions we get more severe restrictions.

Although we have 56 alternatives, in a first approximation it suffices to consider the first 8 SCAs for covering all basic SD variables. In this case, for CA1.1 we have:

$$\{X_{1.1.1}, X_{1.4.4}, X_{2.2.6}, X_{2.3.10}\}, \quad \{X_{1.1.16}, X_{1.4.13}, X_{2.2.11}, X_{2.3.7}\},$$
$$(4.8a)$$

$$\{X_{1.4.4}, X_{1.1.1}, X_{2.3.7}, X_{2.2.11}\}, \quad \{X_{1.4.13}, X_{1.1.16}, X_{2.3.10}, X_{2.2.6}\},$$
$$(4.8b)$$

$$\{X_{2.2.6}, X_{2.3.7}, X_{1.1.1}, X_{1.4.13}\}, \quad \{X_{2.2.11}, X_{2.3.10}, X_{1.1.16}, X_{1.4.4}\},$$
$$(4.8c)$$

$$\{X_{2.3.7}, X_{2.2.6}, X_{1.4.4}, X_{1.1.16}\}, \quad \{X_{2.3.10}, X_{2.2.11}, X_{1.4.13}, X_{1.1.1}\}.$$
$$(4.8d)$$

---

[1] In [AULETTA 2015] I erroneously computed 64 CAs for each of the 16 CAs.

Figure 4.1.   The 16 boxes go from the left to the right according to the columns a,b,c ... p (Table 4.1). The white box means no 1 present, while the black box four 1s are present. The three grey boxes represent one, two, and three 1s.

It may be noted that the left column displays a-a-a-b ordering, while the right one b-b-b-a ordering. Moreover, in each column the first element in each row is determined by the order of the first row, apart from the last row, where there is exchange left–right. Of course, a 4D representation by means of Venn diagram is not possible. However, we can provide for a kind of representation, as shown in Figure 4.1 for the SCA $\{X_{1.1.1}, X_{1.4.4}, X_{2.2.6}, X_{2.3.10}\}$.

With such permutations, we get the basic quadruplets in Table 4.7.

It is instructive now to compare CA1.1 with other CAs. I consider first an example of each CA1.x $(1 \leq x \leq 8)$, starting with CA1.2 and making a similar choice as before. The advantage is that in this way we cover all 128 SD variables and let each one be the first element of a code alphabet. Thus, for CA1.2,

$$\{X_{1.1.2}, X_{1.4.3}, X_{2.2.5}, X_{2.3.9}\}, \quad \{X_{1.1.15}, X_{1.4.14}, X_{2.2.12}, X_{2.3.8}\},$$
$$(4.9a)$$

$$\{X_{1.4.3}, X_{1.1.2}, X_{2.3.8}, X_{2.2.12}\}, \quad \{X_{1.4.14}, X_{1.1.15}, X_{2.3.9}, X_{2.2.5}\},$$
$$(4.9b)$$

$$\{X_{2.2.5}, X_{2.3.8}, X_{1.1.2}, X_{1.4.14}\}, \quad \{X_{2.2.12}, X_{2.3.9}, X_{1.1.15}, X_{1.4.3}\},$$
$$(4.9c)$$

$$\{X_{2.3.8}, X_{2.2.5}, X_{1.4.3}, X_{1.1.15}\}, \quad \{X_{2.3.9}, X_{2.2.12}, X_{1.4.14}, X_{1.1.2}\}.$$
$$(4.9d)$$

This graphic gives the combinations displayed in Table 4.8. Note the characteristic exchange of the two central rows.

Table 4.7.   Quaternary products codified by 8 SCAs $\in$ CA1.1.

1000 0000 0000 0000	$\begin{array}{l}X'_{1.1.1}X'_{1.4.4}X'_{2.2.6}X'_{2.3.10}\\X'_{1.4.4}X'_{1.1.1}X'_{2.3.7}X'_{2.2.11}\\X'_{2.2.6}X'_{2.3.7}X'_{1.1.1}X'_{1.4.13}\\X'_{2.3.7}X'_{2.2.6}X'_{1.4.4}X'_{1.1.16}\end{array}$	$\begin{array}{l}X'_{1.1.16}X'_{1.4.13}X'_{2.2.11}X'_{2.3.7}\\X'_{1.4.13}X'_{1.1.16}X'_{2.3.10}X'_{2.2.6}\\X'_{2.2.11}X'_{2.3.10}X'_{1.1.16}X'_{1.4.4}\\X'_{2.3.10}X'_{2.2.11}X'_{1.4.13}X'_{1.1.1}\end{array}$
0100 0000 0000 0000	$\begin{array}{l}X'_{1.1.1}X'_{1.4.4}X_{2.2.6}X_{2.3.10}\\X'_{1.4.4}X'_{1.1.1}X_{2.3.7}X_{2.2.11}\\X_{2.2.6}X_{2.3.7}X'_{1.1.1}X'_{1.4.13}\\X_{2.3.7}X_{2.2.6}X'_{1.4.4}X'_{1.1.16}\end{array}$	$\begin{array}{l}X'_{1.1.16}X'_{1.4.13}X_{2.2.11}X_{2.3.7}\\X'_{1.4.13}X'_{1.1.16}X_{2.3.10}X_{2.2.6}\\X_{2.2.11}X_{2.3.10}X'_{1.1.16}X'_{1.4.4}\\X_{2.3.10}X_{2.2.11}X'_{1.4.13}X'_{1.1.1}\end{array}$
0010 0000 0000 0000	$\begin{array}{l}X'_{1.1.1}X_{1.4.4}X'_{2.2.6}X_{2.3.10}\\X_{1.4.4}X'_{1.1.1}X_{2.3.7}X'_{2.2.11}\\X'_{2.2.6}X_{2.3.7}X'_{1.1.1}X_{1.4.13}\\X_{2.3.7}X'_{2.2.6}X_{1.4.4}X'_{1.1.16}\end{array}$	$\begin{array}{l}X'_{1.1.16}X_{1.4.13}X'_{2.2.11}X_{2.3.7}\\X_{1.4.13}X'_{1.1.16}X_{2.3.10}X'_{2.2.6}\\X'_{2.2.11}X_{2.3.10}X'_{1.1.16}X_{1.4.4}\\X_{2.3.10}X'_{2.2.11}X_{1.4.13}X'_{1.1.1}\end{array}$
0001 0000 0000 0000	$\begin{array}{l}X'_{1.1.1}X_{1.4.4}X_{2.2.6}X'_{2.3.10}\\X_{1.4.4}X'_{1.1.1}X'_{2.3.7}X_{2.2.11}\\X_{2.2.6}X'_{2.3.7}X'_{1.1.1}X_{1.4.13}\\X'_{2.3.7}X_{2.2.6}X_{1.4.4}X'_{1.1.16}\end{array}$	$\begin{array}{l}X'_{1.1.16}X_{1.4.13}X_{2.2.11}X'_{2.3.7}\\X_{1.4.13}X'_{1.1.16}X'_{2.3.10}X_{2.2.6}\\X_{2.2.11}X'_{2.3.10}X'_{1.1.16}X_{1.4.4}\\X'_{2.3.10}X_{2.2.11}X_{1.4.13}X'_{1.1.1}\end{array}$
0000 1000 0000 0000	$\begin{array}{l}X'_{1.1.1}X'_{1.4.4}X_{2.2.6}X'_{2.3.10}\\X'_{1.4.4}X'_{1.1.1}X'_{2.3.7}X_{2.2.11}\\X_{2.2.6}X'_{2.3.7}X'_{1.1.1}X_{1.4.13}\\X'_{2.3.7}X_{2.2.6}X'_{1.4.4}X_{1.1.16}\end{array}$	$\begin{array}{l}X_{1.1.16}X_{1.4.13}X_{2.2.11}X'_{2.3.7}\\X_{1.4.13}X_{1.1.16}X_{2.3.10}X'_{2.2.6}\\X_{2.2.11}X_{2.3.10}X_{1.1.16}X'_{1.4.4}\\X_{2.3.10}X_{2.2.11}X_{1.4.13}X'_{1.1.1}\end{array}$
0000 0100 0000 0000	$\begin{array}{l}X'_{1.1.1}X'_{1.4.4}X_{2.2.6}X'_{2.3.10}\\X'_{1.4.4}X'_{1.1.1}X_{2.3.7}X'_{2.2.11}\\X_{2.2.6}X_{2.3.7}X'_{1.1.1}X_{1.4.13}\\X_{2.3.7}X_{2.2.6}X'_{1.4.4}X_{1.1.16}\end{array}$	$\begin{array}{l}X_{1.1.16}X_{1.4.13}X'_{2.2.11}X_{2.3.7}\\X_{1.4.13}X_{1.1.16}X'_{2.3.10}X_{2.2.6}\\X'_{2.2.11}X'_{2.3.10}X_{1.1.16}X'_{1.4.4}\\X'_{2.3.10}X'_{2.2.11}X_{1.4.13}X'_{1.1.1}\end{array}$
0000 0010 0000 0000	$\begin{array}{l}X'_{1.1.1}X_{1.4.4}X'_{2.2.6}X'_{2.3.10}\\X_{1.4.4}X'_{1.1.1}X_{2.3.7}X_{2.2.11}\\X'_{2.2.6}X_{2.3.7}X'_{1.1.1}X'_{1.4.13}\\X_{2.3.7}X'_{2.2.6}X_{1.4.4}X_{1.1.16}\end{array}$	$\begin{array}{l}X_{1.1.16}X'_{1.4.13}X'_{2.2.11}X'_{2.3.7}\\X'_{1.4.13}X_{1.1.16}X'_{2.3.10}X'_{2.2.6}\\X_{2.2.11}X'_{2.3.10}X_{1.1.16}X_{1.4.4}\\X'_{2.3.10}X_{2.2.11}X'_{1.4.13}X'_{1.1.1}\end{array}$
0000 0001 0000 0000	$\begin{array}{l}X'_{1.1.1}X_{1.4.4}X_{2.2.6}X_{2.3.10}\\X_{1.4.4}X'_{1.1.1}X'_{2.3.7}X'_{2.2.11}\\X_{2.2.6}X'_{2.3.7}X'_{1.1.1}X'_{1.4.13}\\X'_{2.3.7}X_{2.2.6}X_{1.4.4}X_{1.1.16}\end{array}$	$\begin{array}{l}X_{1.1.16}X'_{1.4.13}X'_{2.2.11}X'_{2.3.7}\\X'_{1.4.13}X_{1.1.16}X_{2.3.10}X_{2.2.6}\\X_{2.2.11}X_{2.3.10}X_{1.1.16}X_{1.4.4}\\X_{2.3.10}X_{2.2.11}X'_{1.4.13}X'_{1.1.1}\end{array}$
0000 0000 1000 0000	$\begin{array}{l}X_{1.1.1}X'_{1.4.4}X'_{2.2.6}X'_{2.3.10}\\X'_{1.4.4}X_{1.1.1}X_{2.3.7}X_{2.2.11}\\X'_{2.2.6}X_{2.3.7}X_{1.1.1}X_{1.4.13}\\X_{2.3.7}X'_{2.2.6}X'_{1.4.4}X'_{1.1.16}\end{array}$	$\begin{array}{l}X'_{1.1.16}X_{1.4.13}X_{2.2.11}X_{2.3.7}\\X_{1.4.13}X'_{1.1.16}X'_{2.3.10}X'_{2.2.6}\\X_{2.2.11}X'_{2.3.10}X'_{1.1.16}X'_{1.4.4}\\X'_{2.3.10}X_{2.2.11}X_{1.4.13}X_{1.1.1}\end{array}$
0000 0000 0100 0000	$\begin{array}{l}X_{1.1.1}X'_{1.4.4}X_{2.2.6}X'_{2.3.10}\\X'_{1.4.4}X_{1.1.1}X'_{2.3.7}X'_{2.2.11}\\X_{2.2.6}X'_{2.3.7}X_{1.1.1}X_{1.4.13}\\X'_{2.3.7}X_{2.2.6}X'_{1.4.4}X'_{1.1.16}\end{array}$	$\begin{array}{l}X'_{1.1.16}X_{1.4.13}X_{2.2.11}X'_{2.3.7}\\X_{1.4.13}X'_{1.1.16}X_{2.3.10}X_{2.2.6}\\X'_{2.2.11}X_{2.3.10}X'_{1.1.16}X'_{1.4.4}\\X_{2.3.10}X'_{2.2.11}X_{1.4.13}X_{1.1.1}\end{array}$

Table 4.7. *(Continued)*

0000 0000 0010 0000	$X_{1.1.1}X_{1.4.4}X'_{2.2.6}X_{2.3.10}$ $X'_{1.4.4}X_{1.1.1}X'_{2.3.7}X_{2.2.11}$ $X'_{2.2.6}X'_{2.3.7}X_{1.1.1}X'_{1.4.13}$ $X'_{2.3.7}X'_{2.2.6}X_{1.4.4}X'_{1.1.16}$	$X'_{1.1.16}X'_{1.4.13}X_{2.2.11}X'_{2.3.7}$ $X'_{1.4.13}X'_{1.1.16}X_{2.3.10}X'_{2.2.6}$ $X_{2.2.11}X_{2.3.10}X'_{1.1.16}X_{1.4.4}$ $X_{2.3.10}X_{2.2.11}X'_{1.4.13}X_{1.1.1}$
0000 0000 0001 0000	$X_{1.1.1}X_{1.4.4}X_{2.2.6}X'_{2.3.10}$ $X_{1.4.4}X_{1.1.1}X_{2.3.7}X'_{2.2.11}$ $X_{2.2.6}X_{2.3.7}X_{1.1.1}X'_{1.4.13}$ $X_{2.3.7}X_{2.2.6}X_{1.4.4}X'_{1.1.16}$	$X'_{1.1.16}X'_{1.4.13}X'_{2.2.11}X_{2.3.7}$ $X'_{1.4.13}X'_{1.1.16}X'_{2.3.10}X_{2.2.6}$ $X'_{2.2.11}X'_{2.3.10}X'_{1.1.16}X_{1.4.4}$ $X'_{2.3.10}X'_{2.2.11}X'_{1.4.13}X_{1.1.1}$
0000 0000 0000 1000	$X_{1.1.1}X'_{1.4.4}X'_{2.2.6}X_{2.3.10}$ $X'_{1.4.4}X_{1.1.1}X_{2.3.7}X'_{2.2.11}$ $X'_{2.2.6}X_{2.3.7}X_{1.1.1}X'_{1.4.13}$ $X_{2.3.7}X'_{2.2.6}X'_{1.4.4}X_{1.1.16}$	$X_{1.1.16}X'_{1.4.13}X'_{2.2.11}X_{2.3.7}$ $X'_{1.4.13}X_{1.1.16}X_{2.3.10}X'_{2.2.6}$ $X'_{2.2.11}X_{2.3.10}X_{1.1.16}X'_{1.4.4}$ $X_{2.3.10}X'_{2.2.11}X'_{1.4.13}X_{1.1.1}$
0000 0000 0000 0100	$X_{1.1.1}X'_{1.4.4}X_{2.2.6}X'_{2.3.10}$ $X'_{1.4.4}X_{1.1.1}X'_{2.3.7}X_{2.2.11}$ $X_{2.2.6}X'_{2.3.7}X_{1.1.1}X'_{1.4.13}$ $X'_{2.3.7}X_{2.2.6}X'_{1.4.4}X_{1.1.16}$	$X_{1.1.16}X'_{1.4.13}X_{2.2.11}X'_{2.3.7}$ $X'_{1.4.13}X_{1.1.16}X'_{2.3.10}X_{2.2.6}$ $X_{2.2.11}X'_{2.3.10}X_{1.1.16}X'_{1.4.4}$ $X'_{2.3.10}X_{2.2.11}X'_{1.4.13}X_{1.1.1}$
0000 0000 0000 0010	$X_{1.1.1}X_{1.4.4}X'_{2.2.6}X'_{2.3.10}$ $X_{1.4.4}X_{1.1.1}X'_{2.3.7}X'_{2.2.11}$ $X'_{2.2.6}X'_{2.3.7}X_{1.1.1}X_{1.4.13}$ $X'_{2.3.7}X'_{2.2.6}X_{1.4.4}X_{1.1.16}$	$X_{1.1.16}X_{1.4.13}X'_{2.2.11}X'_{2.3.7}$ $X_{1.4.13}X_{1.1.16}X'_{2.3.10}X'_{2.2.6}$ $X'_{2.2.11}X'_{2.3.10}X_{1.1.16}X_{1.4.4}$ $X'_{2.3.10}X'_{2.2.11}X_{1.4.13}X_{1.1.1}$
0000 0000 0000 0001	$X_{1.1.1}X_{1.4.4}X_{2.2.6}X_{2.3.10}$ $X_{1.4.4}X_{1.1.1}X_{2.3.7}X_{2.2.11}$ $X_{2.2.6}X_{2.3.7}X_{1.1.1}X_{1.4.13}$ $X_{2.3.7}X_{2.2.6}X_{1.4.4}X_{1.1.16}$	$X_{1.1.16}X_{1.4.13}X_{2.2.11}X_{2.3.7}$ $X_{1.4.13}X_{1.1.16}X_{2.3.10}X_{2.2.6}$ $X_{2.2.11}X_{2.3.10}X_{1.1.16}X_{1.4.4}$ $X_{2.3.10}X_{2.2.11}X_{1.4.13}X_{1.1.1}$

Let us now consider the similar 8 SCAs for CA1.3:

$$\{X_{1.1.3}, X_{1.4.2}, X_{2.2.8}, X_{2.3.12}\}, \quad \{X_{1.1.14}, X_{1.4.15}, X_{2.2.9}, X_{2.3.5}\},$$
$$(4.10a)$$

$$\{X_{1.4.2}, X_{1.1.3}, X_{2.3.5}, X_{2.2.9}\}, \quad \{X_{1.4.15}, X_{1.1.14}, X_{2.3.12}, X_{2.2.8}\},$$
$$(4.10b)$$

$$\{X_{2.2.8}, X_{2.3.5}, X_{1.1.3}, X_{1.4.15}\}, \quad \{X_{2.2.9}, X_{2.3.12}, X_{1.1.14}, X_{1.4.2}\},$$
$$(4.10c)$$

$$\{X_{2.3.5}, X_{2.2.8}, X_{1.4.2}, X_{1.1.14}\}, \quad \{X_{2.3.12}, X_{2.2.9}, X_{1.4.15}, X_{1.1.3}\}.$$
$$(4.10d)$$

Table 4.8. Quaternary products codified by 8 SCAs $\in$ CA1.2.

Code	Product 1	Product 2
1000 0000 0000 0000	$X'_{1.1.2}X'_{1.4.3}X'_{2.2.5}X'_{2.3.9}$ $X'_{1.4.3}X'_{1.1.2}X'_{2.3.8}X'_{2.2.12}$ $X'_{2.2.5}X'_{2.3.8}X'_{1.1.2}X'_{1.4.14}$ $X'_{2.3.8}X'_{2.2.5}X'_{1.4.3}X'_{1.1.15}$	$X'_{1.1.15}X'_{1.4.14}X'_{2.2.12}X'_{2.3.8}$ $X'_{1.4.14}X'_{1.1.15}X'_{2.3.9}X'_{2.2.5}$ $X'_{2.2.12}X'_{2.3.9}X'_{1.1.15}X'_{1.4.3}$ $X'_{2.3.9}X'_{2.2.12}X'_{1.4.14}X'_{1.1.2}$
0100 0000 0000 0000	$X'_{1.1.2}X'_{1.4.3}X_{2.2.5}X_{2.3.9}$ $X'_{1.4.3}X'_{1.1.2}X_{2.3.8}X_{2.2.12}$ $X_{2.2.5}X_{2.3.8}X'_{1.1.2}X'_{1.4.14}$ $X_{2.3.8}X_{2.2.5}X'_{1.4.3}X'_{1.1.15}$	$X'_{1.1.15}X'_{1.4.14}X_{2.2.12}X_{2.3.8}$ $X'_{1.4.14}X'_{1.1.15}X_{2.3.9}X_{2.2.5}$ $X_{2.2.12}X_{2.3.9}X'_{1.1.15}X'_{1.4.3}$ $X_{2.3.9}X_{2.2.12}X'_{1.4.14}X'_{1.1.2}$
0010 0000 0000 0000	$X'_{1.1.2}X_{1.4.3}X'_{2.2.5}X_{2.3.9}$ $X_{1.4.3}X'_{1.1.2}X_{2.3.8}X'_{2.2.12}$ $X'_{2.2.5}X_{2.3.8}X'_{1.1.2}X_{1.4.14}$ $X_{2.3.8}X'_{2.2.5}X_{1.4.3}X'_{1.1.15}$	$X'_{1.1.15}X_{1.4.14}X'_{2.2.12}X_{2.3.8}$ $X_{1.4.14}X'_{1.1.15}X_{2.3.9}X'_{2.2.5}$ $X'_{2.2.12}X_{2.3.9}X'_{1.1.15}X_{1.4.3}$ $X_{2.3.9}X'_{2.2.12}X_{1.4.14}X'_{1.1.2}$
0001 0000 0000 0000	$X'_{1.1.2}X_{1.4.3}X_{2.2.5}X'_{2.3.9}$ $X_{1.4.3}X'_{1.1.2}X'_{2.3.8}X_{2.2.12}$ $X_{2.2.5}X'_{2.3.8}X'_{1.1.2}X_{1.4.14}$ $X'_{2.3.8}X_{2.2.5}X_{1.4.3}X'_{1.1.15}$	$X'_{1.1.15}X_{1.4.14}X_{2.2.12}X'_{2.3.8}$ $X_{1.4.14}X'_{1.1.15}X'_{2.3.9}X_{2.2.5}$ $X_{2.2.12}X'_{2.3.9}X'_{1.1.15}X_{1.4.3}$ $X'_{2.3.9}X_{2.2.12}X_{1.4.14}X'_{1.1.2}$
0000 1000 0000 0000	$X'_{1.1.2}X'_{1.4.3}X_{2.2.5}X'_{2.3.9}$ $X'_{1.4.3}X'_{1.1.2}X'_{2.3.8}X_{2.2.12}$ $X'_{2.2.5}X'_{2.3.8}X'_{1.1.2}X_{1.4.14}$ $X'_{2.3.8}X'_{2.2.5}X'_{1.4.3}X_{1.1.15}$	$X_{1.1.15}X_{1.4.14}X_{2.2.12}X'_{2.3.8}$ $X_{1.4.14}X_{1.1.15}X_{2.3.9}X'_{2.2.5}$ $X_{2.2.12}X_{2.3.9}X_{1.1.15}X'_{1.4.3}$ $X_{2.3.9}X_{2.2.12}X_{1.4.14}X'_{1.1.2}$
0000 0100 0000 0000	$X'_{1.1.2}X'_{1.4.3}X_{2.2.5}X'_{2.3.9}$ $X'_{1.4.3}X'_{1.1.2}X_{2.3.8}X'_{2.2.12}$ $X_{2.2.5}X_{2.3.8}X'_{1.1.2}X_{1.4.14}$ $X_{2.3.8}X_{2.2.5}X'_{1.4.3}X_{1.1.15}$	$X_{1.1.15}X_{1.4.14}X'_{2.2.12}X_{2.3.8}$ $X_{1.4.14}X_{1.1.15}X'_{2.3.9}X_{2.2.5}$ $X'_{2.2.12}X'_{2.3.9}X_{1.1.15}X'_{1.4.3}$ $X'_{2.3.9}X'_{2.2.12}X_{1.4.14}X'_{1.1.2}$
0000 0010 0000 0000	$X'_{1.1.2}X_{1.4.3}X'_{2.2.5}X_{2.3.9}$ $X_{1.4.3}X'_{1.1.2}X_{2.3.8}X_{2.2.12}$ $X'_{2.2.5}X_{2.3.8}X'_{1.1.2}X'_{1.4.14}$ $X_{2.3.8}X'_{2.2.5}X_{1.4.3}X_{1.1.15}$	$X_{1.1.15}X'_{1.4.14}X_{2.2.12}X_{2.3.8}$ $X'_{1.4.14}X_{1.1.15}X'_{2.3.9}X'_{2.2.5}$ $X_{2.2.12}X'_{2.3.9}X_{1.1.15}X_{1.4.3}$ $X'_{2.3.9}X_{2.2.12}X'_{1.4.14}X_{1.1.2}$
0000 0001 0000 0000	$X_{1.1.2}X'_{1.4.3}X'_{2.2.5}X'_{2.3.9}$ $X'_{1.4.3}X_{1.1.2}X_{2.3.8}X_{2.2.12}$ $X'_{2.2.5}X_{2.3.8}X_{1.1.2}X_{1.4.14}$ $X_{2.3.8}X'_{2.2.5}X'_{1.4.3}X_{1.1.15}$	$X'_{1.1.15}X_{1.4.14}X_{2.2.12}X_{2.3.8}$ $X_{1.4.14}X'_{1.1.15}X'_{2.3.9}X_{2.2.5}$ $X_{2.2.12}X'_{2.3.9}X'_{1.1.15}X'_{1.4.3}$ $X_{2.3.9}X_{2.2.12}X_{1.4.14}X_{1.1.2}$
0000 0000 1000 0000	$X'_{1.1.2}X_{1.4.3}X_{2.2.5}X_{2.3.9}$ $X_{1.4.3}X'_{1.1.2}X'_{2.3.8}X'_{2.2.12}$ $X_{2.2.5}X'_{2.3.8}X'_{1.1.2}X'_{1.4.14}$ $X_{2.3.8}X_{2.2.5}X_{1.4.3}X_{1.1.15}$	$X_{1.1.15}X'_{1.4.14}X'_{2.2.12}X'_{2.3.8}$ $X'_{1.4.14}X_{1.1.15}X_{2.3.9}X_{2.2.5}$ $X'_{2.2.12}X_{2.3.9}X_{1.1.15}X_{1.4.3}$ $X_{2.3.9}X'_{2.2.12}X'_{1.4.14}X'_{1.1.2}$
0000 0000 0100 0000	$X_{1.1.2}X'_{1.4.3}X_{2.2.5}X_{2.3.9}$ $X'_{1.4.3}X_{1.1.2}X'_{2.3.8}X'_{2.2.12}$ $X_{2.2.5}X'_{2.3.8}X_{1.1.2}X'_{1.4.14}$ $X'_{2.3.8}X_{2.2.5}X'_{1.4.3}X_{1.1.15}$	$X'_{1.1.15}X_{1.4.14}X'_{2.2.12}X'_{2.3.8}$ $X_{1.4.14}X'_{1.1.15}X_{2.3.9}X_{2.2.5}$ $X'_{2.2.12}X_{2.3.9}X'_{1.1.15}X_{1.4.3}$ $X_{2.3.9}X'_{2.2.12}X_{1.4.14}X_{1.1.2}$

Table 4.8. (*Continued*)

0000 0000 0010 0000	$X_{1.1.2}X_{1.4.3}X'_{2.2.5}X_{2.3.9}$ $X'_{1.1.15}X'_{1.4.14}X_{2.2.12}X'_{2.3.8}$ $X_{1.4.3}X_{1.1.2}X'_{2.3.8}X_{2.2.12}$ $X'_{1.4.14}X'_{1.1.15}X_{2.3.9}X'_{2.2.5}$ $X'_{2.2.5}X'_{2.3.8}X_{1.1.2}X'_{1.4.14}$ $X_{2.2.12}X_{2.3.9}X'_{1.1.15}X_{1.4.3}$ $X'_{2.3.8}X'_{2.2.5}X_{1.4.3}X'_{1.1.15}$ $X_{2.3.9}X_{2.2.12}X'_{1.4.14}X_{1.1.2}$
0000 0000 0001 0000	$X_{1.1.2}X_{1.4.3}X_{2.2.5}X'_{2.3.9}$ $X'_{1.1.15}X'_{1.4.14}X'_{2.2.12}X_{2.3.8}$ $X_{1.4.3}X_{1.1.2}X_{2.3.8}X'_{2.2.12}$ $X'_{1.4.14}X'_{1.1.15}X'_{2.3.9}X_{2.2.5}$ $X_{2.2.5}X_{2.3.8}X_{1.1.2}X'_{1.4.14}$ $X'_{2.2.12}X'_{2.3.9}X'_{1.1.15}X_{1.4.3}$ $X_{2.3.8}X_{2.2.5}X_{1.4.3}X'_{1.1.15}$ $X'_{2.3.9}X'_{2.2.12}X'_{1.4.14}X_{1.1.2}$
0000 0000 0000 1000	$X_{1.1.2}X'_{1.4.3}X'_{2.2.5}X_{2.3.9}$ $X_{1.1.15}X'_{1.4.14}X'_{2.2.12}X_{2.3.8}$ $X'_{1.4.3}X_{1.1.2}X_{2.3.8}X'_{2.2.12}$ $X'_{1.4.14}X_{1.1.15}X_{2.3.9}X'_{2.2.5}$ $X'_{2.2.5}X_{2.3.8}X_{1.1.2}X'_{1.4.14}$ $X'_{2.2.12}X_{2.3.9}X_{1.1.15}X'_{1.4.3}$ $X_{2.3.8}X'_{2.2.5}X'_{1.4.3}X_{1.1.15}$ $X_{2.3.9}X'_{2.2.12}X'_{1.4.14}X_{1.1.2}$
0000 0000 0000 0100	$X_{1.1.2}X'_{1.4.3}X_{2.2.5}X'_{2.3.9}$ $X_{1.1.15}X'_{1.4.14}X_{2.2.12}X'_{2.3.8}$ $X'_{1.4.3}X_{1.1.2}X'_{2.3.8}X_{2.2.12}$ $X'_{1.4.14}X_{1.1.15}X'_{2.3.9}X_{2.2.5}$ $X_{2.2.5}X'_{2.3.8}X_{1.1.2}X'_{1.4.14}$ $X_{2.2.12}X'_{2.3.9}X_{1.1.15}X'_{1.4.3}$ $X'_{2.3.8}X_{2.2.5}X'_{1.4.3}X_{1.1.15}$ $X'_{2.3.9}X_{2.2.12}X'_{1.4.14}X_{1.1.2}$
0000 0000 0000 0010	$X_{1.1.2}X_{1.4.3}X'_{2.2.5}X'_{2.3.9}$ $X_{1.1.15}X_{1.4.14}X'_{2.2.12}X'_{2.3.8}$ $X_{1.4.3}X_{1.1.2}X'_{2.3.8}X'_{2.2.12}$ $X_{1.4.14}X_{1.1.15}X'_{2.3.9}X'_{2.2.5}$ $X'_{2.2.5}X'_{2.3.8}X_{1.1.2}X_{1.4.14}$ $X'_{2.2.12}X'_{2.3.9}X_{1.1.15}X_{1.4.3}$ $X'_{2.3.8}X'_{2.2.5}X_{1.4.3}X_{1.1.15}$ $X'_{2.3.9}X'_{2.2.12}X_{1.4.14}X_{1.1.2}$
0000 0000 0000 0001	$X_{1.1.2}X_{1.4.3}X_{2.2.5}X_{2.3.9}$ $X_{1.1.15}X_{1.4.14}X_{2.2.12}X_{2.3.8}$ $X_{1.4.3}X_{1.1.2}X_{2.3.8}X_{2.2.12}$ $X_{1.4.14}X_{1.1.15}X_{2.3.9}X_{2.2.5}$ $X_{2.2.5}X_{2.3.8}X_{1.1.2}X_{1.4.14}$ $X_{2.2.12}X_{2.3.9}X_{1.1.15}X_{1.4.3}$ $X_{2.3.8}X_{2.2.5}X_{1.4.3}X_{1.1.15}$ $X_{2.3.9}X_{2.2.12}X_{1.4.14}X_{1.1.2}$

This graphic gives the combinations displayed in Table 4.9. Note that, relative to Table 4.7, we have exchange of the seventh and tenth rows.

For CA1.4 we obtain:

$$\{X_{1.1.4}, X_{1.4.1}, X_{2.2.7}, X_{2.3.11}\}, \quad \{X_{1.1.13}, X_{1.4.16}, X_{2.2.10}, X_{2.3.6}\}, \tag{4.11a}$$

$$\{X_{1.4.1}, X_{1.1.4}, X_{2.3.6}, X_{2.2.10}\}, \quad \{X_{1.4.16}, X_{1.1.13}, X_{2.3.11}, X_{2.2.7}\}, \tag{4.11b}$$

$$\{X_{2.2.7}, X_{2.3.6}, X_{1.1.4}, X_{1.4.16}\}, \quad \{X_{2.2.10}, X_{2.3.11}, X_{1.1.13}, X_{1.4.1}\}, \tag{4.11c}$$

$$\{X_{2.3.6}, X_{2.2.7}, X_{1.4.1}, X_{1.1.13}\}, \quad \{X_{2.3.11}, X_{2.2.10}, X_{1.4.16}, X_{1.1.4}\}. \tag{4.11d}$$

Table 4.9.   Quaternary products codified by 8 SCAs $\in$ CA1.3.

1000 0000 0000 0000	$X'_{1.1.3}X'_{1.4.2}X'_{2.2.8}X'_{2.3.12}$ $X'_{1.4.2}X'_{1.1.3}X'_{2.3.5}X'_{2.2.9}$ $X'_{2.2.8}X'_{2.3.5}X'_{1.1.3}X'_{1.4.15}$ $X'_{2.3.5}X'_{2.2.8}X'_{1.4.2}X'_{1.1.14}$	$X'_{1.1.14}X'_{1.4.15}X'_{2.2.9}X'_{2.3.5}$ $X'_{1.4.15}X'_{1.1.14}X'_{2.3.12}X'_{2.2.8}$ $X'_{2.2.9}X'_{2.3.12}X'_{1.1.14}X'_{1.4.2}$ $X'_{2.3.12}X'_{2.2.9}X'_{1.4.15}X'_{1.1.3}$
0100 0000 0000 0000	$X'_{1.1.3}X'_{1.4.2}X_{2.2.8}X_{2.3.12}$ $X'_{1.4.2}X'_{1.1.3}X_{2.3.5}X_{2.2.9}$ $X_{2.2.8}X_{2.3.5}X'_{1.1.3}X'_{1.4.15}$ $X_{2.3.5}X_{2.2.8}X'_{1.4.2}X'_{1.1.14}$	$X'_{1.1.14}X'_{1.4.15}X_{2.2.9}X_{2.3.5}$ $X'_{1.4.15}X'_{1.1.14}X_{2.3.12}X_{2.2.8}$ $X_{2.2.9}X_{2.3.12}X'_{1.1.14}X'_{1.4.2}$ $X_{2.3.12}X_{2.2.9}X'_{1.4.15}X'_{1.1.3}$
0010 0000 0000 0000	$X'_{1.1.3}X_{1.4.2}X'_{2.2.8}X_{2.3.12}$ $X_{1.4.2}X'_{1.1.3}X_{2.3.5}X'_{2.2.9}$ $X'_{2.2.8}X_{2.3.5}X'_{1.1.3}X_{1.4.15}$ $X_{2.3.5}X'_{2.2.8}X_{1.4.2}X'_{1.1.14}$	$X'_{1.1.14}X_{1.4.15}X'_{2.2.9}X_{2.3.5}$ $X_{1.4.15}X'_{1.1.14}X_{2.3.12}X'_{2.2.8}$ $X'_{2.2.9}X_{2.3.12}X'_{1.1.14}X_{1.4.2}$ $X_{2.3.12}X'_{2.2.9}X_{1.4.15}X'_{1.1.3}$
0001 0000 0000 0000	$X'_{1.1.3}X_{1.4.2}X_{2.2.8}X'_{2.3.12}$ $X_{1.4.2}X'_{1.1.3}X'_{2.3.5}X_{2.2.9}$ $X_{2.2.8}X'_{2.3.5}X'_{1.1.3}X_{1.4.15}$ $X'_{2.3.5}X_{2.2.8}X_{1.4.2}X'_{1.1.14}$	$X'_{1.1.14}X_{1.4.15}X_{2.2.9}X'_{2.3.5}$ $X_{1.4.15}X'_{1.1.14}X'_{2.3.12}X_{2.2.8}$ $X_{2.2.9}X'_{2.3.12}X'_{1.1.14}X_{1.4.2}$ $X'_{2.3.12}X_{2.2.9}X_{1.4.15}X'_{1.1.3}$
0000 1000 0000 0000	$X'_{1.1.3}X'_{1.4.2}X_{2.2.8}X'_{2.3.12}$ $X'_{1.4.2}X'_{1.1.3}X'_{2.3.5}X_{2.2.9}$ $X'_{2.2.8}X'_{2.3.5}X'_{1.1.3}X_{1.4.15}$ $X'_{2.3.5}X'_{2.2.8}X'_{1.4.2}X_{1.1.14}$	$X_{1.1.14}X_{1.4.15}X_{2.2.9}X'_{2.3.5}$ $X_{1.4.15}X_{1.1.14}X_{2.3.12}X'_{2.2.8}$ $X_{2.2.9}X_{2.3.12}X_{1.1.14}X_{1.4.2}$ $X_{2.3.12}X_{2.2.9}X_{1.4.15}X'_{1.1.3}$
0000 0100 0000 0000	$X'_{1.1.3}X'_{1.4.2}X_{2.2.8}X'_{2.3.12}$ $X'_{1.4.2}X'_{1.1.3}X_{2.3.5}X'_{2.2.9}$ $X_{2.2.8}X_{2.3.5}X'_{1.1.3}X_{1.4.15}$ $X_{2.3.5}X_{2.2.8}X'_{1.4.2}X_{1.1.14}$	$X_{1.1.14}X_{1.4.15}X'_{2.2.9}X_{2.3.5}$ $X_{1.4.15}X_{1.1.14}X'_{2.3.12}X_{2.2.8}$ $X'_{2.2.9}X'_{2.3.12}X_{1.1.14}X'_{1.4.2}$ $X'_{2.3.12}X'_{2.2.9}X_{1.4.15}X'_{1.1.3}$
0000 0010 0000 0000	$X_{1.1.3}X'_{1.4.2}X_{2.2.8}X_{2.3.12}$ $X'_{1.4.2}X_{1.1.3}X'_{2.3.5}X'_{2.2.9}$ $X_{2.2.8}X'_{2.3.5}X_{1.1.3}X_{1.4.15}$ $X'_{2.3.5}X_{2.2.8}X'_{1.4.2}X'_{1.1.14}$	$X'_{1.1.14}X_{1.4.15}X'_{2.2.9}X'_{2.3.5}$ $X_{1.4.15}X'_{1.1.14}X_{2.3.12}X_{2.2.8}$ $X'_{2.2.9}X_{2.3.12}X'_{1.1.14}X'_{1.4.2}$ $X_{2.3.12}X'_{2.2.9}X_{1.4.15}X_{1.1.3}$
0000 0001 0000 0000	$X'_{1.1.3}X_{1.4.2}X_{2.2.8}X_{2.3.12}$ $X_{1.4.2}X'_{1.1.3}X'_{2.3.5}X'_{2.2.9}$ $X_{2.2.8}X'_{2.3.5}X'_{1.1.3}X'_{1.4.15}$ $X'_{2.3.5}X_{2.2.8}X_{1.4.2}X_{1.1.14}$	$X_{1.1.14}X'_{1.4.15}X'_{2.2.9}X'_{2.3.5}$ $X'_{1.4.15}X_{1.1.14}X_{2.3.12}X_{2.2.8}$ $X'_{2.2.9}X_{2.3.12}X_{1.1.14}X_{1.4.2}$ $X_{2.3.12}X'_{2.2.9}X'_{1.4.15}X'_{1.1.3}$
0000 0000 1000 0000	$X_{1.1.3}X'_{1.4.2}X'_{2.2.8}X'_{2.3.12}$ $X'_{1.4.2}X_{1.1.3}X_{2.3.5}X_{2.2.9}$ $X'_{2.2.8}X_{2.3.5}X_{1.1.3}X_{1.4.15}$ $X_{2.3.5}X'_{2.2.8}X'_{1.4.2}X'_{1.1.14}$	$X'_{1.1.14}X_{1.4.15}X_{2.2.9}X_{2.3.5}$ $X_{1.4.15}X'_{1.1.14}X'_{2.3.12}X'_{2.2.8}$ $X_{2.2.9}X'_{2.3.12}X'_{1.1.14}X'_{1.4.2}$ $X'_{2.3.12}X_{2.2.9}X_{1.4.15}X_{1.1.3}$
0000 0000 0100 0000	$X'_{1.1.3}X_{1.4.2}X'_{2.2.8}X'_{2.3.12}$ $X_{1.4.2}X'_{1.1.3}X_{2.3.5}X_{2.2.9}$ $X'_{2.2.8}X_{2.3.5}X'_{1.1.3}X'_{1.4.15}$ $X_{2.3.5}X'_{2.2.8}X_{1.4.2}X_{1.1.14}$	$X_{1.1.14}X'_{1.4.15}X_{2.2.9}X_{2.3.5}$ $X'_{1.4.15}X_{1.1.14}X'_{2.3.12}X'_{2.2.8}$ $X_{2.2.9}X_{2.3.12}X_{1.1.14}X_{1.4.2}$ $X'_{2.3.12}X_{2.2.9}X'_{1.4.15}X_{1.1.3}$

Table 4.9. (*Continued*)

0000 0000 0010 0000	$X_{1.1.3}X_{1.4.2}X'_{2.2.8}X_{2.3.12}$ $X_{1.4.2}X_{1.1.3}X'_{2.3.5}X_{2.2.9}$ $X'_{2.2.8}X'_{2.3.5}X_{1.1.3}X'_{1.4.15}$ $X'_{2.3.5}X'_{2.2.8}X_{1.4.2}X'_{1.1.14}$	$X'_{1.1.14}X'_{1.4.15}X_{2.2.9}X'_{2.3.5}$ $X'_{1.4.15}X'_{1.1.14}X_{2.3.12}X'_{2.2.8}$ $X_{2.2.9}X_{2.3.12}X'_{1.1.14}X_{1.4.2}$ $X_{2.3.12}X_{2.2.9}X'_{1.4.15}X_{1.1.3}$
0000 0000 0001 0000	$X_{1.1.3}X_{1.4.2}X_{2.2.8}X'_{2.3.12}$ $X_{1.4.2}X_{1.1.3}X_{2.3.5}X'_{2.2.9}$ $X_{2.2.8}X_{2.3.5}X_{1.1.3}X'_{1.4.15}$ $X_{2.3.5}X_{2.2.8}X_{1.4.2}X'_{1.1.14}$	$X'_{1.1.14}X'_{1.4.15}X'_{2.2.9}X_{2.3.5}$ $X'_{1.4.15}X'_{1.1.14}X'_{2.3.12}X_{2.2.8}$ $X'_{2.2.9}X'_{2.3.12}X'_{1.1.14}X_{1.4.2}$ $X'_{2.3.12}X'_{2.2.9}X'_{1.4.15}X_{1.1.3}$
0000 0000 0000 1000	$X_{1.1.3}X'_{1.4.2}X'_{2.2.8}X_{2.3.12}$ $X'_{1.4.2}X_{1.1.3}X_{2.3.5}X'_{2.2.9}$ $X'_{2.2.8}X_{2.3.5}X_{1.1.3}X'_{1.4.15}$ $X_{2.3.5}X'_{2.2.8}X'_{1.4.2}X_{1.1.14}$	$X_{1.1.14}X'_{1.4.15}X'_{2.2.9}X_{2.3.5}$ $X'_{1.4.15}X_{1.1.14}X_{2.3.12}X'_{2.2.8}$ $X'_{2.2.9}X_{2.3.12}X_{1.1.14}X'_{1.4.2}$ $X_{2.3.12}X'_{2.2.9}X'_{1.4.15}X_{1.1.3}$
0000 0000 0000 0100	$X_{1.1.3}X'_{1.4.2}X_{2.2.8}X'_{2.3.12}$ $X'_{1.4.2}X_{1.1.3}X'_{2.3.5}X_{2.2.9}$ $X_{2.2.8}X'_{2.3.5}X_{1.1.3}X'_{1.4.15}$ $X'_{2.3.5}X_{2.2.8}X'_{1.4.2}X_{1.1.14}$	$X_{1.1.14}X'_{1.4.15}X_{2.2.9}X'_{2.3.5}$ $X'_{1.4.15}X_{1.1.14}X'_{2.3.12}X_{2.2.8}$ $X_{2.2.9}X'_{2.3.12}X_{1.1.14}X'_{1.4.2}$ $X'_{2.3.12}X_{2.2.9}X'_{1.4.15}X_{1.1.3}$
0000 0000 0000 0010	$X_{1.1.3}X_{1.4.2}X'_{2.2.8}X'_{2.3.12}$ $X_{1.4.2}X_{1.1.3}X'_{2.3.5}X'_{2.2.9}$ $X'_{2.2.8}X'_{2.3.5}X_{1.1.3}X_{1.4.15}$ $X'_{2.3.5}X'_{2.2.8}X_{1.4.2}X_{1.1.14}$	$X_{1.1.14}X_{1.4.15}X'_{2.2.9}X'_{2.3.5}$ $X_{1.4.15}X_{1.1.14}X'_{2.3.12}X'_{2.2.8}$ $X'_{2.2.9}X'_{2.3.12}X_{1.1.14}X_{1.4.2}$ $X'_{2.3.12}X'_{2.2.9}X_{1.4.15}X_{1.1.3}$
0000 0000 0000 0001	$X_{1.1.3}X_{1.4.2}X_{2.2.8}X_{2.3.12}$ $X_{1.4.2}X_{1.1.3}X_{2.3.5}X_{2.2.9}$ $X_{2.2.8}X_{2.3.5}X_{1.1.3}X_{1.4.15}$ $X_{2.3.5}X_{2.2.8}X_{1.4.2}X_{1.1.14}$	$X_{1.1.14}X_{1.4.15}X_{2.2.9}X_{2.3.5}$ $X_{1.4.15}X_{1.1.14}X_{2.3.12}X_{2.2.8}$ $X_{2.2.9}X_{2.3.12}X_{1.1.14}X_{1.4.2}$ $X_{2.3.12}X_{2.2.9}X_{1.4.15}X_{1.1.3}$

This graphic gives the combinations displayed in Table 4.10. Note that, relative to Table 4.7, we exchange both the seventh with the tenth row and the eighth with the ninth row (combining the two previous exchanges).

For CA1.5 we obtain:

$$\{X_{1.1.5}, X_{1.4.8}, X_{2.2.2}, X_{2.3.14}\}, \quad \{X_{1.1.12}, X_{1.4.9}, X_{2.2.15}, X_{2.3.3}\},$$
$$(4.12\text{a})$$

$$\{X_{1.4.8}, X_{1.1.5}, X_{2.3.3}, X_{2.2.15}\}, \quad \{X_{1.4.9}, X_{1.1.12}, X_{2.3.14}, X_{2.2.2}\},$$
$$(4.12\text{b})$$

$$\{X_{2.2.2}, X_{2.3.3}, X_{1.1.5}, X_{1.4.9}\}, \quad \{X_{2.2.15}, X_{2.3.14}, X_{1.1.12}, X_{1.4.8}\},$$
$$(4.12\text{c})$$

$$\{X_{2.3.3}, X_{2.2.2}, X_{1.4.8}, X_{1.1.12}\}, \quad \{X_{2.3.14}, X_{2.2.15}, X_{1.4.9}, X_{1.1.5}\}.$$
$$(4.12\text{d})$$

Table 4.10.   Quaternary products codified by 8 SCAs $\in$ CA1.4.

1000 0000 0000 0000	$X'_{1.1.4}X'_{1.4.1}X'_{2.2.7}X'_{2.3.11}$ $X'_{1.4.1}X'_{1.1.4}X'_{2.3.6}X'_{2.2.10}$ $X'_{2.2.7}X'_{2.3.6}X'_{1.1.4}X'_{1.4.16}$ $X'_{2.3.6}X'_{2.2.7}X'_{1.4.1}X'_{1.1.13}$	$X'_{1.1.13}X'_{1.4.16}X'_{2.2.10}X'_{2.3.6}$ $X'_{1.4.16}X'_{1.1.13}X'_{2.3.11}X'_{2.2.7}$ $X'_{2.2.10}X'_{2.3.11}X'_{1.1.13}X'_{1.4.1}$ $X'_{2.3.11}X'_{2.2.10}X'_{1.4.16}X'_{1.1.4}$
0100 0000 0000 0000	$X'_{1.1.4}X'_{1.4.1}X'_{2.2.7}X'_{2.3.11}$ $X'_{1.4.1}X'_{1.1.4}X'_{2.3.6}X'_{2.2.10}$ $X_{2.2.7}X_{2.3.6}X'_{1.1.4}X'_{1.4.16}$ $X_{2.3.6}X_{2.2.7}X'_{1.4.1}X'_{1.1.13}$	$X'_{1.1.13}X'_{1.4.16}X_{2.2.10}X_{2.3.6}$ $X'_{1.4.16}X'_{1.1.13}X_{2.3.11}X_{2.2.7}$ $X_{2.2.10}X_{2.3.11}X'_{1.1.13}X'_{1.4.1}$ $X_{2.3.11}X_{2.2.10}X'_{1.4.16}X'_{1.1.4}$
0010 0000 0000 0000	$X'_{1.1.4}X_{1.4.1}X'_{2.2.7}X'_{2.3.11}$ $X_{1.4.1}X'_{1.1.4}X'_{2.3.6}X'_{2.2.10}$ $X'_{2.2.7}X'_{2.3.6}X'_{1.1.4}X_{1.4.16}$ $X'_{2.3.6}X_{2.2.7}X'_{1.4.1}X'_{1.1.13}$	$X'_{1.1.13}X'_{1.4.16}X'_{2.2.10}X'_{2.3.6}$ $X'_{1.4.16}X'_{1.1.13}X'_{2.3.11}X'_{2.2.7}$ $X'_{2.2.10}X'_{2.3.11}X'_{1.1.13}X_{1.4.1}$ $X'_{2.3.11}X'_{2.2.10}X_{1.4.16}X'_{1.1.4}$
0001 0000 0000 0000	$X'_{1.1.4}X'_{1.4.1}X'_{2.2.7}X'_{2.3.11}$ $X_{1.4.1}X'_{1.1.4}X'_{2.3.6}X_{2.2.10}$ $X_{2.2.7}X'_{2.3.6}X'_{1.1.4}X_{1.4.16}$ $X'_{2.3.6}X_{2.2.7}X'_{1.4.1}X'_{1.1.13}$	$X'_{1.1.13}X'_{1.4.16}X_{2.2.10}X'_{2.3.6}$ $X'_{1.4.16}X'_{1.1.13}X_{2.3.11}X_{2.2.7}$ $X_{2.2.10}X'_{2.3.11}X'_{1.1.13}X_{1.4.1}$ $X'_{2.3.11}X_{2.2.10}X'_{1.4.16}X'_{1.1.4}$
0000 1000 0000 0000	$X'_{1.1.4}X'_{1.4.1}X'_{2.2.7}X_{2.3.11}$ $X'_{1.4.1}X'_{1.1.4}X'_{2.3.6}X_{2.2.10}$ $X'_{2.2.7}X'_{2.3.6}X'_{1.1.4}X_{1.4.16}$ $X'_{2.3.6}X'_{2.2.7}X'_{1.4.1}X_{1.1.13}$	$X_{1.1.13}X_{1.4.16}X_{2.2.10}X_{2.3.6}$ $X_{1.4.16}X_{1.1.13}X_{2.3.11}X'_{2.2.7}$ $X_{2.2.10}X_{2.3.11}X_{1.1.13}X'_{1.4.1}$ $X_{2.3.11}X_{2.2.10}X_{1.4.16}X'_{1.1.4}$
0000 0100 0000 0000	$X'_{1.1.4}X'_{1.4.1}X_{2.2.7}X'_{2.3.11}$ $X'_{1.4.1}X'_{1.1.4}X_{2.3.6}X'_{2.2.10}$ $X_{2.2.7}X_{2.3.6}X'_{1.1.4}X_{1.4.16}$ $X_{2.3.6}X_{2.2.7}X'_{1.4.1}X_{1.1.13}$	$X'_{1.1.13}X_{1.4.16}X'_{2.2.10}X_{2.3.6}$ $X_{1.4.16}X'_{1.1.13}X_{2.3.11}X_{2.2.7}$ $X'_{2.2.10}X'_{2.3.11}X_{1.1.13}X'_{1.4.1}$ $X'_{2.3.11}X'_{2.2.10}X_{1.4.16}X'_{1.1.4}$
0000 0010 0000 0000	$X_{1.1.4}X'_{1.4.1}X'_{2.2.7}X'_{2.3.11}$ $X'_{1.4.1}X_{1.1.4}X'_{2.3.6}X'_{2.2.10}$ $X_{2.2.7}X'_{2.3.6}X_{1.1.4}X_{1.4.16}$ $X'_{2.3.6}X_{2.2.7}X'_{1.4.1}X'_{1.1.13}$	$X'_{1.1.13}X'_{1.4.16}X'_{2.2.10}X'_{2.3.6}$ $X'_{1.4.16}X'_{1.1.13}X_{2.3.11}X_{2.2.7}$ $X'_{2.2.10}X_{2.3.11}X'_{1.1.13}X'_{1.4.1}$ $X_{2.3.11}X'_{2.2.10}X_{1.4.16}X_{1.1.4}$
0000 0001 0000 0000	$X_{1.1.4}X'_{1.4.1}X'_{2.2.7}X'_{2.3.11}$ $X'_{1.4.1}X_{1.1.4}X_{2.3.6}X_{2.2.10}$ $X'_{2.2.7}X_{2.3.6}X_{1.1.4}X_{1.4.16}$ $X_{2.3.6}X'_{2.2.7}X'_{1.4.1}X_{1.1.13}$	$X'_{1.1.13}X_{1.4.16}X_{2.2.10}X_{2.3.6}$ $X_{1.4.16}X'_{1.1.13}X'_{2.3.11}X'_{2.2.7}$ $X_{2.2.10}X'_{2.3.11}X'_{1.1.13}X'_{1.4.1}$ $X'_{2.3.11}X_{2.2.10}X_{1.4.16}X_{1.1.4}$
0000 0000 1000 0000	$X'_{1.1.4}X_{1.4.1}X'_{2.2.7}X_{2.3.11}$ $X_{1.4.1}X'_{1.1.4}X_{2.3.6}X'_{2.2.10}$ $X'_{2.2.7}X'_{2.3.6}X'_{1.1.4}X_{1.4.16}$ $X'_{2.3.6}X_{2.2.7}X_{1.4.1}X_{1.1.13}$	$X'_{1.1.13}X_{1.4.16}X_{2.2.10}X_{2.3.6}$ $X_{1.4.16}X_{1.1.13}X_{2.3.11}X'_{2.2.7}$ $X'_{2.2.10}X'_{2.3.11}X'_{1.1.13}X_{1.4.1}$ $X_{2.3.11}X'_{2.2.10}X'_{1.4.16}X'_{1.1.4}$
0000 0000 0100 0000	$X'_{1.1.4}X_{1.4.1}X'_{2.2.7}X'_{2.3.11}$ $X_{1.4.1}X'_{1.1.4}X_{2.3.6}X_{2.2.10}$ $X'_{2.2.7}X_{2.3.6}X'_{1.1.4}X'_{1.4.16}$ $X_{2.3.6}X'_{2.2.7}X_{1.4.1}X_{1.1.13}$	$X'_{1.1.13}X'_{1.4.16}X_{2.2.10}X_{2.3.6}$ $X'_{1.4.16}X_{1.1.13}X'_{2.3.11}X'_{2.2.7}$ $X_{2.2.10}X'_{2.3.11}X'_{1.1.13}X_{1.4.1}$ $X'_{2.3.11}X_{2.2.10}X'_{1.4.16}X'_{1.1.4}$

Table 4.10. (*Continued*)

0000 0000 0010 0000	$X_{1.1.4}X_{1.4.1}X'_{2.2.7}X_{2.3.11}$ $X_{1.4.1}X_{1.1.4}X'_{2.3.6}X_{2.2.10}$ $X'_{2.2.7}X'_{2.3.6}X_{1.1.4}X'_{1.4.16}$ $X'_{2.3.6}X'_{2.2.7}X_{1.4.1}X'_{1.1.13}$	$X'_{1.1.13}X'_{1.4.16}X_{2.2.10}X'_{2.3.6}$ $X'_{1.4.16}X'_{1.1.13}X_{2.3.11}X'_{2.2.7}$ $X_{2.2.10}X_{2.3.11}X'_{1.1.13}X_{1.4.1}$ $X_{2.3.11}X_{2.2.10}X'_{1.4.16}X_{1.1.4}$
0000 0000 0001 0000	$X_{1.1.4}X_{1.4.1}X_{2.2.7}X'_{2.3.11}$ $X_{1.4.1}X_{1.1.4}X_{2.3.6}X'_{2.2.10}$ $X_{2.2.7}X_{2.3.6}X_{1.1.4}X'_{1.4.16}$ $X_{2.3.6}X_{2.2.7}X_{1.4.1}X'_{1.1.13}$	$X'_{1.1.13}X'_{1.4.16}X'_{2.2.10}X_{2.3.6}$ $X'_{1.4.16}X'_{1.1.13}X'_{2.3.11}X_{2.2.7}$ $X'_{2.2.10}X'_{2.3.11}X'_{1.1.13}X_{1.4.1}$ $X'_{2.3.11}X'_{2.2.10}X'_{1.4.16}X_{1.1.4}$
0000 0000 0000 1000	$X_{1.1.4}X'_{1.4.1}X'_{2.2.7}X_{2.3.11}$ $X'_{1.4.1}X_{1.1.4}X_{2.3.6}X'_{2.2.10}$ $X'_{2.2.7}X_{2.3.6}X_{1.1.4}X'_{1.4.16}$ $X_{2.3.6}X'_{2.2.7}X'_{1.4.1}X_{1.1.13}$	$X_{1.1.13}X'_{1.4.16}X'_{2.2.10}X_{2.3.6}$ $X'_{1.4.16}X_{1.1.13}X_{2.3.11}X'_{2.2.7}$ $X'_{2.2.10}X_{2.3.11}X_{1.1.13}X'_{1.4.1}$ $X_{2.3.11}X'_{2.2.10}X'_{1.4.16}X_{1.1.4}$
0000 0000 0000 0100	$X_{1.1.4}X'_{1.4.1}X_{2.2.7}X'_{2.3.11}$ $X'_{1.4.1}X_{1.1.4}X'_{2.3.6}X_{2.2.10}$ $X_{2.2.7}X'_{2.3.6}X_{1.1.4}X'_{1.4.16}$ $X'_{2.3.6}X_{2.2.7}X'_{1.4.1}X_{1.1.13}$	$X_{1.1.13}X'_{1.4.16}X_{2.2.10}X'_{2.3.6}$ $X'_{1.4.16}X_{1.1.13}X'_{2.3.11}X_{2.2.7}$ $X_{2.2.10}X'_{2.3.11}X_{1.1.13}X'_{1.4.1}$ $X'_{2.3.11}X_{2.2.10}X'_{1.4.16}X_{1.1.4}$
0000 0000 0000 0010	$X_{1.1.4}X_{1.4.1}X'_{2.2.7}X'_{2.3.11}$ $X_{1.4.1}X_{1.1.4}X'_{2.3.6}X'_{2.2.10}$ $X'_{2.2.7}X'_{2.3.6}X_{1.1.4}X_{1.4.16}$ $X'_{2.3.6}X'_{2.2.7}X_{1.4.1}X_{1.1.13}$	$X_{1.1.13}X_{1.4.16}X'_{2.2.10}X'_{2.3.6}$ $X_{1.4.16}X_{1.1.13}X'_{2.3.11}X'_{2.2.7}$ $X'_{2.2.10}X'_{2.3.11}X_{1.1.13}X_{1.4.1}$ $X'_{2.3.11}X'_{2.2.10}X_{1.4.16}X_{1.1.4}$
0000 0000 0000 0001	$X_{1.1.4}X_{1.4.1}X_{2.2.7}X_{2.3.11}$ $X_{1.4.1}X_{1.1.4}X_{2.3.6}X_{2.2.10}$ $X_{2.2.7}X_{2.3.6}X_{1.1.4}X_{1.4.16}$ $X_{2.3.6}X_{2.2.7}X_{1.4.1}X_{1.1.13}$	$X_{1.1.13}X_{1.4.16}X_{2.2.10}X_{2.3.6}$ $X_{1.4.16}X_{1.1.13}X_{2.3.11}X_{2.2.7}$ $X_{2.2.10}X_{2.3.11}X_{1.1.13}X_{1.4.1}$ $X_{2.3.11}X_{2.2.10}X_{1.4.16}X_{1.1.4}$

This graphic gives the results displayed in Table 4.11. Note that, relative to Table 4.7, we exchange the sixth and the eleventh rows.

Let us consider a similar choice for CA1.6:

$$\{X_{1.1.6}, X_{1.4.7}, X_{2.2.1}, X_{2.3.13}\}, \quad \{X_{1.1.11}, X_{1.4.10}, X_{2.2.16}, X_{2.3.4}\},$$
$$(4.13a)$$

$$\{X_{1.4.7}, X_{1.1.6}, X_{2.3.4}, X_{2.2.16}\}, \quad \{X_{1.4.10}, X_{1.1.11}, X_{2.3.13}, X_{2.2.1}\},$$
$$(4.13b)$$

$$\{X_{2.2.1}, X_{2.3.4}, X_{1.1.6}, X_{1.4.10}\}, \quad \{X_{2.2.16}, X_{2.3.13}, X_{1.1.11}, X_{1.4.7}\},$$
$$(4.13c)$$

$$\{X_{2.3.4}, X_{2.2.1}, X_{1.4.7}, X_{1.1.11}\}, \quad \{X_{2.3.13}, X_{2.2.16}, X_{1.4.10}, X_{1.1.6}\}.$$
$$(4.13d)$$

Table 4.11.   Quaternary products codified by 8 SCAs $\in$ CA1.5.

1000 0000 0000 0000	$X'_{1.1.5}X'_{1.4.8}X'_{2.2.2}X'_{2.3.14}$ $X'_{1.4.8}X'_{1.1.5}X'_{2.3.3}X'_{2.2.15}$ $X'_{2.2.2}X'_{2.3.3}X'_{1.1.5}X'_{1.4.9}$ $X'_{2.3.3}X'_{2.2.2}X'_{1.4.8}X'_{1.1.12}$	$X'_{1.1.12}X'_{1.4.9}X'_{2.2.15}X'_{2.3.3}$ $X'_{1.4.9}X'_{1.1.12}X'_{2.3.14}X'_{2.2.2}$ $X'_{2.2.15}X'_{2.3.14}X'_{1.1.12}X'_{1.4.8}$ $X'_{2.3.14}X'_{2.2.15}X'_{1.4.9}X'_{1.1.5}$
0100 0000 0000 0000	$X'_{1.1.5}X'_{1.4.8}X_{2.2.2}X_{2.3.14}$ $X'_{1.4.8}X'_{1.1.5}X_{2.3.3}X_{2.2.15}$ $X_{2.2.2}X_{2.3.3}X'_{1.1.5}X'_{1.4.9}$ $X_{2.3.3}X_{2.2.2}X'_{1.4.8}X'_{1.1.12}$	$X'_{1.1.12}X'_{1.4.9}X_{2.2.15}X_{2.3.3}$ $X'_{1.4.9}X'_{1.1.12}X_{2.3.14}X_{2.2.2}$ $X_{2.2.15}X_{2.3.14}X'_{1.1.12}X'_{1.4.8}$ $X_{2.3.14}X_{2.2.15}X'_{1.4.9}X'_{1.1.5}$
0010 0000 0000 0000	$X'_{1.1.5}X_{1.4.8}X'_{2.2.2}X_{2.3.14}$ $X_{1.4.8}X'_{1.1.5}X_{2.3.3}X'_{2.2.15}$ $X'_{2.2.2}X_{2.3.3}X'_{1.1.5}X_{1.4.9}$ $X_{2.3.3}X'_{2.2.2}X_{1.4.8}X'_{1.1.12}$	$X'_{1.1.12}X_{1.4.9}X'_{2.2.15}X_{2.3.3}$ $X_{1.4.9}X'_{1.1.12}X_{2.3.14}X'_{2.2.2}$ $X'_{2.2.15}X_{2.3.14}X'_{1.1.12}X_{1.4.8}$ $X_{2.3.14}X'_{2.2.15}X_{1.4.9}X'_{1.1.5}$
0001 0000 0000 0000	$X'_{1.1.5}X_{1.4.8}X_{2.2.2}X'_{2.3.14}$ $X_{1.4.8}X'_{1.1.5}X'_{2.3.3}X_{2.2.15}$ $X_{2.2.2}X'_{2.3.3}X'_{1.1.5}X_{1.4.9}$ $X'_{2.3.3}X_{2.2.2}X_{1.4.8}X'_{1.1.12}$	$X'_{1.1.12}X_{1.4.9}X_{2.2.15}X'_{2.3.3}$ $X_{1.4.9}X'_{1.1.12}X'_{2.3.14}X_{2.2.2}$ $X_{2.2.15}X'_{2.3.14}X'_{1.1.12}X_{1.4.8}$ $X'_{2.3.14}X_{2.2.15}X_{1.4.9}X'_{1.1.5}$
0000 1000 0000 0000	$X'_{1.1.5}X'_{1.4.8}X'_{2.2.2}X_{2.3.14}$ $X'_{1.4.8}X'_{1.1.5}X'_{2.3.3}X_{2.2.15}$ $X'_{2.2.2}X'_{2.3.3}X'_{1.1.5}X_{1.4.9}$ $X'_{2.3.3}X'_{2.2.2}X'_{1.4.8}X_{1.1.12}$	$X_{1.1.12}X_{1.4.9}X_{2.2.15}X'_{2.3.3}$ $X_{1.4.9}X_{1.1.12}X_{2.3.14}X'_{2.2.2}$ $X_{2.2.15}X_{2.3.14}X_{1.1.12}X'_{1.4.8}$ $X_{2.3.14}X_{2.2.15}X_{1.4.9}X'_{1.1.5}$
0000 0100 0000 0000	$X_{1.1.5}X_{1.4.8}X'_{2.2.2}X_{2.3.14}$ $X_{1.4.8}X_{1.1.5}X'_{2.3.3}X_{2.2.15}$ $X'_{2.2.2}X'_{2.3.3}X_{1.1.5}X'_{1.4.9}$ $X'_{2.3.3}X'_{2.2.2}X_{1.4.8}X'_{1.1.12}$	$X'_{1.1.12}X'_{1.4.9}X_{2.2.15}X'_{2.3.3}$ $X'_{1.4.9}X'_{1.1.12}X_{2.3.14}X'_{2.2.2}$ $X_{2.2.15}X_{2.3.14}X'_{1.1.12}X_{1.4.8}$ $X_{2.3.14}X_{2.2.15}X'_{1.4.9}X_{1.1.5}$
0000 0010 0000 0000	$X'_{1.1.5}X_{1.4.8}X'_{2.2.2}X'_{2.3.14}$ $X_{1.4.8}X'_{1.1.5}X_{2.3.3}X_{2.2.15}$ $X'_{2.2.2}X_{2.3.3}X'_{1.1.5}X'_{1.4.9}$ $X_{2.3.3}X'_{2.2.2}X_{1.4.8}X_{1.1.12}$	$X_{1.1.12}X'_{1.4.9}X_{2.2.15}X_{2.3.3}$ $X'_{1.4.9}X_{1.1.12}X'_{2.3.14}X'_{2.2.2}$ $X_{2.2.15}X'_{2.3.14}X_{1.1.12}X_{1.4.8}$ $X'_{2.3.14}X_{2.2.15}X'_{1.4.9}X'_{1.1.5}$
0000 0001 0000 0000	$X'_{1.1.5}X_{1.4.8}X_{2.2.2}X_{2.3.14}$ $X_{1.4.8}X'_{1.1.5}X'_{2.3.3}X'_{2.2.15}$ $X_{2.2.2}X'_{2.3.3}X'_{1.1.5}X'_{1.4.9}$ $X'_{2.3.3}X_{2.2.2}X_{1.4.8}X_{1.1.12}$	$X_{1.1.12}X'_{1.4.9}X'_{2.2.15}X'_{2.3.3}$ $X'_{1.4.9}X_{1.1.12}X_{2.3.14}X_{2.2.2}$ $X'_{2.2.15}X_{2.3.14}X_{1.1.12}X_{1.4.8}$ $X_{2.3.14}X'_{2.2.15}X_{1.4.9}X'_{1.1.5}$
0000 0000 1000 0000	$X_{1.1.5}X'_{1.4.8}X'_{2.2.2}X'_{2.3.14}$ $X'_{1.4.8}X_{1.1.5}X_{2.3.3}X_{2.2.15}$ $X'_{2.2.2}X_{2.3.3}X_{1.1.5}X_{1.4.9}$ $X_{2.3.3}X'_{2.2.2}X'_{1.4.8}X'_{1.1.12}$	$X'_{1.1.12}X_{1.4.9}X_{2.2.15}X_{2.3.3}$ $X_{1.4.9}X'_{1.1.12}X'_{2.3.14}X'_{2.2.2}$ $X_{2.2.15}X'_{2.3.14}X'_{1.1.12}X'_{1.4.8}$ $X'_{2.3.14}X_{2.2.15}X_{1.4.9}X_{1.1.5}$
0000 0000 0100 0000	$X_{1.1.5}X'_{1.4.8}X_{2.2.2}X_{2.3.14}$ $X'_{1.4.8}X_{1.1.5}X'_{2.3.3}X'_{2.2.15}$ $X_{2.2.2}X'_{2.3.3}X_{1.1.5}X_{1.4.9}$ $X'_{2.3.3}X_{2.2.2}X'_{1.4.8}X'_{1.1.12}$	$X'_{1.1.12}X_{1.4.9}X'_{2.2.15}X'_{2.3.3}$ $X_{1.4.9}X'_{1.1.12}X_{2.3.14}X_{2.2.2}$ $X'_{2.2.15}X_{2.3.14}X'_{1.1.12}X'_{1.4.8}$ $X_{2.3.14}X'_{2.2.15}X_{1.4.9}X_{1.1.5}$

Table 4.11. (*Continued*)

0000 0000 0010 0000	$X'_{1.1.5}X'_{1.4.8}X_{2.2.2}X'_{2.3.14}$ $X'_{1.4.8}X'_{1.1.5}X_{2.3.3}X'_{2.2.15}$ $X_{2.2.2}X_{2.3.3}X'_{1.1.5}X_{1.4.9}$ $X_{2.3.3}X_{2.2.2}X'_{1.4.8}X_{1.1.12}$ $X_{1.1.12}X_{1.4.9}X'_{2.2.15}X_{2.3.3}$ $X_{1.4.9}X_{1.1.12}X'_{2.3.14}X_{2.2.2}$ $X'_{2.2.15}X'_{2.3.14}X_{1.1.12}X'_{1.4.8}$ $X'_{2.3.14}X'_{2.2.15}X_{1.4.9}X'_{1.1.5}$
0000 0000 0001 0000	$X_{1.1.5}X_{1.4.8}X_{2.2.2}X'_{2.3.14}$ $X_{1.4.8}X_{1.1.5}X_{2.3.3}X'_{2.2.15}$ $X_{2.2.2}X_{2.3.3}X_{1.1.5}X'_{1.4.9}$ $X_{2.3.3}X_{2.2.2}X_{1.4.8}X'_{1.1.12}$ $X'_{1.1.12}X'_{1.4.9}X'_{2.2.15}X_{2.3.3}$ $X'_{1.4.9}X'_{1.1.12}X'_{2.3.14}X_{2.2.2}$ $X'_{2.2.15}X'_{2.3.14}X'_{1.1.12}X_{1.4.8}$ $X'_{2.3.14}X'_{2.2.15}X'_{1.4.9}X_{1.1.5}$
0000 0000 0000 1000	$X_{1.1.5}X'_{1.4.8}X'_{2.2.2}X_{2.3.14}$ $X'_{1.4.8}X_{1.1.5}X_{2.3.3}X'_{2.2.15}$ $X'_{2.2.2}X_{2.3.3}X_{1.1.5}X'_{1.4.9}$ $X_{2.3.3}X'_{2.2.2}X'_{1.4.8}X_{1.1.12}$ $X_{1.1.12}X'_{1.4.9}X'_{2.2.15}X_{2.3.3}$ $X'_{1.4.9}X_{1.1.12}X_{2.3.14}X'_{2.2.2}$ $X'_{2.2.15}X_{2.3.14}X_{1.1.12}X'_{1.4.8}$ $X_{2.3.14}X'_{2.2.15}X'_{1.4.9}X_{1.1.5}$
0000 0000 0000 0100	$X_{1.1.5}X'_{1.4.8}X_{2.2.2}X'_{2.3.14}$ $X'_{1.4.8}X_{1.1.5}X'_{2.3.3}X_{2.2.15}$ $X_{2.2.2}X'_{2.3.3}X_{1.1.5}X'_{1.4.9}$ $X'_{2.3.3}X_{2.2.2}X'_{1.4.8}X_{1.1.12}$ $X_{1.1.12}X'_{1.4.9}X_{2.2.15}X'_{2.3.3}$ $X'_{1.4.9}X_{1.1.12}X'_{2.3.14}X_{2.2.2}$ $X_{2.2.15}X'_{2.3.14}X_{1.1.12}X'_{1.4.8}$ $X'_{2.3.14}X_{2.2.15}X'_{1.4.9}X_{1.1.5}$
0000 0000 0000 0010	$X_{1.1.5}X_{1.4.8}X'_{2.2.2}X'_{2.3.14}$ $X_{1.4.8}X_{1.1.5}X'_{2.3.3}X'_{2.2.15}$ $X'_{2.2.2}X'_{2.3.3}X_{1.1.5}X_{1.4.9}$ $X'_{2.3.3}X'_{2.2.2}X_{1.4.8}X_{1.1.12}$ $X_{1.1.12}X_{1.4.9}X'_{2.2.15}X'_{2.3.3}$ $X_{1.4.9}X_{1.1.12}X'_{2.3.14}X'_{2.2.2}$ $X'_{2.2.15}X'_{2.3.14}X_{1.1.12}X_{1.4.8}$ $X'_{2.3.14}X'_{2.2.15}X_{1.4.9}X_{1.1.5}$
0000 0000 0000 0001	$X_{1.1.5}X_{1.4.8}X_{2.2.2}X_{2.3.14}$ $X_{1.4.8}X_{1.1.5}X_{2.3.3}X_{2.2.15}$ $X_{2.2.2}X_{2.3.3}X_{1.1.5}X_{1.4.9}$ $X_{2.3.3}X_{2.2.2}X_{1.4.8}X_{1.1.12}$ $X_{1.1.12}X_{1.4.9}X_{2.2.15}X_{2.3.3}$ $X_{1.4.9}X_{1.1.12}X_{2.3.14}X_{2.2.2}$ $X_{2.2.15}X_{2.3.14}X_{1.1.12}X_{1.4.8}$ $X_{2.3.14}X_{2.2.15}X_{1.4.9}X_{1.1.5}$

This graphic gives the combinations of Table 4.12. Relative to Table 4.7, we have both the exchange of the eighth with the ninth row (according to Table 4.8) and of the sixth with the eleventh row (according to Table 4.11).

Let us consider a similar choice for CA1.7:

$$\{X_{1.1.7}, X_{1.4.6}, X_{2.2.4}, X_{2.3.16}\}, \quad \{X_{1.1.10}, X_{1.4.11}, X_{2.2.13}, X_{2.3.1}\},$$
(4.14a)

$$\{X_{1.4.6}, X_{1.1.7}, X_{2.3.1}, X_{2.2.13}\}, \quad \{X_{1.4.11}, X_{1.1.10}, X_{2.3.16}, X_{2.2.4}\},$$
(4.14b)

$$\{X_{2.2.4}, X_{2.3.1}, X_{1.1.7}, X_{1.4.11}\}, \quad \{X_{2.2.13}, X_{2.3.16}, X_{1.1.10}, X_{1.4.6}\},$$
(4.14c)

$$\{X_{2.3.1}, X_{2.2.4}, X_{1.4.6}, X_{1.1.10}\}, \quad \{X_{2.3.16}, X_{2.2.13}, X_{1.4.11}, X_{1.1.7}\}.$$
(4.14d)

Table 4.12.   Quaternary products codified by 8 SCAs $\in$ CA1.6.

1000 0000 0000 0000	$X'_{1.1.6}X'_{1.4.7}X'_{2.2.1}X'_{2.3.13}$ $X'_{1.4.7}X'_{1.1.6}X'_{2.3.4}X'_{2.2.16}$ $X'_{2.2.1}X'_{2.3.4}X'_{1.1.6}X'_{1.4.10}$ $X'_{2.3.4}X'_{2.2.1}X'_{1.4.7}X'_{1.1.11}$	$X'_{1.1.11}X'_{1.4.10}X'_{2.2.16}X'_{2.3.4}$ $X'_{1.4.10}X'_{1.1.11}X'_{2.3.13}X'_{2.2.1}$ $X'_{2.2.16}X'_{2.3.13}X'_{1.1.11}X'_{1.4.7}$ $X'_{2.3.13}X'_{2.2.16}X'_{1.4.10}X'_{1.1.6}$
0100 0000 0000 0000	$X'_{1.1.6}X'_{1.4.7}X_{2.2.1}X_{2.3.13}$ $X'_{1.4.7}X'_{1.1.6}X_{2.3.4}X_{2.2.16}$ $X_{2.2.1}X_{2.3.4}X'_{1.1.6}X'_{1.4.10}$ $X_{2.3.4}X_{2.2.1}X'_{1.4.7}X'_{1.1.11}$	$X'_{1.1.11}X'_{1.4.10}X_{2.2.16}X_{2.3.4}$ $X'_{1.4.10}X'_{1.1.11}X_{2.3.13}X_{2.2.1}$ $X_{2.2.16}X_{2.3.13}X'_{1.1.11}X'_{1.4.7}$ $X_{2.3.13}X_{2.2.16}X'_{1.4.10}X'_{1.1.6}$
0010 0000 0000 0000	$X'_{1.1.6}X_{1.4.7}X'_{2.2.1}X_{2.3.13}$ $X_{1.4.7}X'_{1.1.6}X_{2.3.4}X'_{2.2.16}$ $X'_{2.2.1}X_{2.3.4}X'_{1.1.6}X_{1.4.10}$ $X_{2.3.4}X'_{2.2.1}X_{1.4.7}X'_{1.1.11}$	$X'_{1.1.11}X_{1.4.10}X'_{2.2.16}X_{2.3.4}$ $X_{1.4.10}X'_{1.1.11}X_{2.3.13}X'_{2.2.1}$ $X'_{2.2.16}X_{2.3.13}X'_{1.1.11}X_{1.4.7}$ $X_{2.3.13}X'_{2.2.16}X_{1.4.10}X'_{1.1.6}$
0001 0000 0000 0000	$X'_{1.1.6}X_{1.4.7}X_{2.2.1}X'_{2.3.13}$ $X_{1.4.7}X'_{1.1.6}X'_{2.3.4}X_{2.2.16}$ $X_{2.2.1}X'_{2.3.4}X'_{1.1.6}X_{1.4.10}$ $X'_{2.3.4}X_{2.2.1}X_{1.4.7}X'_{1.1.11}$	$X'_{1.1.11}X_{1.4.10}X_{2.2.16}X'_{2.3.4}$ $X_{1.4.10}X'_{1.1.11}X'_{2.3.13}X_{2.2.1}$ $X_{2.2.16}X'_{2.3.13}X'_{1.1.11}X_{1.4.7}$ $X'_{2.3.13}X_{2.2.16}X_{1.4.10}X'_{1.1.6}$
0000 1000 0000 0000	$X'_{1.1.6}X'_{1.4.7}X'_{2.2.1}X_{2.3.13}$ $X'_{1.4.7}X'_{1.1.6}X'_{2.3.4}X_{2.2.16}$ $X'_{2.2.1}X'_{2.3.4}X'_{1.1.6}X_{1.4.10}$ $X'_{2.3.4}X'_{2.2.1}X'_{1.4.7}X_{1.1.11}$	$X_{1.1.11}X_{1.4.10}X_{2.2.16}X'_{2.3.4}$ $X_{1.4.10}X_{1.1.11}X_{2.3.13}X'_{2.2.1}$ $X_{2.2.16}X_{2.3.13}X_{1.1.11}X'_{1.4.7}$ $X_{2.3.13}X_{2.2.16}X_{1.4.10}X'_{1.1.6}$
0000 0100 0000 0000	$X_{1.1.6}X_{1.4.7}X'_{2.2.1}X_{2.3.13}$ $X_{1.4.7}X_{1.1.6}X'_{2.3.4}X_{2.2.16}$ $X'_{2.2.1}X'_{2.3.4}X_{1.1.6}X'_{1.4.10}$ $X'_{2.3.4}X'_{2.2.1}X_{1.4.7}X'_{1.1.11}$	$X'_{1.1.11}X'_{1.4.10}X_{2.2.16}X'_{2.3.4}$ $X'_{1.4.10}X'_{1.1.11}X_{2.3.13}X'_{2.2.1}$ $X_{2.2.16}X_{2.3.13}X'_{1.1.11}X_{1.4.7}$ $X_{2.3.13}X_{2.2.16}X'_{1.4.10}X_{1.1.6}$
0000 0010 0000 0000	$X'_{1.1.6}X_{1.4.7}X'_{2.2.1}X'_{2.3.13}$ $X_{1.4.7}X'_{1.1.6}X_{2.3.4}X_{2.2.16}$ $X'_{2.2.1}X_{2.3.4}X'_{1.1.6}X_{1.4.10}$ $X_{2.3.4}X'_{2.2.1}X_{1.4.7}X_{1.1.11}$	$X'_{1.1.11}X_{1.4.10}X_{2.2.16}X_{2.3.4}$ $X'_{1.4.10}X_{1.1.11}X'_{2.3.13}X'_{2.2.1}$ $X_{2.2.16}X'_{2.3.13}X_{1.1.11}X_{1.4.7}$ $X'_{2.3.13}X_{2.2.16}X'_{1.4.10}X'_{1.1.6}$
0000 0001 0000 0000	$X_{1.1.6}X'_{1.4.7}X'_{2.2.1}X'_{2.3.13}$ $X'_{1.4.7}X_{1.1.6}X_{2.3.4}X_{2.2.16}$ $X'_{2.2.1}X_{2.3.4}X_{1.1.6}X_{1.4.10}$ $X_{2.3.4}X'_{2.2.1}X'_{1.4.7}X'_{1.1.11}$	$X'_{1.1.11}X_{1.4.10}X_{2.2.16}X_{2.3.4}$ $X_{1.4.10}X'_{1.1.11}X'_{2.3.13}X'_{2.2.1}$ $X_{2.2.16}X'_{2.3.13}X'_{1.1.11}X'_{1.4.7}$ $X'_{2.3.13}X_{2.2.16}X_{1.4.10}X_{1.1.6}$
0000 0000 1000 0000	$X'_{1.1.6}X'_{1.4.7}X_{2.2.1}X_{2.3.13}$ $X_{1.4.7}X'_{1.1.6}X'_{2.3.4}X'_{2.2.16}$ $X_{2.2.1}X'_{2.3.4}X'_{1.1.6}X'_{1.4.10}$ $X'_{2.3.4}X_{2.2.1}X_{1.4.7}X_{1.1.11}$	$X'_{1.1.11}X_{1.4.10}X'_{2.2.16}X'_{2.3.4}$ $X'_{1.4.10}X_{1.1.11}X_{2.3.13}X_{2.2.1}$ $X'_{2.2.16}X_{2.3.13}X_{1.1.11}X_{1.4.7}$ $X_{2.3.13}X'_{2.2.16}X'_{1.4.10}X'_{1.1.6}$
0000 0000 0100 0000	$X'_{1.1.6}X'_{1.4.7}X_{2.2.1}X_{2.3.13}$ $X'_{1.4.7}X_{1.1.6}X'_{2.3.4}X'_{2.2.16}$ $X_{2.2.1}X'_{2.3.4}X_{1.1.6}X_{1.4.10}$ $X_{2.3.4}X_{2.2.1}X'_{1.4.7}X'_{1.1.11}$	$X'_{1.1.11}X_{1.4.10}X'_{2.2.16}X'_{2.3.4}$ $X_{1.4.10}X'_{1.1.11}X_{2.3.13}X_{2.2.1}$ $X'_{2.2.16}X_{2.3.13}X_{1.1.11}X'_{1.4.7}$ $X_{2.3.13}X'_{2.2.16}X_{1.4.10}X_{1.1.6}$

Table 4.12.   (*Continued*)

0000 0000 0010 0000	$X'_{1.1.6}X'_{1.4.7}X_{2.2.1}X'_{2.3.13}$ $X'_{1.4.7}X'_{1.1.6}X_{2.3.4}X'_{2.2.16}$ $X_{2.2.1}X_{2.3.4}X'_{1.1.6}X_{1.4.10}$ $X_{2.3.4}X_{2.2.1}X'_{1.4.7}X_{1.1.11}$	$X_{1.1.11}X_{1.4.10}X'_{2.2.16}X_{2.3.4}$ $X_{1.4.10}X_{1.1.11}X'_{2.3.13}X_{2.2.1}$ $X'_{2.2.16}X'_{2.3.13}X_{1.1.11}X'_{1.4.7}$ $X'_{2.3.13}X'_{2.2.16}X_{1.4.10}X'_{1.1.6}$
0000 0000 0001 0000	$X_{1.1.6}X_{1.4.7}X_{2.2.1}X'_{2.3.13}$ $X_{1.4.7}X_{1.1.6}X_{2.3.4}X'_{2.2.16}$ $X_{2.2.1}X_{2.3.4}X_{1.1.6}X'_{1.4.10}$ $X_{2.3.4}X_{2.2.1}X_{1.4.7}X'_{1.1.11}$	$X'_{1.1.11}X'_{1.4.10}X'_{2.2.16}X_{2.3.4}$ $X'_{1.4.10}X'_{1.1.11}X'_{2.3.13}X_{2.2.1}$ $X'_{2.2.16}X'_{2.3.13}X'_{1.1.11}X_{1.4.7}$ $X'_{2.3.13}X'_{2.2.16}X'_{1.4.10}X_{1.1.6}$
0000 0000 0000 1000	$X_{1.1.6}X'_{1.4.7}X'_{2.2.1}X_{2.3.13}$ $X'_{1.4.7}X_{1.1.6}X_{2.3.4}X'_{2.2.16}$ $X'_{2.2.1}X_{2.3.4}X_{1.1.6}X'_{1.4.10}$ $X_{2.3.4}X'_{2.2.1}X'_{1.4.7}X_{1.1.11}$	$X_{1.1.11}X'_{1.4.10}X'_{2.2.16}X_{2.3.4}$ $X'_{1.4.10}X_{1.1.11}X_{2.3.13}X'_{2.2.1}$ $X'_{2.2.16}X_{2.3.13}X_{1.1.11}X'_{1.4.7}$ $X_{2.3.13}X'_{2.2.16}X'_{1.4.10}X_{1.1.6}$
0000 0000 0000 0100	$X_{1.1.6}X'_{1.4.7}X_{2.2.1}X'_{2.3.13}$ $X'_{1.4.7}X_{1.1.6}X'_{2.3.4}X_{2.2.16}$ $X_{2.2.1}X'_{2.3.4}X_{1.1.6}X'_{1.4.10}$ $X'_{2.3.4}X_{2.2.1}X'_{1.4.7}X_{1.1.11}$	$X_{1.1.11}X'_{1.4.10}X_{2.2.16}X'_{2.3.4}$ $X'_{1.4.10}X_{1.1.11}X'_{2.3.13}X_{2.2.1}$ $X_{2.2.16}X'_{2.3.13}X_{1.1.11}X'_{1.4.7}$ $X'_{2.3.13}X_{2.2.16}X'_{1.4.10}X_{1.1.6}$
0000 0000 0000 0001	$X_{1.1.6}X_{1.4.7}X'_{2.2.1}X'_{2.3.13}$ $X_{1.4.7}X_{1.1.6}X'_{2.3.4}X'_{2.2.16}$ $X'_{2.2.1}X'_{2.3.4}X_{1.1.6}X_{1.4.10}$ $X'_{2.3.4}X'_{2.2.1}X_{1.4.7}X_{1.1.11}$	$X_{1.1.11}X_{1.4.10}X'_{2.2.16}X'_{2.3.4}$ $X_{1.4.10}X_{1.1.11}X'_{2.3.13}X'_{2.2.1}$ $X'_{2.2.16}X'_{2.3.13}X_{1.1.11}X_{1.4.7}$ $X'_{2.3.13}X'_{2.2.16}X_{1.4.10}X_{1.1.6}$
0000 0000 0000 0001	$X_{1.1.6}X_{1.4.7}X_{2.2.1}X_{2.3.13}$ $X_{1.4.7}X_{1.1.6}X_{2.3.4}X_{2.2.16}$ $X_{2.2.1}X_{2.3.4}X_{1.1.6}X_{1.4.10}$ $X_{2.3.4}X_{2.2.1}X_{1.4.7}X_{1.1.11}$	$X_{1.1.11}X_{1.4.10}X_{2.2.16}X_{2.3.4}$ $X_{1.4.10}X_{1.1.11}X_{2.3.13}X_{2.2.1}$ $X_{2.2.16}X_{2.3.13}X_{1.1.11}X_{1.4.7}$ $X_{2.3.13}X_{2.2.16}X_{1.4.10}X_{1.1.6}$

This graphic gives the combinations of Table 4.13. Relative to Table 4.7, we have both the exchange of the seventh with the tenth row (according to Table 4.9) and of the sixth with the eleventh row (according to Table 4.11).

Finally, let us consider CA1.8:

$$\{X_{1.1.8}, X_{1.4.5}, X_{2.2.3}, X_{2.3.15}\}, \quad \{X_{1.1.9}, X_{1.4.12}, X_{2.2.14}, X_{2.3.2}\},$$
$$(4.15a)$$

$$\{X_{1.4.5}, X_{1.1.8}, X_{2.3.2}, X_{2.2.14}\}, \quad \{X_{1.4.12}, X_{1.1.9}, X_{2.3.15}, X_{2.2.3}\},$$
$$(4.15b)$$

$$\{X_{2.2.3}, X_{2.3.2}, X_{1.1.8}, X_{1.4.12}\}, \quad \{X_{2.2.14}, X_{2.3.15}, X_{1.1.9}, X_{1.4.5}\},$$
$$(4.15c)$$

$$\{X_{2.3.2}, X_{2.2.3}, X_{1.4.5}, X_{1.1.9}\}, \quad \{X_{2.3.15}, X_{2.2.14}, X_{1.4.12}, X_{1.1.8}\}.$$
$$(4.15d)$$

Table 4.13.   Quaternary products codified by 8 SCAs $\in$ CA1.7.

Code	Products (I)	Products (II)
1000 0000 0000 0000	$X'_{1.1.7}X'_{1.4.6}X'_{2.2.4}X'_{2.3.16}$ $X'_{1.4.6}X'_{1.1.7}X'_{2.3.1}X'_{2.2.13}$ $X'_{2.2.4}X'_{2.3.1}X'_{1.1.7}X'_{1.4.11}$ $X'_{2.3.1}X'_{2.2.4}X'_{1.4.6}X'_{1.1.10}$	$X'_{1.1.10}X'_{1.4.11}X'_{2.2.13}X'_{2.3.1}$ $X'_{1.4.11}X'_{1.1.10}X'_{2.3.16}X'_{2.2.4}$ $X'_{2.2.13}X'_{2.3.16}X'_{1.1.10}X'_{1.4.6}$ $X'_{2.3.16}X'_{2.2.13}X'_{1.4.11}X'_{1.1.7}$
0100 0000 0000 0000	$X'_{1.1.7}X'_{1.4.6}X_{2.2.4}X_{2.3.16}$ $X'_{1.4.6}X'_{1.1.7}X_{2.3.1}X_{2.2.13}$ $X_{2.2.4}X_{2.3.1}X'_{1.1.7}X'_{1.4.11}$ $X_{2.3.1}X_{2.2.4}X'_{1.4.6}X'_{1.1.10}$	$X'_{1.1.10}X'_{1.4.11}X_{2.2.13}X_{2.3.1}$ $X'_{1.4.11}X'_{1.1.10}X_{2.3.16}X_{2.2.4}$ $X_{2.2.13}X_{2.3.16}X'_{1.1.10}X'_{1.4.6}$ $X_{2.3.16}X_{2.2.13}X'_{1.4.11}X'_{1.1.7}$
0010 0000 0000 0000	$X'_{1.1.7}X_{1.4.6}X'_{2.2.4}X_{2.3.16}$ $X_{1.4.6}X'_{1.1.7}X_{2.3.1}X'_{2.2.13}$ $X'_{2.2.4}X_{2.3.1}X'_{1.1.7}X_{1.4.11}$ $X_{2.3.1}X'_{2.2.4}X_{1.4.6}X'_{1.1.10}$	$X'_{1.1.10}X_{1.4.11}X'_{2.2.13}X_{2.3.1}$ $X_{1.4.11}X'_{1.1.10}X_{2.3.16}X'_{2.2.4}$ $X'_{2.2.13}X_{2.3.16}X'_{1.1.10}X_{1.4.6}$ $X_{2.3.16}X'_{2.2.13}X_{1.4.11}X'_{1.1.7}$
0001 0000 0000 0000	$X'_{1.1.7}X_{1.4.6}X_{2.2.4}X'_{2.3.16}$ $X_{1.4.6}X'_{1.1.7}X'_{2.3.1}X_{2.2.13}$ $X_{2.2.4}X'_{2.3.1}X'_{1.1.7}X_{1.4.11}$ $X'_{2.3.1}X_{2.2.4}X_{1.4.6}X'_{1.1.10}$	$X'_{1.1.10}X_{1.4.11}X_{2.2.13}X'_{2.3.1}$ $X_{1.4.11}X'_{1.1.10}X'_{2.3.16}X_{2.2.4}$ $X_{2.2.13}X'_{2.3.16}X'_{1.1.10}X_{1.4.6}$ $X'_{2.3.16}X_{2.2.13}X_{1.4.11}X'_{1.1.7}$
0000 1000 0000 0000	$X'_{1.1.7}X'_{1.4.6}X'_{2.2.4}X_{2.3.16}$ $X'_{1.4.6}X'_{1.1.7}X'_{2.3.1}X_{2.2.13}$ $X'_{2.2.4}X'_{2.3.1}X'_{1.1.7}X_{1.4.11}$ $X'_{2.3.1}X'_{2.2.4}X'_{1.4.6}X_{1.1.10}$	$X_{1.1.10}X_{1.4.11}X_{2.2.13}X'_{2.3.1}$ $X_{1.4.11}X_{1.1.10}X_{2.3.16}X'_{2.2.4}$ $X_{2.2.13}X_{2.3.16}X_{1.1.10}X'_{1.4.6}$ $X_{2.3.16}X_{2.2.13}X_{1.4.11}X'_{1.1.7}$
0000 0100 0000 0000	$X_{1.1.7}X'_{1.4.6}X'_{2.2.4}X_{2.3.16}$ $X_{1.4.6}X_{1.1.7}X'_{2.3.1}X_{2.2.13}$ $X'_{2.2.4}X'_{2.3.1}X_{1.1.7}X'_{1.4.11}$ $X'_{2.3.1}X'_{2.2.4}X_{1.4.6}X'_{1.1.10}$	$X'_{1.1.10}X'_{1.4.11}X_{2.2.13}X_{2.3.1}$ $X'_{1.4.11}X'_{1.1.10}X_{2.3.16}X_{2.2.4}$ $X_{2.2.13}X_{2.3.16}X'_{1.1.10}X_{1.4.6}$ $X_{2.3.16}X_{2.2.13}X'_{1.4.11}X_{1.1.7}$
0000 0010 0000 0000	$X_{1.1.7}X'_{1.4.6}X_{2.2.4}X_{2.3.16}$ $X'_{1.4.6}X_{1.1.7}X'_{2.3.1}X'_{2.2.13}$ $X_{2.2.4}X'_{2.3.1}X_{1.1.7}X_{1.4.11}$ $X'_{2.3.1}X_{2.2.4}X'_{1.4.6}X'_{1.1.10}$	$X'_{1.1.10}X_{1.4.11}X'_{2.2.13}X'_{2.3.1}$ $X_{1.4.11}X'_{1.1.10}X_{2.3.16}X_{2.2.4}$ $X'_{2.2.13}X_{2.3.16}X'_{1.1.10}X'_{1.4.6}$ $X_{2.3.16}X'_{2.2.13}X_{1.4.11}X_{1.1.7}$
0000 0001 0000 0000	$X'_{1.1.7}X_{1.4.6}X_{2.2.4}X_{2.3.16}$ $X_{1.4.6}X'_{1.1.7}X'_{2.3.1}X'_{2.2.13}$ $X_{2.2.4}X'_{2.3.1}X'_{1.1.7}X'_{1.4.11}$ $X'_{2.3.1}X_{2.2.4}X_{1.4.6}X_{1.1.10}$	$X_{1.1.10}X'_{1.4.11}X'_{2.2.13}X'_{2.3.1}$ $X'_{1.4.11}X_{1.1.10}X_{2.3.16}X_{2.2.4}$ $X'_{2.2.13}X_{2.3.16}X_{1.1.10}X_{1.4.6}$ $X_{2.3.16}X'_{2.2.13}X'_{1.4.11}X'_{1.1.7}$
0000 0000 1000 0000	$X'_{1.1.7}X_{1.4.6}X'_{2.2.4}X'_{2.3.16}$ $X_{1.4.6}X'_{1.1.7}X_{2.3.1}X_{2.2.13}$ $X'_{2.2.4}X_{2.3.1}X_{1.1.7}X_{1.4.11}$ $X_{2.3.1}X'_{2.2.4}X'_{1.4.6}X'_{1.1.10}$	$X_{1.1.10}X'_{1.4.11}X_{2.2.13}X_{2.3.1}$ $X_{1.4.11}X'_{1.1.10}X'_{2.3.16}X'_{2.2.4}$ $X_{2.2.13}X'_{2.3.16}X'_{1.1.10}X_{1.4.6}$ $X'_{2.3.16}X_{2.2.13}X_{1.4.11}X_{1.1.7}$
0000 0000 0100 0000	$X'_{1.1.7}X_{1.4.6}X'_{2.2.4}X'_{2.3.16}$ $X_{1.4.6}X'_{1.1.7}X_{2.3.1}X_{2.2.13}$ $X'_{2.2.4}X_{2.3.1}X'_{1.1.7}X'_{1.4.11}$ $X_{2.3.1}X'_{2.2.4}X_{1.4.6}X_{1.1.10}$	$X_{1.1.10}X'_{1.4.11}X_{2.2.13}X_{2.3.1}$ $X'_{1.4.11}X_{1.1.10}X'_{2.3.16}X'_{2.2.4}$ $X_{2.2.13}X'_{2.3.16}X_{1.1.10}X_{1.4.6}$ $X'_{2.3.16}X_{2.2.13}X'_{1.4.11}X'_{1.1.7}$

Table 4.13. (*Continued*)

0000 0000 0010 0000	$X'_{1.1.7}X'_{1.4.6}X_{2.2.4}X'_{2.3.16}$ $X'_{1.4.6}X'_{1.1.7}X_{2.3.1}X'_{2.2.13}$ $X_{2.2.4}X_{2.3.1}X'_{1.1.7}X_{1.4.11}$ $X_{2.3.1}X_{2.2.4}X'_{1.4.6}X_{1.1.10}$	$X_{1.1.10}X_{1.4.11}X'_{2.2.13}X_{2.3.1}$ $X_{1.4.11}X_{1.1.10}X'_{2.3.16}X_{2.2.4}$ $X'_{2.2.13}X'_{2.3.16}X_{1.1.10}X'_{1.4.6}$ $X'_{2.3.16}X'_{2.2.13}X_{1.4.11}X'_{1.1.7}$
0000 0000 0001 0000	$X_{1.1.7}X_{1.4.6}X_{2.2.4}X'_{2.3.16}$ $X_{1.4.6}X_{1.1.7}X_{2.3.1}X'_{2.2.13}$ $X_{2.2.4}X_{2.3.1}X_{1.1.7}X'_{1.4.11}$ $X_{2.3.1}X_{2.2.4}X_{1.4.6}X'_{1.1.10}$	$X'_{1.1.10}X'_{1.4.11}X'_{2.2.13}X_{2.3.1}$ $X'_{1.4.11}X'_{1.1.10}X'_{2.3.16}X_{2.2.4}$ $X'_{2.2.13}X'_{2.3.16}X'_{1.1.10}X_{1.4.6}$ $X'_{2.3.16}X'_{2.2.13}X'_{1.4.11}X_{1.1.7}$
0000 0000 0000 1000	$X_{1.1.7}X'_{1.4.6}X'_{2.2.4}X_{2.3.16}$ $X'_{1.4.6}X_{1.1.7}X_{2.3.1}X'_{2.2.13}$ $X'_{2.2.4}X_{2.3.1}X_{1.1.7}X'_{1.4.11}$ $X_{2.3.1}X'_{2.2.4}X'_{1.4.6}X_{1.1.10}$	$X_{1.1.10}X'_{1.4.11}X'_{2.2.13}X_{2.3.1}$ $X_{1.4.11}X_{1.1.10}X_{2.3.16}X'_{2.2.4}$ $X'_{2.2.13}X_{2.3.16}X_{1.1.10}X'_{1.4.6}$ $X_{2.3.16}X'_{2.2.13}X'_{1.4.11}X_{1.1.7}$
0000 0000 0000 0100	$X_{1.1.7}X'_{1.4.6}X_{2.2.4}X'_{2.3.16}$ $X'_{1.4.6}X_{1.1.7}X'_{2.3.1}X_{2.2.13}$ $X_{2.2.4}X'_{2.3.1}X_{1.1.7}X'_{1.4.11}$ $X'_{2.3.1}X_{2.2.4}X'_{1.4.6}X_{1.1.10}$	$X_{1.1.10}X'_{1.4.11}X_{2.2.13}X'_{2.3.1}$ $X'_{1.4.11}X_{1.1.10}X'_{2.3.16}X_{2.2.4}$ $X_{2.2.13}X'_{2.3.16}X_{1.1.10}X'_{1.4.6}$ $X'_{2.3.16}X_{2.2.13}X'_{1.4.11}X_{1.1.7}$
0000 0000 0000 0010	$X_{1.1.7}X_{1.4.6}X'_{2.2.4}X'_{2.3.16}$ $X_{1.4.6}X_{1.1.7}X'_{2.3.1}X'_{2.2.13}$ $X'_{2.2.4}X'_{2.3.1}X_{1.1.7}X_{1.4.11}$ $X'_{2.3.1}X'_{2.2.4}X_{1.4.6}X_{1.1.10}$	$X_{1.1.10}X_{1.4.11}X'_{2.2.13}X'_{2.3.1}$ $X_{1.4.11}X_{1.1.10}X'_{2.3.16}X'_{2.2.4}$ $X'_{2.2.13}X'_{2.3.16}X_{1.1.10}X_{1.4.6}$ $X'_{2.3.16}X'_{2.2.13}X_{1.4.11}X_{1.1.7}$
0000 0000 0000 0001	$X_{1.1.7}X_{1.4.6}X_{2.2.4}X_{2.3.16}$ $X_{1.4.6}X_{1.1.7}X_{2.3.1}X_{2.2.13}$ $X_{2.2.4}X_{2.3.1}X_{1.1.7}X_{1.4.11}$ $X_{2.3.1}X_{2.2.4}X_{1.4.6}X_{1.1.10}$	$X_{1.1.10}X_{1.4.11}X_{2.2.13}X_{2.3.1}$ $X_{1.4.11}X_{1.1.10}X_{2.3.16}X_{2.2.4}$ $X_{2.2.13}X_{2.3.16}X_{1.1.10}X_{1.4.6}$ $X_{2.3.16}X_{2.2.13}X_{1.4.11}X_{1.1.7}$

This graphic gives the combinations of Table 4.14. Note the inversion of the six central rows. It combines all three exchanges so far considered.

We proceed in a similar way for CA2.1 and get following SCAs:

$$\{X_{1.2.1}, X_{1.3.4}, X_{2.1.6}, X_{2.4.10}\}, \quad \{X_{1.2.16}, X_{1.3.13}, X_{2.1.11}, X_{2.4.7}\},$$
$$(4.16a)$$

$$\{X_{1.3.4}, X_{1.2.1}, X_{2.4.7}, X_{2.1.11}\}, \quad \{X_{1.3.13}, X_{1.2.16}, X_{2.4.10}, X_{2.1.6}\},$$
$$(4.16b)$$

$$\{X_{2.1.6}, X_{2.4.7}, X_{1.2.1}, X_{1.3.13}\}, \quad \{X_{2.1.11}, X_{2.4.10}, X_{1.2.16}, X_{1.3.4}\},$$
$$(4.16c)$$

$$\{X_{2.4.7}, X_{2.1.6}, X_{1.3.4}, X_{1.2.16}\}, \quad \{X_{2.4.10}, X_{2.1.11}, X_{1.3.13}, X_{1.2.1}\}.$$
$$(4.16d)$$

Table 4.14.   Quaternary products codified by 8 SCAs $\in$ CA1.8.

1000 0000 0000 0000	$X'_{1.1.8}X'_{1.4.5}X'_{2.2.3}X'_{2.3.15}$ $X'_{1.4.5}X'_{1.1.8}X'_{2.3.2}X'_{2.2.14}$ $X'_{2.2.3}X'_{2.3.2}X'_{1.1.8}X'_{1.4.12}$ $X'_{2.3.2}X'_{2.2.3}X'_{1.4.5}X'_{1.1.9}$	$X'_{1.1.9}X'_{1.4.12}X'_{2.2.14}X'_{2.3.2}$ $X'_{1.4.12}X'_{1.1.9}X'_{2.3.15}X'_{2.2.3}$ $X'_{2.2.14}X'_{2.3.15}X'_{1.1.9}X'_{1.4.5}$ $X'_{2.3.15}X'_{2.2.14}X'_{1.4.12}X'_{1.1.8}$
0100 0000 0000 0000	$X'_{1.1.8}X'_{1.4.5}X_{2.2.3}X_{2.3.15}$ $X'_{1.4.5}X'_{1.1.8}X_{2.3.2}X_{2.2.14}$ $X_{2.2.3}X_{2.3.2}X'_{1.1.8}X'_{1.4.12}$ $X_{2.3.2}X_{2.2.3}X'_{1.4.5}X'_{1.1.9}$	$X'_{1.1.9}X'_{1.4.12}X_{2.2.14}X_{2.3.2}$ $X'_{1.4.12}X'_{1.1.9}X_{2.3.15}X_{2.2.3}$ $X_{2.2.14}X_{2.3.15}X'_{1.1.9}X'_{1.4.5}$ $X_{2.3.15}X_{2.2.14}X'_{1.4.12}X'_{1.1.8}$
0010 0000 0000 0000	$X'_{1.1.8}X_{1.4.5}X'_{2.2.3}X_{2.3.15}$ $X_{1.4.5}X'_{1.1.8}X_{2.3.2}X'_{2.2.14}$ $X'_{2.2.3}X_{2.3.2}X'_{1.1.8}X_{1.4.12}$ $X_{2.3.2}X'_{2.2.3}X_{1.4.5}X'_{1.1.9}$	$X'_{1.1.9}X_{1.4.12}X'_{2.2.14}X_{2.3.2}$ $X_{1.4.12}X'_{1.1.9}X_{2.3.15}X'_{2.2.3}$ $X'_{2.2.14}X_{2.3.15}X'_{1.1.9}X_{1.4.5}$ $X_{2.3.15}X'_{2.2.14}X_{1.4.12}X'_{1.1.8}$
0001 0000 0000 0000	$X'_{1.1.8}X_{1.4.5}X_{2.2.3}X'_{2.3.15}$ $X_{1.4.5}X'_{1.1.8}X'_{2.3.2}X_{2.2.14}$ $X_{2.2.3}X'_{2.3.2}X'_{1.1.8}X_{1.4.12}$ $X'_{2.3.2}X_{2.2.3}X_{1.4.5}X'_{1.1.9}$	$X'_{1.1.9}X_{1.4.12}X_{2.2.14}X'_{2.3.2}$ $X_{1.4.12}X'_{1.1.9}X'_{2.3.15}X_{2.2.3}$ $X_{2.2.14}X'_{2.3.15}X'_{1.1.9}X_{1.4.5}$ $X'_{2.3.15}X_{2.2.14}X_{1.4.12}X'_{1.1.8}$
0000 1000 0000 0000	$X'_{1.1.8}X'_{1.4.5}X'_{2.2.3}X_{2.3.15}$ $X'_{1.4.5}X'_{1.1.8}X'_{2.3.2}X_{2.2.14}$ $X'_{2.2.3}X'_{2.3.2}X'_{1.1.8}X_{1.4.12}$ $X'_{2.3.2}X'_{2.2.3}X'_{1.4.5}X_{1.1.9}$	$X_{1.1.9}X_{1.4.12}X_{2.2.14}X'_{2.3.2}$ $X_{1.4.12}X_{1.1.9}X_{2.3.15}X'_{2.2.3}$ $X_{2.2.14}X_{2.3.15}X_{1.1.9}X'_{1.4.5}$ $X_{2.3.15}X_{2.2.14}X_{1.4.12}X'_{1.1.8}$
0000 0100 0000 0000	$X_{1.1.8}X_{1.4.5}X'_{2.2.3}X_{2.3.15}$ $X_{1.4.5}X_{1.1.8}X'_{2.3.2}X_{2.2.14}$ $X'_{2.2.3}X'_{2.3.2}X_{1.1.8}X'_{1.4.12}$ $X'_{2.3.2}X'_{2.2.3}X_{1.4.5}X'_{1.1.9}$	$X'_{1.1.9}X'_{1.4.12}X_{2.2.14}X'_{2.3.2}$ $X'_{1.4.12}X'_{1.1.9}X_{2.3.15}X'_{2.2.3}$ $X_{2.2.14}X_{2.3.15}X'_{1.1.9}X_{1.4.5}$ $X_{2.3.15}X_{2.2.14}X'_{1.4.12}X_{1.1.8}$
0000 0010 0000 0000	$X_{1.1.8}X'_{1.4.5}X_{2.2.3}X_{2.3.15}$ $X'_{1.4.5}X_{1.1.8}X'_{2.3.2}X'_{2.2.14}$ $X_{2.2.3}X'_{2.3.2}X_{1.1.8}X_{1.4.12}$ $X'_{2.3.2}X_{2.2.3}X'_{1.4.5}X'_{1.1.9}$	$X'_{1.1.9}X_{1.4.12}X'_{2.2.14}X'_{2.3.2}$ $X_{1.4.12}X'_{1.1.9}X_{2.3.15}X_{2.2.3}$ $X'_{2.2.14}X_{2.3.15}X'_{1.1.9}X'_{1.4.5}$ $X_{2.3.15}X'_{2.2.14}X_{1.4.12}X_{1.1.8}$
0000 0001 0000 0000	$X_{1.1.8}X'_{1.4.5}X'_{2.2.3}X'_{2.3.15}$ $X'_{1.4.5}X_{1.1.8}X_{2.3.2}X_{2.2.14}$ $X'_{2.2.3}X_{2.3.2}X_{1.1.8}X_{1.4.12}$ $X_{2.3.2}X'_{2.2.3}X'_{1.4.5}X'_{1.1.9}$	$X'_{1.1.9}X_{1.4.12}X_{2.2.14}X_{2.3.2}$ $X_{1.4.12}X'_{1.1.9}X'_{2.3.15}X'_{2.2.3}$ $X_{2.2.14}X'_{2.3.15}X'_{1.1.9}X'_{1.4.5}$ $X'_{2.3.15}X_{2.2.14}X_{1.4.12}X_{1.1.8}$
0000 0000 1000 0000	$X'_{1.1.8}X_{1.4.5}X_{2.2.3}X_{2.3.15}$ $X_{1.4.5}X'_{1.1.8}X'_{2.3.2}X'_{2.2.14}$ $X_{2.2.3}X'_{2.3.2}X'_{1.1.8}X'_{1.4.12}$ $X'_{2.3.2}X_{2.2.3}X_{1.4.5}X_{1.1.9}$	$X_{1.1.9}X'_{1.4.12}X'_{2.2.14}X'_{2.3.2}$ $X'_{1.4.12}X_{1.1.9}X_{2.3.15}X_{2.2.3}$ $X'_{2.2.14}X_{2.3.15}X_{1.1.9}X_{1.4.5}$ $X_{2.3.15}X'_{2.2.14}X'_{1.4.12}X'_{1.1.8}$
0000 0000 0100 0000	$X'_{1.1.8}X_{1.4.5}X'_{2.2.3}X'_{2.3.15}$ $X_{1.4.5}X'_{1.1.8}X_{2.3.2}X_{2.2.14}$ $X'_{2.2.3}X_{2.3.2}X'_{1.1.8}X_{1.4.12}$ $X_{2.3.2}X'_{2.2.3}X'_{1.4.5}X_{1.1.9}$	$X_{1.1.9}X'_{1.4.12}X_{2.2.14}X_{2.3.2}$ $X'_{1.4.12}X_{1.1.9}X'_{2.3.15}X'_{2.2.3}$ $X_{2.2.14}X'_{2.3.15}X_{1.1.9}X_{1.4.5}$ $X'_{2.3.15}X_{2.2.14}X'_{1.4.12}X'_{1.1.8}$

Table 4.14. (*Continued*)

0000 0000 0010 0000	$X'_{1.1.8}X'_{1.4.5}X_{2.2.3}X'_{2.3.15}$ $X'_{1.4.5}X'_{1.1.8}X_{2.3.2}X'_{2.2.14}$ $X_{2.2.3}X_{2.3.2}X'_{1.1.8}X_{1.4.12}$ $X_{2.3.2}X_{2.2.3}X'_{1.4.5}X_{1.1.9}$	$X_{1.1.9}X_{1.4.12}X'_{2.2.14}X_{2.3.2}$ $X_{1.4.12}X_{1.1.9}X'_{2.3.15}X_{2.2.3}$ $X'_{2.2.14}X'_{2.3.15}X_{1.1.9}X'_{1.4.5}$ $X'_{2.3.15}X'_{2.2.14}X_{1.4.12}X'_{1.1.8}$
0000 0000 0001 0000	$X_{1.1.8}X_{1.4.5}X_{2.2.3}X'_{2.3.15}$ $X_{1.4.5}X_{1.1.8}X_{2.3.2}X'_{2.2.14}$ $X_{2.2.3}X_{2.3.2}X_{1.1.8}X'_{1.4.12}$ $X_{2.3.2}X_{2.2.3}X_{1.4.5}X'_{1.1.9}$	$X'_{1.1.9}X'_{1.4.12}X'_{2.2.14}X_{2.3.2}$ $X'_{1.4.12}X'_{1.1.9}X'_{2.3.15}X_{2.2.3}$ $X'_{2.2.14}X'_{2.3.15}X'_{1.1.9}X_{1.4.5}$ $X'_{2.3.15}X'_{2.2.14}X'_{1.4.12}X_{1.1.8}$
0000 0000 0000 1000	$X_{1.1.8}X'_{1.4.5}X'_{2.2.3}X_{2.3.15}$ $X'_{1.4.5}X_{1.1.8}X_{2.3.2}X'_{2.2.14}$ $X'_{2.2.3}X_{2.3.2}X_{1.1.8}X'_{1.4.12}$ $X_{2.3.2}X'_{2.2.3}X'_{1.4.5}X_{1.1.9}$	$X_{1.1.9}X'_{1.4.12}X'_{2.2.14}X_{2.3.2}$ $X'_{1.4.12}X_{1.1.9}X_{2.3.15}X'_{2.2.3}$ $X'_{2.2.14}X_{2.3.15}X_{1.1.9}X'_{1.4.5}$ $X_{2.3.15}X'_{2.2.14}X'_{1.4.12}X_{1.1.8}$
0000 0000 0000 0100	$X_{1.1.8}X'_{1.4.5}X_{2.2.3}X'_{2.3.15}$ $X'_{1.4.5}X_{1.1.8}X'_{2.3.2}X_{2.2.14}$ $X_{2.2.3}X'_{2.3.2}X_{1.1.8}X'_{1.4.12}$ $X'_{2.3.2}X_{2.2.3}X'_{1.4.5}X_{1.1.9}$	$X_{1.1.9}X'_{1.4.12}X_{2.2.14}X'_{2.3.2}$ $X'_{1.4.12}X_{1.1.9}X'_{2.3.15}X_{2.2.3}$ $X_{2.2.14}X'_{2.3.15}X_{1.1.9}X'_{1.4.5}$ $X'_{2.3.15}X_{2.2.14}X'_{1.4.12}X_{1.1.8}$
0000 0000 0000 0010	$X_{1.1.8}X_{1.4.5}X'_{2.2.3}X'_{2.3.15}$ $X_{1.4.5}X_{1.1.8}X'_{2.3.2}X'_{2.2.14}$ $X'_{2.2.3}X'_{2.3.2}X_{1.1.8}X_{1.4.12}$ $X'_{2.3.2}X'_{2.2.3}X_{1.4.5}X_{1.1.9}$	$X_{1.1.9}X_{1.4.12}X'_{2.2.14}X'_{2.3.2}$ $X_{1.4.12}X_{1.1.9}X'_{2.3.15}X'_{2.2.3}$ $X'_{2.2.14}X'_{2.3.15}X_{1.1.9}X_{1.4.5}$ $X'_{2.3.15}X'_{2.2.14}X_{1.4.12}X_{1.1.8}$
0000 0000 0000 0001	$X_{1.1.8}X_{1.4.5}X_{2.2.3}X_{2.3.15}$ $X_{1.4.5}X_{1.1.8}X_{2.3.2}X_{2.2.14}$ $X_{2.2.3}X_{2.3.2}X_{1.1.8}X_{1.4.12}$ $X_{2.3.2}X_{2.2.3}X_{1.4.5}X_{1.1.9}$	$X_{1.1.9}X_{1.4.12}X_{2.2.14}X_{2.3.2}$ $X_{1.4.12}X_{1.1.9}X_{2.3.15}X_{2.2.3}$ $X_{2.2.14}X_{2.3.15}X_{1.1.9}X_{1.4.5}$ $X_{2.3.15}X_{2.2.14}X_{1.4.12}X_{1.1.8}$

This graphic gives the combinations displayed in Table 4.15. Note the characteristic exchange between rows 4 and 13 relative to CA1.1.

This inversion is the "signature" when passing from CAs 1 to CAs 2. The sequence of the CAs of the second group (right column of Table 4.3) follows a similar pattern.

4.3.   Some Examples of Nodes

4.3.1.   *2–chunks*

We can give rise to the 65,536 nodes of $\mathscr{B}_4$ by making use of the double chunks of $\mathscr{B}_3$, similar to what was done in Table 3.5 for the latter, where I recall that $256 \times 256 = 65536$. Note that here and in the following discussion, to avoid confusion with a single chunk with 4 digits (that I still use), I call the 2 chunks with 4 digits (each constituting a half of a 4D SD variable) 2-chunk, the 4 chunks of 4 digits each

Table 4.15.   Quaternary products codified by 8 SCAs $\in$ CA2.1.

Code	Products (left)	Products (right)
1000 0000 0000 0000	$X'_{1.2.1}X'_{1.3.4}X'_{2.1.6}X'_{2.4.10}$ $X'_{1.3.4}X'_{1.2.1}X'_{2.4.7}X'_{2.1.11}$ $X'_{2.1.6}X'_{2.4.7}X'_{1.2.1}X'_{1.3.13}$ $X'_{2.4.7}X'_{2.1.6}X'_{1.3.4}X'_{1.2.16}$	$X'_{1.2.16}X'_{1.3.13}X'_{2.1.11}X'_{2.4.7}$ $X'_{1.3.13}X'_{1.2.16}X'_{2.4.10}X'_{2.1.6}$ $X'_{2.1.11}X'_{2.4.10}X'_{1.2.16}X'_{1.3.4}$ $X'_{2.4.10}X'_{2.1.11}X'_{1.3.13}X'_{1.2.1}$
0100 0000 0000 0000	$X'_{1.2.1}X'_{1.3.4}X_{2.1.6}X_{2.4.10}$ $X'_{1.3.4}X'_{1.2.1}X_{2.4.7}X_{2.1.11}$ $X_{2.1.6}X_{2.4.7}X'_{1.2.1}X'_{1.3.13}$ $X_{2.4.7}X_{2.1.6}X'_{1.3.4}X'_{1.2.16}$	$X'_{1.2.16}X'_{1.3.13}X_{2.1.11}X_{2.4.7}$ $X'_{1.3.13}X'_{1.2.16}X_{2.4.10}X_{2.1.6}$ $X_{2.1.11}X_{2.4.10}X'_{1.2.16}X'_{1.3.4}$ $X_{2.4.10}X_{2.1.11}X'_{1.3.13}X'_{1.2.1}$
0010 0000 0000 0000	$X'_{1.2.1}X_{1.3.4}X'_{2.1.6}X_{2.4.10}$ $X_{1.3.4}X'_{1.2.1}X_{2.4.7}X'_{2.1.11}$ $X'_{2.1.6}X_{2.4.7}X'_{1.2.1}X_{1.3.13}$ $X_{2.4.7}X'_{2.1.6}X_{1.3.4}X'_{1.2.16}$	$X'_{1.2.16}X_{1.3.13}X'_{2.1.11}X_{2.4.7}$ $X_{1.3.13}X'_{1.2.16}X_{2.4.10}X'_{2.1.6}$ $X'_{2.1.11}X_{2.4.10}X'_{1.2.16}X_{1.3.4}$ $X_{2.4.10}X'_{2.1.11}X_{1.3.13}X'_{1.2.1}$
0001 0000 0000 0000	$X_{1.2.1}X'_{1.3.4}X'_{2.1.6}X_{2.4.10}$ $X'_{1.3.4}X_{1.2.1}X_{2.4.7}X'_{2.1.11}$ $X'_{2.1.6}X_{2.4.7}X_{1.2.1}X'_{1.3.13}$ $X_{2.4.7}X'_{2.1.6}X'_{1.3.4}X_{1.2.16}$	$X_{1.2.16}X'_{1.3.13}X'_{2.1.11}X_{2.4.7}$ $X'_{1.3.13}X_{1.2.16}X_{2.4.10}X'_{2.1.6}$ $X'_{2.1.11}X_{2.4.10}X_{1.2.16}X'_{1.3.4}$ $X_{2.4.10}X'_{2.1.11}X'_{1.3.13}X_{1.2.1}$
0000 1000 0000 0000	$X'_{1.2.1}X'_{1.3.4}X'_{2.1.6}X_{2.4.10}$ $X'_{1.3.4}X'_{1.2.1}X'_{2.4.7}X_{2.1.11}$ $X'_{2.1.6}X'_{2.4.7}X'_{1.2.1}X_{1.3.13}$ $X'_{2.4.7}X'_{2.1.6}X'_{1.3.4}X_{1.2.16}$	$X_{1.2.16}X_{1.3.13}X'_{2.1.11}X'_{2.4.7}$ $X_{1.3.13}X_{1.2.16}X'_{2.4.10}X'_{2.1.6}$ $X_{2.1.11}X_{2.4.10}X'_{1.2.16}X'_{1.3.4}$ $X_{2.4.10}X_{2.1.11}X'_{1.3.13}X'_{1.2.1}$
0000 0100 0000 0000	$X'_{1.2.1}X'_{1.3.4}X_{2.1.6}X'_{2.4.10}$ $X'_{1.3.4}X'_{1.2.1}X_{2.4.7}X'_{2.1.11}$ $X_{2.1.6}X_{2.4.7}X'_{1.2.1}X_{1.3.13}$ $X_{2.4.7}X_{2.1.6}X'_{1.3.4}X_{1.2.16}$	$X_{1.2.16}X_{1.3.13}X'_{2.1.11}X_{2.4.7}$ $X_{1.3.13}X_{1.2.16}X'_{2.4.10}X_{2.1.6}$ $X'_{2.1.11}X'_{2.4.10}X_{1.2.16}X'_{1.3.4}$ $X'_{2.4.10}X'_{2.1.11}X_{1.3.13}X'_{1.2.1}$
0000 0010 0000 0000	$X'_{1.2.1}X_{1.3.4}X'_{2.1.6}X_{2.4.10}$ $X_{1.3.4}X'_{1.2.1}X_{2.4.7}X'_{2.1.11}$ $X'_{2.1.6}X_{2.4.7}X'_{1.2.1}X_{1.3.13}$ $X_{2.4.7}X'_{2.1.6}X_{1.3.4}X'_{1.2.16}$	$X'_{1.2.16}X_{1.3.13}X'_{2.1.11}X_{2.4.7}$ $X_{1.3.13}X_{1.2.16}X_{2.4.10}X'_{2.1.6}$ $X_{2.1.11}X_{2.4.10}X'_{1.2.16}X_{1.3.4}$ $X'_{2.4.10}X_{2.1.11}X_{1.3.13}X'_{1.2.1}$
0000 0001 0000 0000	$X_{1.2.1}X'_{1.3.4}X_{2.1.6}X'_{2.4.10}$ $X'_{1.3.4}X_{1.2.1}X'_{2.4.7}X'_{2.1.11}$ $X_{2.1.6}X'_{2.4.7}X_{1.2.1}X'_{1.3.13}$ $X'_{2.4.7}X_{2.1.6}X'_{1.3.4}X_{1.2.16}$	$X_{1.2.16}X'_{1.3.13}X'_{2.1.11}X_{2.4.7}$ $X'_{1.3.13}X_{1.2.16}X_{2.4.10}X_{2.1.6}$ $X'_{2.1.11}X_{2.4.10}X_{1.2.16}X'_{1.3.4}$ $X_{2.4.10}X'_{2.1.11}X'_{1.3.13}X_{1.2.1}$
0000 0000 1000 0000	$X_{1.2.1}X'_{1.3.4}X'_{2.1.6}X'_{2.4.10}$ $X'_{1.3.4}X_{1.2.1}X_{2.4.7}X_{2.1.11}$ $X'_{2.1.6}X_{2.4.7}X_{1.2.1}X_{1.3.13}$ $X_{2.4.7}X'_{2.1.6}X'_{1.3.4}X'_{1.2.16}$	$X'_{1.2.16}X_{1.3.13}X_{2.1.11}X_{2.4.7}$ $X_{1.3.13}X'_{1.2.16}X'_{2.4.10}X'_{2.1.6}$ $X_{2.1.11}X'_{2.4.10}X_{1.2.16}X_{1.3.4}$ $X'_{2.4.10}X_{2.1.11}X_{1.3.13}X_{1.2.1}$
0000 0000 0100 0000	$X_{1.2.1}X'_{1.3.4}X_{2.1.6}X'_{2.4.10}$ $X'_{1.3.4}X_{1.2.1}X'_{2.4.7}X'_{2.1.11}$ $X_{2.1.6}X'_{2.4.7}X_{1.2.1}X_{1.3.13}$ $X'_{2.4.7}X_{2.1.6}X'_{1.3.4}X_{1.2.16}$	$X'_{1.2.16}X_{1.3.13}X_{2.1.11}X'_{2.4.7}$ $X_{1.3.13}X'_{1.2.16}X_{2.4.10}X_{2.1.6}$ $X_{2.1.11}X_{2.4.10}X'_{1.2.16}X_{1.3.4}$ $X_{2.4.10}X'_{2.1.11}X_{1.3.13}X_{1.2.1}$

Table 4.15. (*Continued*)

0000 0000 0010 0000	$X_{1.2.1}X_{1.3.4}X'_{2.1.6}X_{2.4.10}$ $X_{1.3.4}X_{1.2.1}X'_{2.4.7}X_{2.1.11}$ $X'_{2.1.6}X'_{2.4.7}X_{1.2.1}X'_{1.3.13}$ $X'_{2.4.7}X'_{2.1.6}X_{1.3.4}X'_{1.2.16}$	$X'_{1.2.16}X'_{1.3.13}X_{2.1.11}X'_{2.4.7}$ $X'_{1.3.13}X'_{1.2.16}X_{2.4.10}X'_{2.1.6}$ $X_{2.1.11}X_{2.4.10}X'_{1.2.16}X_{1.3.4}$ $X_{2.4.10}X_{2.1.11}X'_{1.3.13}X_{1.2.1}$
0000 0000 0001 0000	$X_{1.2.1}X_{1.3.4}X_{2.1.6}X'_{2.4.10}$ $X_{1.3.4}X_{1.2.1}X_{2.4.7}X'_{2.1.11}$ $X_{2.1.6}X_{2.4.7}X_{1.2.1}X'_{1.3.13}$ $X_{2.4.7}X_{2.1.6}X_{1.3.4}X'_{1.2.16}$	$X'_{1.2.16}X'_{1.3.13}X'_{2.1.11}X_{2.4.7}$ $X'_{1.3.13}X'_{1.2.16}X'_{2.4.10}X_{2.1.6}$ $X'_{2.1.11}X'_{2.4.10}X'_{1.2.16}X_{1.3.4}$ $X'_{2.4.10}X'_{2.1.11}X'_{1.3.13}X_{1.2.1}$
0000 0000 0000 1000	$X'_{1.2.1}X_{1.3.4}X_{2.1.6}X'_{2.4.10}$ $X_{1.3.4}X'_{1.2.1}X'_{2.4.7}X_{2.1.11}$ $X_{2.1.6}X'_{2.4.7}X'_{1.2.1}X_{1.3.13}$ $X'_{2.4.7}X_{2.1.6}X_{1.3.4}X'_{1.2.16}$	$X'_{1.2.16}X_{1.3.13}X_{2.1.11}X'_{2.4.7}$ $X_{1.3.13}X'_{1.2.16}X'_{2.4.10}X_{2.1.6}$ $X_{2.1.11}X'_{2.4.10}X'_{1.2.16}X_{1.3.4}$ $X'_{2.4.10}X_{2.1.11}X_{1.3.13}X'_{1.2.1}$
0000 0000 0000 0100	$X_{1.2.1}X'_{1.3.4}X_{2.1.6}X'_{2.4.10}$ $X'_{1.3.4}X_{1.2.1}X'_{2.4.7}X_{2.1.11}$ $X_{2.1.6}X'_{2.4.7}X_{1.2.1}X'_{1.3.13}$ $X'_{2.4.7}X_{2.1.6}X'_{1.3.4}X_{1.2.16}$	$X_{1.2.16}X'_{1.3.13}X_{2.1.11}X'_{2.4.7}$ $X'_{1.3.13}X_{1.2.16}X'_{2.4.10}X_{2.1.6}$ $X_{2.1.11}X'_{2.4.10}X_{1.2.16}X'_{1.3.4}$ $X'_{2.4.10}X_{2.1.11}X'_{1.3.13}X_{1.2.1}$
0000 0000 0000 0010	$X_{1.2.1}X_{1.3.4}X'_{2.1.6}X'_{2.4.10}$ $X_{1.3.4}X_{1.2.1}X'_{2.4.7}X'_{2.1.11}$ $X'_{2.1.6}X'_{2.4.7}X_{1.2.1}X_{1.3.13}$ $X'_{2.4.7}X'_{2.1.6}X_{1.3.4}X_{1.2.16}$	$X_{1.2.16}X_{1.3.13}X'_{2.1.11}X'_{2.4.7}$ $X_{1.3.13}X_{1.2.16}X'_{2.4.10}X'_{2.1.6}$ $X'_{2.1.11}X'_{2.4.10}X_{1.2.16}X_{1.3.4}$ $X'_{2.4.10}X'_{2.1.11}X_{1.3.13}X_{1.2.1}$
0000 0000 0000 0001	$X_{1.2.1}X_{1.3.4}X_{2.1.6}X_{2.4.10}$ $X_{1.3.4}X_{1.2.1}X_{2.4.7}X_{2.1.11}$ $X_{2.1.6}X_{2.4.7}X_{1.2.1}X_{1.3.13}$ $X_{2.4.7}X_{2.1.6}X_{1.3.4}X_{1.2.16}$	$X_{1.2.16}X_{1.3.13}X_{2.1.11}X_{2.4.7}$ $X_{1.3.13}X_{1.2.16}X_{2.4.10}X_{2.1.6}$ $X_{2.1.11}X_{2.4.10}X_{1.2.16}X_{1.3.4}$ $X_{2.4.10}X_{2.1.11}X_{1.3.13}X_{1.2.1}$

constituting a 5D SD variable 4-chunk, and so on (in some cases, to avoid confusion I call the single chunk 1-chunk).

Using the information contained in Table 4.3, we can distribute the 2-chunks in the 16 main CAs, as displayed in Table 4.16. Note the following patterns:

- CA1.1, CA2.2, CA2.3, CA1.4, CA2.5, CA1.6, CA1.7, CA2.8 show endogenous pairing, since they display relations that are only endogenous to each code alphabet (all nodes pertain to levels already present in 3D). In particular, CA1.1, CA1.4, CA1.6, CA1.7 combine chunks that come from the 3D CA1, while CA2.2, CA2.3, CA2.5, CA2.8 combine chunks that come from the 3D CA2. Both CA1.1 and CA2.8 keep the same behaviour of CA1, that is, the only repeats (or substitutions) are on the same row (identity of the 3 indices labelling the 4D SD variables). CA2.2

Table 4.16.   2-chunks of 4D SD variables distributed in the main CAs. Note that only in CA1.1 and CA2.8 we have the four 3D SD variables of CA1 and CA2, respectively.

CA1.1				CA2.1			
0	$X_{1.1.1.A} = X'_{1.1.1.B}$ 0000 0000	$X_{1.1.1.B} = X'_{1.1.1.A}$ 1111 1111	8	1	$X_{1.2.1.A} = X_{1.1.9.B}$ 0001 0000	$X_{1.1.9.B} = X'_{1.2.1.A}$ 1110 1111	7
4c	$X_{1.1.16.A} = X_{1.1.16.B}$ 0000 1111	$X'_{1.1.16.B} = X'_{1.1.16.A}$ 1111 0000	4c	5b	$X_{1.2.16.A} = X_{1.1.8.B}$ 0001 1111	$X'_{1.1.8.B} = X'_{1.2.16.A}$ 1110 0000	3b
4a	$X_{1.4.4.A} = X_{1.4.4.B}$ 0011 0011	$X'_{1.4.4.B} = X'_{1.4.4.A}$ 1100 1100	4a	3a	$X_{1.3.4.A} = X_{1.4.12.B}$ 0010 0011	$X_{1.4.12.B} = X'_{1.3.4.A}$ 1101 1100	5a
4a	$X_{1.4.13.A} = X'_{1.4.13.B}$ 0011 1100	$X_{1.4.13.B} = X'_{1.4.13.A}$ 1100 0011	4a	3a	$X_{1.3.13.A} = X'_{1.4.5.B}$ 0010 1100	$X_{1.4.5.B} = X'_{1.3.13.A}$ 1101 0011	5a
4a	$X_{2.2.6.A} = X_{2.2.6.B}$ 0101 0101	$X_{2.2.6.B} = X'_{2.2.6.A}$ 1010 1010	4a	3a	$X_{2.1.6.A} = X_{2.2.14.B}$ 0100 1100	$X'_{2.2.14.B} = X'_{2.1.6.A}$ 1011 1010	5a
4a	$X_{2.2.11.A} = X'_{2.2.11.B}$ 0101 1010	$X_{2.2.11.B} = X'_{2.2.11.A}$ 1010 0101	4a	3a	$X_{2.1.11.A} = X_{2.2.3.B}$ 0100 1010	$X_{2.2.3.B} = X'_{2.1.11.A}$ 1011 0101	5a
4a	$X_{2.3.7.A} = X'_{2.3.7.B}$ 0110 0110	$X_{2.3.7.B} = X'_{2.3.7.A}$ 1001 1001	4a	5a	$X_{2.4.7.A} = X_{2.3.15.B}$ 0111 0110	$X_{2.3.15.B} = X'_{2.4.7.A}$ 1000 1001	3a
4a	$X_{2.3.10.A} = X_{2.3.10.B}$ 0110 1001	$X'_{2.3.10.B} = X'_{2.3.10.A}$ 1001 0110	4a	5a	$X_{2.4.10.A} = X_{2.3.2.B}$ 0111 1001	$X'_{2.3.2.B} = X'_{2.4.10.A}$ 1000 0110	3a

Table 4.16. (Continued)

	CA1.2						CA2.2				
1	$X_{1.1.2.A} = X'_{2.4.16.B}$ 0000 0001		$X'_{2.4.16.B} = X'_{1.1.2.A}$ 1111 1110		7	2b	$X_{1.2.2.A} = X_{2.4.8.B}$ 0001 0001		$X'_{2.4.8.B} = X'_{1.2.2.A}$ 1110 1110		6b
3b	$X_{1.1.15.A} = X'_{2.4.1.B}$ 0000 1110		$X_{2.4.1.B} = X'_{1.1.15.A}$ 1111 0001		5b	4b	$X_{1.2.15.A} = X'_{2.4.9.B}$ 0001 1110		$X_{2.4.9.B} = X'_{1.2.15.A}$ 1110 0001		4b
3a	$X_{1.4.3.A} = X'_{2.1.13.B}$ 0011 0010		$X_{2.1.13.B} = X'_{1.4.3.A}$ 1100 1101		5a	2b	$X_{1.3.3.A} = X'_{2.1.5.B}$ 0010 0010		$X_{2.1.5.B} = X'_{1.3.3.A}$ 1101 1101		6b
5a	$X_{1.4.14.A} = X_{2.1.4.B}$ 0011 1101		$X'_{2.1.4.B} = X'_{1.4.14.A}$ 1100 0010		3a	4b	$X_{1.3.14.A} = X_{2.1.12.B}$ 0010 1101		$X'_{2.1.12.B} = X'_{1.3.14.A}$ 1101 0010		4b
3a	$X_{2.2.5.A} = X'_{1.3.11.B}$ 0101 0100		$X_{1.3.11.B} = X'_{2.2.5.A}$ 1010 1011		5a	2b	$X_{2.1.5.A} = X'_{1.3.3.B}$ 0100 0100		$X_{1.3.3.B} = X'_{2.1.5.A}$ 1011 1011		6b
5a	$X_{2.2.12.A} = X_{1.3.6.B}$ 0101 1011		$X'_{1.3.6.B} = X'_{2.2.12.A}$ 1010 0100		3a	4b	$X_{2.1.12.A} = X_{1.3.14.B}$ 0100 1011		$X'_{1.3.14.B} = X'_{2.1.12.A}$ 1011 0100		4b
5a	$X_{2.3.8.A} = X_{1.2.10.B}$ 0110 0111		$X'_{1.2.10.B} = X'_{2.3.8.A}$ 1001 1000		3a	6b	$X_{2.4.8.A} = X_{1.2.2.B}$ 0111 0111		$X'_{1.2.2.B} = X'_{2.4.8.A}$ 1000 1000		2b
3a	$X_{2.3.9.A} = X'_{1.2.7.B}$ 0110 1000		$X_{1.2.7.B} = X'_{2.3.9.A}$ 1001 0111		5a	4b	$X_{2.4.9.A} = X'_{1.2.15.B}$ 0111 1000		$X_{1.2.15.B} = X'_{2.4.9.A}$ 1000 0111		4b

(Continued)

Table 4.16.   (*Continued*)

	CA1.3				CA2.3		
1	$X_{1.1.3.A} = X'_{2.1.1.B}$ 0000 0010	$X_{2.1.1.B} = X'_{1.1.3.A}$ 1111 1101	7	2b	$X_{1.2.3.A} = X'_{2.1.9.B}$ 0001 0010	$X_{2.1.9.B} = X'_{1.2.3.A}$ 1110 1101	6b
3b	$X_{1.1.14.A} = X_{2.1.16.B}$ 0000 1101	$X'_{2.1.16.B} = X'_{1.1.14.A}$ 1111 0010	5b	4b	$X_{1.2.14.A} = X_{2.1.8.B}$ 0001 1101	$X_{2.1.8.B} = X'_{1.2.14.A}$ 1110 0010	4b
3a	$X_{1.4.2.A} = X_{2.4.4.B}$ 0011 0001	$X_{2.4.4.B} = X'_{1.4.2.A}$ 1100 1110	5a	2b	$X_{1.3.2.A} = X_{2.4.12.B}$ 0010 0001	$X_{2.4.12.B} = X'_{1.3.2.A}$ 1101 1110	6b
5a	$X_{1.4.15.A} = X_{2.4.13.B}$ 0011 1110	$X_{2.4.13.B} = X'_{1.4.15.A}$ 1100 0001	3a	4b	$X_{1.3.15.A} = X_{2.4.5.B}$ 0010 1110	$X_{2.4.5.B} = X'_{1.3.15.A}$ 1101 0001	4b
5a	$X_{2.2.8.A} = X_{1.2.6.B}$ 0101 0111	$X'_{1.2.6.B} = X'_{2.2.8.A}$ 1010 1000	3a	4b	$X_{2.1.8.A} = X_{1.2.14.B}$ 0100 0111	$X_{1.2.14.B} = X'_{2.1.8.A}$ 1011 1000	4b
3a	$X_{2.2.9.A} = X'_{1.2.11.B}$ 0101 1000	$X_{1.2.11.B} = X_{2.2.9.A}$ 1010 0111	5a	2b	$X_{2.1.9.A} = X'_{1.2.3.B}$ 0100 1000	$X_{1.2.3.B} = X'_{2.1.9.A}$ 1011 0111	6b
3a	$X_{2.3.5.A} = X'_{1.3.7.B}$ 0110 0100	$X_{1.3.7.B} = X'_{2.3.5.A}$ 1001 1011	5a	4b	$X_{2.4.5.A} = X'_{1.3.15.B}$ 0111 0100	$X_{1.3.15.B} = X'_{2.4.5.A}$ 1000 1011	4b
5a	$X_{2.3.12.A} = X_{1.3.10.B}$ 0110 1011	$X'_{1.3.10.B} = X'_{2.3.12.A}$ 1001 0100	3a	6b	$X_{2.4.12.A} = X_{1.3.2.B}$ 0111 1011	$X'_{1.3.2.B} = X'_{2.4.12.A}$ 1000 0100	2b

Table 4.16. (Continued)

CA1.4				CA2.4			
2a	$X_{1.1.4.A} = X_{1.4.16.B}$ 0000 · 0011	$X'_{1.4.16.B} = X'_{1.1.4.A}$ 1111 · 1100	6a	3a	$X_{1.2.4.A} = X_{1.4.8.B}$ 0001 · 0011	$X_{1.4.8.B} = X'_{1.2.4.A}$ 1110 · 1100	5a
2a	$X_{1.1.13.A} = X'_{1.4.1.B}$ 0000 · 1100	$X_{1.4.1.B} = X'_{1.1.13.A}$ 1111 · 0011	6a	3a	$X_{1.2.13.A} = X'_{1.4.9.B}$ 0001 · 1100	$X_{1.4.9.B} = X'_{1.2.13.A}$ 1110 · 0011	5a
2a	$X_{1.4.1.A} = X'_{1.1.13.B}$ 0011 · 0000	$X_{1.1.13.B} = X'_{1.4.1.A}$ 1100 · 1111	6a	1	$X_{1.3.1.A} = X'_{1.1.5.B}$ 0010 · 0000	$X_{1.1.5.B} = X'_{1.3.1.A}$ 1101 · 1111	7
6a	$X_{1.4.16.A} = X_{1.1.4.B}$ 0011 · 1111	$X'_{1.1.4.B} = X'_{1.4.16.A}$ 1100 · 0000	2a	5b	$X_{1.3.16.A} = X_{1.1.12.B}$ 0010 · 1111	$X'_{1.1.12.B} = X'_{1.3.16.A}$ 1101 · 0000	3b
4a	$X_{2.2.7.A} = X'_{2.3.11.B}$ 0101 · 0110	$X_{2.3.11.B} = X'_{2.2.7.A}$ 1010 · 1001	4a	3a	$X_{2.1.7.A} = X'_{2.3.3.B}$ 0100 · 0110	$X_{2.3.3.B} = X'_{2.1.7.A}$ 1011 · 1001	5a
4a	$X_{2.2.10.A} = X_{2.3.6.B}$ 0101 · 1001	$X_{2.3.6.B} = X'_{2.2.10.A}$ 1010 · 0110	4a	3a	$X_{2.1.10.A} = X_{2.3.14.B}$ 0100 · 1001	$X_{2.3.14.B} = X'_{2.1.10.A}$ 1011 · 0110	5a
4a	$X_{2.3.6.A} = X_{2.2.10.B}$ 0110 · 0101	$X_{2.2.10.B} = X'_{2.3.6.A}$ 1001 · 1010	4a	5a	$X_{2.4.6.A} = X_{2.2.2.B}$ 0111 · 0101	$X_{2.2.2.B} = X'_{2.4.6.A}$ 1000 · 1010	3a
4a	$X_{2.3.11.A} = X'_{2.2.7.B}$ 0110 · 1010	$X_{2.2.7.B} = X'_{2.3.11.A}$ 1001 · 0101	4a	5a	$X_{2.4.11.A} = X_{2.2.15.B}$ 0111 · 1010	$X_{2.2.15.B} = X'_{2.4.11.A}$ 1000 · 0101	3a

(Continued)

Table 4.16.   (*Continued*)

CA1.5				CA2.5			
1	$X_{1.1.5.A} = X'_{1.3.1.B}$ 0000 / 0100	$X_{1.3.1.B} = X'_{1.1.5.A}$ 1111 / 1011	7	2b	$X_{1.2.5.A} = X'_{1.3.9.B}$ 0001 / 0100	$X_{1.3.9.B} = X'_{1.2.5.A}$ 1110 / 1011	6b
3b	$X_{1.1.12.A} = X'_{1.3.16.B}$ 0000 / 1011	$X_{1.3.16.B} = X'_{1.1.12.A}$ 1111 / 0100	5b	4b	$X_{1.2.12.A} = X'_{1.3.8.B}$ 0001 / 1011	$X_{1.3.8.B} = X'_{1.2.12.A}$ 1110 / 0100	4b
5a	$X_{1.4.8.A} = X_{1.2.4.B}$ 0011 / 0111	$X_{1.2.4.B} = X'_{1.4.8.A}$ 1100 / 1000	3a	4b	$X_{1.3.8.A} = X'_{1.2.12.B}$ 0010 / 0111	$X_{1.2.12.B} = X'_{1.3.8.A}$ 1101 / 1000	4b
3a	$X_{1.4.9.A} = X'_{1.2.13.B}$ 0011 / 1000	$X_{1.2.13.B} = X'_{1.4.9.A}$ 1100 / 0111	5a	2b	$X_{1.3.9.A} = X'_{1.2.5.B}$ 0010 / 1000	$X_{1.2.5.B} = X'_{1.3.9.A}$ 1101 / 0111	6b
3a	$X_{2.2.2.A} = X'_{2.4.6.B}$ 0101 / 0001	$X_{2.4.6.B} = X'_{2.2.2.A}$ 1010 / 1110	5a	2b	$X_{2.2.12.A} = X'_{2.4.14.B}$ 0100 / 0001	$X_{2.4.14.B} = X'_{2.2.12.A}$ 1011 / 1110	6b
3a	$X_{2.2.15.A} = X'_{2.4.11.B}$ 0101 / 1110	$X_{2.4.11.B} = X'_{2.2.15.A}$ 1010 / 0001	3a	4b	$X_{2.1.15.A} = X'_{2.4.3.B}$ 0100 / 1110	$X_{2.4.3.B} = X'_{2.1.15.A}$ 1011 / 0001	4b
5a	$X_{2.3.3.A} = X'_{2.1.7.B}$ 0110 / 0010	$X_{2.1.7.B} = X'_{2.3.3.A}$ 1001 / 1101	5a	4b	$X_{2.4.3.A} = X'_{2.1.15.B}$ 0111 / 0010	$X_{2.1.15.B} = X'_{2.4.3.A}$ 1000 / 1101	4b
5a	$X_{2.3.14.A} = X_{2.1.10.B}$ 0110 / 1101	$X_{2.1.10.B} = X'_{2.3.14.A}$ 1001 / 0010	3a	6b	$X_{2.4.14.A} = X_{2.1.2.B}$ 0111 / 1101	$X_{2.1.2.B} = X'_{2.4.14.A}$ 1000 / 0010	2b

Table 4.16.   (*Continued*)

CA1.6

2a	$X_{1.1.6.A} = X_{2.2.16.B}$ 0000 0101	$X'_{2.2.16.B} = X'_{1.1.6.A}$ 1111 1010	6a
2a	$X_{1.1.11.A} = X_{2.2.1.B}$ 0000 1010	$X'_{2.2.1.B} = X'_{1.1.11.A}$ 1111 0101	6a
4a	$X_{1.4.7.A} = X_{2.3.13.B}$ 0011 0110	$X'_{2.3.13.B} = X'_{1.4.7.A}$ 1100 1001	4a
4a	$X_{1.4.10.A} = X_{2.3.4.B}$ 0011 1001	$X'_{2.3.4.B} = X'_{1.4.10.A}$ 1100 0110	4a
2a	$X_{2.2.1.A} = X_{1.1.11.B}$ 0101 0000	$X'_{1.1.11.B} = X'_{2.2.1.A}$ 1010 1111	6a
6a	$X_{2.2.16.A} = X_{1.1.6.B}$ 0101 1111	$X'_{1.1.6.B} = X'_{2.2.16.A}$ 1010 0000	2a
4a	$X_{2.3.4.A} = X_{1.4.10.B}$ 0110 0011	$X'_{1.4.10.B} = X'_{2.3.4.A}$ 1001 1100	4a
4a	$X_{2.3.13.A} = X_{1.4.7.B}$ 0110 1100	$X'_{1.4.7.B} = X'_{2.3.13.A}$ 1001 0011	4a

CA2.6

3a	$X_{1.2.6.A} = X_{2.2.8.B}$ 0001 0101	$X'_{2.2.8.B} = X'_{1.2.6.A}$ 1110 1010	5a
3a	$X_{1.2.11.A} = X_{2.2.9.B}$ 0001 1010	$X'_{2.2.9.B} = X'_{1.2.11.A}$ 1110 0101	5a
3a	$X_{1.3.7.A} = X_{2.3.5.B}$ 0010 0110	$X'_{2.3.5.B} = X'_{1.3.7.A}$ 1101 1001	5a
3a	$X_{1.3.10.A} = X_{2.3.12.B}$ 0010 1001	$X'_{2.3.12.B} = X'_{1.3.10.A}$ 1101 0110	5a
1	$X_{2.1.1.A} = X_{1.1.3.B}$ 0100 0000	$X'_{1.1.3.B} = X'_{2.1.1.A}$ 1011 1111	7
5b	$X_{2.1.16.A} = X_{1.1.14.B}$ 0100 1111	$X'_{1.1.14.B} = X'_{2.1.16.A}$ 1011 0000	3b
5a	$X_{2.4.4.A} = X_{1.4.2.B}$ 0111 0011	$X'_{1.4.2.B} = X'_{2.4.4.A}$ 1000 1100	3a
5a	$X_{2.4.13.A} = X_{1.4.15.B}$ 0111 1100	$X'_{1.4.15.B} = X'_{2.4.13.A}$ 1000 0011	3a

(*Continued*)

Table 4.16.   (*Continued*)

	CA1.7				CA2.7		
2a	$X_{1.1.7.A} = X'_{2.3.1.B}$ 0000 0110	$X_{2.3.1.B} = X'_{1.1.7.A}$ 1111 1001	6a	3a	$X_{1.2.7.A} = X'_{2.3.9.B}$ 0001 0110	$X_{2.3.9.B} = X'_{1.2.7.A}$ 1110 1001	5a
2a	$X_{1.1.10.A} = X'_{2.3.16.B}$ 0000 1001	$X_{2.3.16.B} = X'_{1.1.10.A}$ 1111 0110	6a	3a	$X_{1.2.10.A} = X'_{2.3.8.B}$ 0001 1001	$X_{2.3.8.B} = X'_{1.2.10.A}$ 1110 0110	5a
4a	$X_{1.4.6.A} = X'_{2.2.4.B}$ 0011 0101	$X_{2.2.4.B} = X'_{1.4.6.A}$ 1100 1010	4a	3a	$X_{1.3.6.A} = X'_{2.2.12.B}$ 0010 1010	$X_{2.2.12.B} = X'_{1.3.6.A}$ 1101 0101	5a
4a	$X_{1.4.11.A} = X'_{2.2.13.B}$ 0011 1010	$X_{2.2.13.B} = X'_{1.4.11.A}$ 1100 0101	4a	3a	$X_{1.3.11.A} = X'_{2.2.5.B}$ 0010 0101	$X_{2.2.5.B} = X'_{1.3.11.A}$ 1101 1010	5a
4a	$X_{2.2.4.A} = X'_{1.4.6.B}$ 0101 0011	$X_{1.4.6.B} = X'_{2.2.4.A}$ 1010 1100	4a	3a	$X_{2.1.4.A} = X'_{1.4.14.B}$ 0100 1100	$X_{1.4.14.B} = X'_{2.1.4.A}$ 1011 0011	5a
4a	$X_{2.2.13.A} = X'_{1.4.11.B}$ 0101 1100	$X_{1.4.11.B} = X'_{2.2.13.A}$ 1010 0011	4a	3a	$X_{2.1.13.A} = X'_{1.4.3.B}$ 0100 0011	$X_{1.4.3.B} = X'_{2.1.13.A}$ 1011 1100	5a
2a	$X_{2.3.1.A} = X'_{1.1.7.B}$ 0110 1111	$X_{1.1.7.B} = X'_{2.3.1.A}$ 1001 0000	6a	3b	$X_{2.4.1.A} = X'_{1.1.15.B}$ 0111 1111	$X_{1.1.15.B} = X'_{2.4.1.A}$ 1000 0000	5b
6a	$X_{2.3.16.A} = X'_{1.1.10.B}$ 0110 1001	$X_{1.1.10.B} = X'_{2.3.16.A}$ 1001 1111	2a	7	$X_{2.4.16.A} = X'_{1.1.2.B}$ 0111 0000	$X_{1.1.2.B} = X'_{2.4.16.A}$ 1000 1111	1

Table 4.16. (Continued)

CA1.8

3b	$X_{1.1.8.A} = X'_{1.2.16.B}$ 0000 / 0111	$X'_{1.2.16.B} = X'_{1.1.8.A}$ 1111 / 1000	5b
1	$X_{1.1.9.A} = X'_{1.2.1.B}$ 0000 / 1000	$X'_{1.2.1.B} = X'_{1.1.9.A}$ 1111 / 0111	7
3a	$X_{1.4.5.A} = X_{1.3.13.B}$ 0011 / 0100	$X_{1.3.13.B} = X'_{1.4.5.A}$ 1100 / 1011	5a
5a	$X_{1.4.12.A} = X_{1.3.4.B}$ 0011 / 1011	$X_{1.3.4.B} = X'_{1.4.12.A}$ 1100 / 0100	3a
3a	$X_{2.2.3.A} = X_{2.1.11.B}$ 0101 / 0010	$X_{2.1.11.B} = X'_{2.2.3.A}$ 1010 / 1101	5a
5a	$X_{2.2.14.A} = X_{2.1.6.B}$ 0101 / 1101	$X'_{2.1.6.B} = X'_{2.2.14.A}$ 1010 / 0010	3a
3a	$X_{2.3.2.A} = X_{2.4.10.B}$ 0110 / 0001	$X'_{2.4.10.B} = X'_{2.3.2.A}$ 1001 / 1110	5a
5a	$X_{2.3.15.A} = X'_{2.4.7.B}$ 0110 / 1110	$X'_{2.4.7.B} = X'_{2.3.15.A}$ 1001 / 0001	3a

CA2.8

4b	$X_{1.2.8.A} = X'_{1.2.8.B}$ 0001 / 0111	$X'_{1.2.8.B} = X'_{1.2.8.A}$ 1110 / 1000	4b
2b	$X_{1.2.9.A} = X'_{1.2.9.B}$ 0001 / 1000	$X_{1.2.9.B} = X'_{1.2.9.A}$ 1110 / 0111	6b
2b	$X_{1.3.5.A} = X'_{1.3.5.B}$ 0010 / 0100	$X_{1.3.5.B} = X'_{1.3.5.A}$ 1101 / 1011	6b
4b	$X_{1.3.12.A} = X'_{1.3.12.B}$ 0010 / 1011	$X_{1.3.12.B} = X'_{1.3.12.A}$ 1101 / 0100	4b
2b	$X_{2.1.3.A} = X'_{2.1.3.B}$ 0100 / 0010	$X_{2.1.3.B} = X'_{2.1.3.A}$ 1011 / 1101	6b
4b	$X_{2.1.14.A} = X'_{2.1.14.B}$ 0100 / 1101	$X'_{2.1.14.B} = X'_{2.1.14.A}$ 1011 / 0010	4b
4b	$X_{2.4.2.A} = X'_{2.4.2.B}$ 0111 / 0001	$X'_{2.4.2.B} = X'_{2.4.2.A}$ 1000 / 1110	4b
6b	$X_{2.4.15.A} = X'_{2.4.15.B}$ 0111 / 1110	$X_{2.4.15.B} = X'_{2.4.15.A}$ 1000 / 0001	2b

behaves as CA2 and shows, in particular, identities between the
3D indices 1.2 and 2.4 as well as between 1.3 and 2.1. Similarly,
CA2.3 behaves as CA2, but with pairings 1.2 with 2.1 and 1.3
with 2.4. CA1.4 shows pairings of the first indices of the SD
variables, that is, 1.1 with 1.4 and 2.2 with 2.3. Similarly, CA2.5
that displays identities between 1.2 and 1.3 as well as between
2.1 and 2.4. Finally, CA1.6 couples 1.1 with 2.2 and 1.4 with 2.3,
while CA1.7 pairs 1.1 and 2.3 and 1.4 with 2.2.

- At the opposite, the other ones may be called exogenous code
  alphabets since each displays relations to another code alphabet,
  and they combine a chunk coming from CA1 and another coming
  from CA2 (all involved nodes pertain to new levels of this
  algebra). The distinction between endogenous and exogenous CAs
  is related to the observed coherence and incoherence of the IDs in
  CA1 and CA2, respectively, of $\mathscr{B}_3$ (Subsection 3.1.1, Table 3.5).
  In particular, we have the couples CA2.1-CA1.8, CA1.2-CA2.7,
  CA1.3-CA2.6, CA2.4-CA.1.5 (note that the sum of the last indices
  here is always 9). Thus, CA2.1 displays identities with CA1.8,
  especially between the first indices of the SD variables, that
  is, between 1.2 and 1.1, between 1.3 and 1.4, between 2.1 and 2.2,
  and between 2.4 and 2.3. CA2.4 behaves similarly and displays
  identities with CA1.5, again between the first indices, but pairing
  the second indices, that is, between 1.2 and 1.4, between 1.3 and
  1.1, between 2.1 and 2.3, and between 2.4 and 2.2. CA1.2 shows
  identities with CA2.7 exchanging the first indices, in particular
  connecting 1.1 with 2.4, 1.4 with 2.1, 2.2 with 1.3, and 2.3 with
  1.2. Similarly, CA1.3 shows identities with CA2.6, in particular
  between the second indices, that is, between 1.1 and 2.1, between
  1.4 and 2.4, between 2.2 and 1.2, and between 2.3 and 1.3. Note
  that each of these 8 code alphabets contains a 2-chunk with a
  single 1, that coupled with 0000 0000 ($X_{1.1.1.A} = X'_{1.1.1.B}$) gives
  rise to the 16 nodes of Level 1/16.

All endogenous odd CAs combine only chunks of CA1, while all
endogenous even CAs combine only chunks of CA2, while all exogenous CAs take one chunk from CA1 and another from CA2, where

here and in the following I call a CA even or odd accordingly to the second index. In the light of this examination, CA2 is more exogenous relative to CA1, since it couples two different halves of the code alphabet.

In the previous table we labelled all 2-chunks. Let us consider them in detail.

- There is one case with eight 0s and it occurs in CA1.1. Let us call it **0**.
- There is a 1 in a 1-chunk. There are 8 cases, one for each exogenous CA. Let us call this form 1.
- There are two 1s in a single 1-chunk. There are 12 2-chunks with this pattern, and are all in odd endogenous CAs. Let us label them 2a.
- There is a 1 in a 1-chunk and another 1 in another. There are 16 2-chunks like this, and are all in even endogenous CAs. Let us call them 2b.
- Then, there are two 1s in a 1-chunk and another 1 in another. There are 48 cases of this form and are all in exogenous CAs. Let us call them 3a.
- There are three 1s in a single 1-chunk. There are 8 cases like this, one for each exogenous CA. Let us call this form 3b.
- There are two 1s in a 1-chunk and two 1s in another. There are 36 cases like this, and are all in endogenous odd CAs. Let us call these 2-chunks 4a.
- Then, there are three 1s in a 1-chunk and another 1 in another 1-chunk. There are 32 cases, and are all in endogenous even CAs. Let us call these cases 4b.
- There are four 1s in a single 1-chunk. There are two 2-chunks like this, and are all in CA1.1. Let us call these 2-chunks 4c.
- There are two 1s in a 1-chunk and three 1s in another. There are 48 2-chunks with this pattern, and are all in exogenous CAs. Let us call this pattern 5a.
- There is a 1 in a 1-chunk and four 1s in another. There are 8 2-chunks like this, one for each exogenous CA. Let us call such a pattern 5b.

- There are two 1s in a 1-chunk and four 1s in another. There are 12 cases with this pattern, and are all in endogenous odd CAs. Let us call them 6a.
- There are three 1s in each of two 1-chunk. There 16 cases with such a pattern, and are all in endogenous even CAs. Let us call them 6b.
- There are three 1s in a 1-chunk and four 1s in another. There are 8 cases with this pattern, and are all in exogenous CAs. Let us call this pattern 7.
- There is one case with eight 1s and is in CA1.1. Let us call it 8.

All 2-chunks with an even number of 1s are in endogenous CAs, while all 2-chunks with an odd number of 1s are in exogenous CAs, in agreement with $\mathscr{B}_3$. In particular, the first 12 products of Level 2/8 are 2a chunks, and therefore in odd endogenous CAs, while the other 12 products are 2b chunks, and therefore in even endogenous CAs (the four ternary equivalences in CA2.8). Similarly for the expressions of Level 6/8. The 32 first expressions of Level 3/8 are subdivided in 8 3b (each for each exogenous CA) and 24 3a patterns distributed three for each exogenous CA, and the same for the other 24 3a chunks representing the second 24 expressions of Level 3/8. Similarly for Level 5/8. The advantage to dealing with chunks now is more evident: the number of combinations is great, and to subdivide IDs in chunks and to consider the number of permutations of 1s (or 0s) in each chunk, as well as the permutations (combinatorics) of the chunks themselves, helps us in the task. In this way, we subdivide the nodes of each level in groups that determine the logical form.

In the following section I show several examples of nodes by using different SCAs of CA1.1. Also in the 4D case, the different combinations show distinctive patterns. Since here column a can be combined with four other columns in order to obtain the correct logical expressions, we have 4 main patterns for each SCA, as displayed by Figure 4.2 for some of them. Of course, there are many other combinations. For instance, in the $\{X_{1.1.1}, X_{1.4.4}, X_{2.2.6}, X_{2.3.10}\}$ SCA the

$\{X_{1.1.1},$ $X_{1.4.4},$ $X_{2.2.6},$ $X_{2.3.10}\}$	a-e, b-f, c-g, d-h, i-m, j-n, k-o, l-p a-i, b-j, c-k, d-l, e-m, f-n, g-o, h-p a-f, b-h, c-e, d-g, i-n, j-p, k-m, l-o a-g, b-f, c-e, d-h, i-o, j-n, k-m, l-p	a-b, c-d, e-f, g-h, i-j, k-l, m-n, o-p a-c, b-d, e-g, f-h, i-k, j-l, m-o, n-p a-e, b-f, c-g, d-h, i-m, j-n, k-o, l-p a-i, b-j, c-k, d-l, e-m, f-n, g-o, h-p	$\{X_{1.1.1},$ $X_{1.1.16},$ $X_{1.4.4},$ $X_{2.2.6}\}$
$\{X_{1.1.16},$ $X_{1.4.4},$ $X_{1.4.13},$ $X_{2.3.7}\}$	a-b, c-d, e-f, g-h, i-j, k-l, m-n, o-p a-n, b-m, c-p, d-o, e-j, f-i, g-l, h-k a-j, b-l, c-i, d-k, e-n, f-p, g-m, h-o a-k, b-i, c-l, d-j, e-o, f-m, g-p, h-n	a-d, b-c, e-h, f-g, i-l, j-k, m-p, n-o a-i, b-j, c-k, d-l, e-m, f-n, g-o, h-p a-k, b-i, c-l, d-j, e-o, f-m, g-p, h-n a-o, b-p, c-m, d-n, e-k, f-l, g-i, h-j	$\{X_{1.1.16},$ $X_{2.2.6},$ $X_{2.3.7},$ $X_{2.3.10}\}$
$\{X_{1.1.1},$ $X_{2.2.6},$ $X_{2.2.11},$ $X_{2.3.10}\}$	a-c, b-d, e-g, f-h, i-k, j-l, m-o, n-p a-f, b-h, c-e, d-g, i-n, j-p, k-m, l-o a-g, b-f, c-e, d-h, i-o, j-n, k-m, l-p a-o, b-p, c-m, d-n, e-k, f-l, g-i, h-j	a-h, b-g, c-f, d-e, i-p, j-o, k-n, l-m a-j, b-l, c-i, d-k, e-n, f-p, g-m, h-o a-k, b-i, c-l, d-j, e-o, f-m, g-p, h-n a-l, b-k, c-j, d-i, e-p, f-o, g-n, h-m	$\{X_{1.1.16},$ $X_{1.4.13},$ $X_{2.2.11},$ $X_{2.3.7}\}$
$\{X_{1.1.1},$ $X_{1.4.13},$ $X_{2.2.11},$ $X_{2.3.10}\}$	a-f, b-h, c-e, d-g, i-n, j-p, k-m, l-o a-g, b-f, c-e, d-h, i-o, j-n, k-m, l-p a-h, b-g, c-f, d-e, i-p, j-o, k-n, l-m a-l, b-k, c-j, d-i, e-p, f-o, g-n, h-m	a-d, b-c, e-h, f-g, i-l, j-k, m-p, n-o a-j, b-l, c-i, d-k, e-n, f-p, g-m, h-o a-l, b-k, c-j, d-i, e-p, f-o, g-n, h-m a-o, b-p, c-m, d-n, e-k, f-l, g-i, h-j	$\{X_{1.1.16},$ $X_{2.2.11},$ $X_{2.3.7},$ $X_{2.3.10}\}$
$\{X_{1.1.1},$ $X_{1.4.4},$ $X_{1.4.13},$ $X_{2.2.6}\}$	a-b, c-d, e-f, g-h, i-j, k-l, m-n, o-p a-e, b-f, c-g, d-h, i-m, j-n, k-o, l-p a-g, b-f, c-e, d-h, i-o, j-n, k-m, l-p a-m, b-n, c-o, d-p, e-i, f-j, g-k, h-l	a-d, b-c, e-h, f-g, i-l, j-k, m-p, n-o a-e, b-f, c-g, d-h, i-m, j-n, k-o, l-p a-f, b-h, c-e, d-g, i-n, j-p, k-m, l-o a-n, b-m, c-p, d-o, e-j, f-i, g-l, h-k	$\{X_{1.1.1},$ $X_{1.4.4},$ $X_{2.3.7},$ $X_{2.3.10}\}$

Figure 4.2. 4D SCA's patterns.

first two lines can be combined crosswise so as to get (i) a-e, b-j, c-g, d-l, i-m, f-n, k-o, h-p and (ii) a-i, b-f, c-k, d-h, e-m, j-n, g-o, h-p. Moreover, both sequences can be cut in the middle and each first half of the first row can be combined with the second half of the second row and vice versa. This is also true for the third and fourth row, and even between the first row and the last two.

Note that we have transformation rules from a SD variable to another in a similar way to what is seen in Subsection 3.1.3 for

3 dimensions. Note that, for example

$$X_{1.1.14} = (X_{1.1.1} + X_{1.1.16})(X_{1.4.4} + X'_{1.4.13}), \qquad (4.17\text{a})$$

$$X_{2.2.7} = (X_{2.2.6} + X_{2.2.11})(X_{2.3.7} + X'_{2.3.10}), \qquad (4.17\text{b})$$

$$X_{1.2.2} = (X_{1.1.1} + X_{1.4.4})(X_{2.2.6} + X'_{2.3.7}), \qquad (4.17\text{c})$$

$$X_{2.4.9} = (X'_{1.1.16} + X_{1.4.13})(X_{2.2.11} + X_{2.3.10}). \qquad (4.17\text{d})$$

Note also, for example, that

$$X_{1.1.1} = X_{1.1.16} \sim (X_{1.4.4} \sim X_{1.4.13}) = X_{1.1.16} \sim (X_{2.2.6} \sim X_{2.2.11})$$

$$= X_{1.1.16} \sim (X_{2.3.7} \sim X_{2.3.10}),$$

$$X_{1.1.16} = X_{1.1.1} \sim (X_{1.4.4} \sim X_{1.4.13}) = X_{1.1.1} \sim (X_{2.2.6} \sim X_{2.2.11})$$

$$= X_{1.1.1} \sim (X_{2.3.7} \sim X_{2.3.10}),$$

$$X_{1.4.4} = X_{1.4.13} \sim (X_{1.1.1} \sim X_{1.1.16}) = X_{1.4.13} \sim (X_{2.2.6} \sim X_{2.2.11})$$

$$= X_{1.4.13} \sim (X_{2.3.7} \sim X_{2.3.10}).$$

For the sake of notation, making use of the second index and of a and b for the third index, I shall use the following convention (but when necessary I shall go back to the full indices): $X_{1a} := X_{1.1.1}$, $X_{1b} := X_{1.1.16}$, $X_{4a} := X_{1.4.4}$, $X_{4b} := X_{1.4.13}$, $X_{2a} := X_{2.2.6}$, $X_{2b} :=$ $X_{2.2.11}$, $X_{3a} := X_{2.3.7}$, $X_{3b} := X_{2.3.10}$.

4.3.2. *Levels 2/16 and 3/16*

We have

$$\binom{16}{2} = \binom{4}{1}\binom{4}{2} + \binom{4}{2}\binom{4}{1}\binom{4}{1}$$

$$= 24 + 96 = 120 \qquad (4.18)$$

nodes for the IDs with two 1s (Level 2/16 corresponding to 1/8): the first term on the right-hand side tells us that we have 6 different ways to arrange two 1s in a chunk (right factor) and 4 different ways to arrange the chunk among four chunks (left factor), what makes

24 cases, while the second term on the right-hand side tells us that we have to multiply the two combinatorics of a 1 in two different chunks (the two right factors) and 6 different ways to distribute these two chunks among 4 chunks (the left factor), what makes 96 cases. Note that each node of level $1/16$ contributes to 15 nodes and each node of Level $2/16$ receives 2 arrows (from two different nodes), so that $(16 \times 15)/2 = 120$.

From the point of view of the IDs, we pair either endogenous CAs with endogenous CAs or exogenous CAs with exogenous CAs, according to Level $2/8$. In particular, the first 24 cases are generated by a combination of 2a (endogenous odd CA) with 0000 0000 (CA1.1), i.e. $2/8 + 0/8$. The last 96 cases are divided in two groups: (i) 64 1+1 nodes ($1/8 + 1/8$) and (ii) 32 2b + 0000 000 nodes (again $2/8 + 0/8$).

Let us consider now the logical expressions. At this level, we have ternary products. All ternary products are either (i) sums of a 2-chunk pertaining to an endogenous odd CA (i.e. 2a) and 0000 0000 ($X_{1.1.1.4}$, element of CA1.1 itself endogenous) or (ii) the sum of two 2-chunks pertaining both to exogenous CAs, with a 1 each. These exogenous CAs need to have the same first index; for instance, the product $X_{1b}X_{4a}X_{2b}$ is the sum of a 2-chunk pertaining to CA1.3 and another pertaining to CA1.2. In any case, the congruence of the CAs is either all endogenous or all exogenous, as already said.

The generation of nodes from previous nodes of less dimensional algebras follows combinatorial patterns that ultimately depend on the generative production of SD variables. Let us consider the relations between the IDs of Level $1/8$ and those of Level $2/16$. Table 4.17 shows some examples. The truth values in bold determine the pattern. Note that in $\mathscr{B}_4$ the 8 central values cover all possibilities of one 1 in a sequence of 4 values for each 1-chunk. All other combinations that derive from combinations in the same row of Table 3.7 give the same results. For instance, $X_{4a}X_{1a}X_{3a} = X_{1a}X_{4a}X_{2a}$, and so on. Moreover, if we take parallel expressions in different main CAs, we get the same result. For instance, the product $X_{1a}X_{4a}X_{2a}$ is paralleled by $X_{1.1.4}X_{1.4.1}X_{2.2.7}$ of CA1.4 as well as by $X_{1.2.1}X_{1.3.4}X_{2.1.6}$ of

Table 4.17. Relations among some nodes of Level 1/8 and some nodes of Level 2/16.

$\mathscr{B}_3$			$\mathscr{B}_4$				
			$X_{1a}X_{4a}X_{2a}$	0000	0000	0001	0001
			$X_{1a}X_{4a}X_{2b}$	0000	0000	0010	0001
			$X_{1a}X_{4b}X_{2a}$	0000	0000	0100	0001
$X_{1.1}X_{1.4}X_{2.2}$	0000	0001	$X_{1a}X_{4b}X_{2b}$	0000	0000	1000	0001
			$X_{1b}X_{4a}X_{2a}$	0000	0001	0000	0001
			$X_{1b}X_{4a}X_{2b}$	0000	0010	0000	0001
			$X_{1b}X_{4b}X_{2a}$	0000	0100	0000	0001
			$X_{1b}X_{4b}X_{2b}$	0000	1000	0000	0001
			$X_{1a}X_{4a}X'_{2a}$	0000	0000	0010	0010
			$X_{1a}X_{4a}X'_{2b}$	0000	0000	0001	0010
			$X_{1a}X_{4b}X'_{2a}$	0000	0000	1000	0010
$X_{1.1}X_{1.4}X'_{2.2}$	0000	0010	$X_{1a}X_{4b}X'_{2b}$	0000	0000	0100	0010
			$X_{1b}X_{4a}X'_{2a}$	0000	0010	0000	0010
			$X_{1b}X_{4a}X'_{2b}$	0000	0001	0000	0010
			$X_{1b}X_{4b}X'_{2a}$	0000	1000	0000	0010
			$X_{1b}X_{4b}X'_{2b}$	0000	0100	0000	0010
			$X'_{1a}X'_{4a}X_{2a}$	0100	0100	0000	0000
			$X'_{1a}X'_{4a}X_{2b}$	0100	1000	0000	0000
			$X'_{1a}X'_{4b}X_{2a}$	0100	0001	0000	0000
$X'_{1.1}X'_{1.4}X_{2.2}$	0100	0000	$X'_{1a}X'_{4b}X_{2b}$	0100	0010	0000	0000
			$X'_{1b}X'_{4a}X_{2a}$	0100	0000	0100	0000
			$X'_{1b}X'_{4a}X_{2b}$	0100	0000	1000	0000
			$X'_{1b}X'_{4b}X_{2a}$	0100	0000	0001	0000
			$X'_{1b}X'_{4b}X_{2b}$	0100	0000	0010	0000

Table 4.18. Examples of quaternary equivalences.

1000	0000	0000	0001	$X_{1a} \sim X_{4a} \sim X_{2a} \sim X_{3b}$
0000	0010	0100	0000	$X_{1a} \sim X'_{4a} \sim X_{2a} \sim X_{3b}$
0001	0000	0000	1000	$X_{1a} \sim X'_{4a} \sim X'_{2a} \sim X_{3b}$

CA2.1 and $X_{1.2.2}X_{1.3.3}X_{2.1.5}$ of CA2.2. All of these products represent $8 \times 8 = 64$ nodes.

In the 2D Level 2/4 we have binary equivalences and in the 3D Level 2/8 we have ternary equivalences. As expected, in Level 2/16 we have quaternary equivalences, as displayed in Table 4.18. Quaternary equivalences are the sum of two 2-chunks with a 1 each pertaining

to exogenous CAs in such a way that they are complementary, like
CA2.1-CA1.8. Keep in mind that, when all terms are negated, we get
the same equivalence. Note that across all nD algebras, increasing the
number of the terms of the equivalence keeps the two 1s as a constant.
We have 8 quaternary equivalences as a whole, which means that we
have 112 ternary products (including other SCAs).

By using the SCA example $\{X_{1a}, X_{1b}, X_{4a}, X_{2a}\}$, we get the
triplets displayed in Table 4.19. A comparison with Tables 3.12–3.13
is helpful. In particular, they derive from the products (in the
same order): $X_{1.1}X_{1.4}$, $X_{1.1}X_{2.2}$, $X_{1.1}X'_{2.2}$, $X_{1.1}X'_{1.4}$, $X'_{1.1}X_{1.4}$. These
expressions are the the the combination of a 2-chunk (with two 1s, i.e. 2a)
pertaining to an endogenous CA and 0000 0000. Similar results are
obtained through the combination of two 2-chunks (with a 1 in each,
i.e. 2b) pertaining to exogenous CAs, as displayed in Table 4.20,
where again a comparison with Table 3.12 is helpful: in the same
order, we have $X'_{1.4}X'_{2.2}$, $X'_{1.4}X_{2.2}$, $X_{1.4}X_{2.2}$, and $X'_{1.4}X'_{2.2}$. It is also
interesting to consider here the 3D triplets (in the same order):
$X'_{1.1}X'_{1.4}X'_{2.2}$, $X'_{1.1}X'_{1.4}X_{2.2}$, $X_{1.1}X_{1.4}X_{2.2}$, $X_{1.1}X'_{1.4}X'_{2.2}$. In such a
case, we need to consider the doubling of the entire 3D 2-chunk.

In summary, we have 64 ternary products, as displayed in
Table 4.17, and 8 quaternary equivalences (with symmetric forms),
according to Table 4.18. We have then 24 nodes with two 1s in a

Table 4.19.   Other triplets of Level 2/16.

0000	0000	0000	0011	$X_{1a}X_{1b}X_{4a}$
0000	0000	0000	0101	$X_{1a}X_{1b}X_{2a}$
0000	0000	0000	1010	$X_{1a}X_{1b}X'_{2a}$
0000	1100	0000	0000	$X'_{1a}X_{1b}X'_{4a}$
0011	0000	0000	0000	$X'_{1a}X'_{1b}X_{4a}$

Table 4.20.   Examples of other triplets for Level 2/16.

1000	0000	1000	0000	$X'_{1b}X'_{4a}X'_{2a}$
0100	0000	0100	0000	$X'_{1b}X'_{4a}X_{2a}$
0000	0001	0000	0001	$X_{1b}X_{4a}X_{2a}$
0000	1000	0000	1000	$X_{1b}X'_{4a}X'_{2a}$

single chunk, and these give rise to triplets expressed with other SCAs; other 24 triplets come from two 1s distributed in two differ- ent 2-chunks, making again use of other SCAs. Note that there is a certain analogy with Level 2/8: there we have binary products and ternary equivalences, and here ternary products and quater- nary equivalences. Although there are some correspondences between Levels 2/8 and 2/4, they are not analogical in the sense that we find here. Note also that, relative to Level 1/8 (of which Level 2/16 is a prosecution), we have an additional logical form as well as Level 2/8 relative to Level 1/4. Of course, the common form of Level 2/16 is represented by a sum of two quaternary products. In such a case, we have $1/16 + 1/16 = 2/16$, and the same for quaternary equivalences.

Finally, four products of ternary products give rise to an expres- sion of Level 1/16, such as:

$$X_{1a}X_{4a}X_{2a}X_{3b}$$

$$= (X_{1a}X_{4a}X_{2a})(X_{1a}X_{4a}X_{3b})(X_{1a}X_{2a}X_{3b})(X_{4a}X_{2a}X_{3b}). \quad (4.19)$$

Level 3/16 is somehow intermediate between Levels 1/8 and 2/8 (summing numerator and denominator of the last two expressions we get 3/16). It is intermediate between Levels 2/16 and 4/16. We have

$$\binom{16}{3} = \binom{4}{1}\binom{4}{3} + 2\binom{4}{2}\binom{4}{2}\binom{4}{1} + \binom{4}{1}\binom{4}{1}\binom{4}{1}\binom{4}{1}$$

$$= 16 + 288 + 256 = 560 \quad (4.20)$$

nodes. In fact, we have three different arrangements: either three 1s grouped in a single chunk or two in one chunk and one in the other or even one 1 in each of three chunks. Both three 1s and one 1 in a chunk give rise to 4 permutations, while two 1s to 6 permutations.

- In the first case (three 1s in a single chunk), we have the 4 per- mutations of the chunk as a whole, so that we have $4 \times 4 = 16$ nodes.

- In the second case (two 1s in a chunk and one 1 in another), we have 6 permutations for the first chunk and 4 for the second ($6 \times 4 = 24$ permutations), plus 6 permutations of the two chunks themselves, but since the two chunks are different, the possible configurations are doubled, so that we have $6 \times 2 = 12$ permutations. This makes $24 \times 12 = 288$ nodes.
- Finally, in the last case (a single 1 in each of three chunks), we have 4 permutations inside a single chunk, which makes $4 \times 4 \times 4 = 64$, and 4 permutations of the 3 chunks as a whole, which makes $64 \times 4 = 256$ nodes.

Let us now consider the IDs. First, note that all nodes for Level 3/8 have a 2-chunk pertaining to an endogenous CA and another pertaining to an exogenous CA:

- The first case (three 1s in a chunk) presents analogies with the first 8 cases of Table 3.15, for which some examples are presented in Table 4.21. Note that the 2-chunk either on the left or on the right contains a whole expression of Level 3/8: the corresponding expressions of Level 3/8 are (in order): $X_{1.1}(X_{1.4} + X_{2.2})$, $X_{1.1}(X_{1.4} + X'_{2.2})$, $X_{1.1}(X'_{1.4} + X_{2.2})$, $X_{1.1}(X'_{1.4} + X'_{2.2})$, $X'_{1.1}(X'_{1.4} + X'_{2.2})$. They are, in fact, a sum of chunk of CA1 (0000 0000) and another (with the form 3b) that pertains to an exogenous CA (1 for each of such CAs): 0/8 + 3/8.
- The other 48 cases of Level 3/8 (two 1s in a chunk and one 1 in another) are now distributed over 4 chunks (instead of 2), so that we generate $48 \times 6 = 288$ IDs. We distinguish two subcases: (i) 96 combinations of 3a (exogenous CAs) and 0000 0000 (again 0/8 + 3/8),

Table 4.21. Nodes of Level 3/16 given by **0** + 3b. Note the superposition of columns o-p (the first ternary product) and n-p (the second ternary product) for the first row.

0000	0000	0000	0111	$X_{1a}X_{1b}(X_{4a} + X_{2a})$
0000	0000	0000	1011	$X_{1a}X_{1b}(X_{4a} + X'_{2a})$
0000	0000	0000	1101	$X_{1a}X_{1b}(X'_{4a} + X_{2a})$
0000	1110	0000	0000	$X'_{1a}X_{1b}(X'_{4a} + X'_{2a})$
1110	0000	0000	0000	$X'_{1a}X'_{1b}(X'_{4a} + X'_{2a})$

Table 4.22.   **0** + 3a and 2a + 1 for Level 3/16.

0000	0000	0011	0001	$X_{1a}X_{4a}(X_{2a} + X_{3b})$
0000	0000	0001	0011	$X_{1a}X_{4a}(X_{2a} + X'_{3b})$
1000	0110	0000	0000	$X'_{1a}X'_{3b}(X'_{4a} + X'_{2a})$
0001	0101	0000	0000	$X'_{1a}X_{2a}(X_{4a} + X'_{3b})$
0011	0001	0000	0000	$X'_{1a}X_{4a}(X_{2a} + X_{3b})$
0000	0000	1100	0001	$X_{1a}X'_{1b}X'_{4a} + X_{1a}X_{1b}X_{4a}X_{2a}$
0011	0000	0001	0000	$X'_{1b}X_{4a}(X'_{1a} + X_{2a})$
0000	0010	0000	0110	$X_{4a}X'_{3b}(X_{1a} + X'_{2a})$
0000	0011	0001	0000	$X'_{1a}X_{1b}X_{4a} + X_{1a}X'_{1b}X_{4a}X_{2a}$
1000	0000	0000	0011	$X_{1a}X_{1b}X_{4a} + X'_{1a}X'_{1b}X'_{4a}X'_{2a}$
0100	0000	0000	1010	$X_{1a}X_{1b}X'_{2a} + X'_{1a}X'_{1b}X'_{4a}X_{2a}$

and (ii) 192 combinations of 2a (endogenous odd CA) and a 2-chunk with a single 1 (2/8 + 1/8). Table 4.22 shows some examples. The first 6 rows display the first subcase, while the last 5 rows the second subcase (in the second SCA). Of course, all expressions of the first group represent the sum of a 2-chunk that is already present as such in the first 32 expressions of Level 3/8 and 0000 0000, so that they have a corresponding behaviour (e.g. the first row corresponds to $X_{1.4}(X'_{1.1} + X_{2.2})$); thus the rules of agreement and disagreement of Level 3/8 also work here. For instance, $X_{1a}X'_{1b}X'_{4a} + X_{1a}X_{1b}X_{4a}X_{2a}$ has a counterpart in $X'_{1.1}X'_{1.4} + X_{1.1}X_{1.4}X_{2.2}$. The second group is different, since we need to distinguish between two subcases. If the 1s are in 1-chunk occupying the same position in any of the 2–chunks, then we have again the rules of Level 3/8. If they show mismatch of the position, no matter whether there is agreement or disagreement between the 1s, we always get a sum of a ternary and a quaternary product.

- Finally, the third case is given by a combination of 2b (endogenous even CAs) and a 1 (2/8 + 1/8). Table 4.23 shows some examples. We have a situation that is similar to the previous one, but with a 1 in each of three chunks. However, the same rule is valid here: do the 1s that are in agreement in a 1-chunk occupy the same position in the two 2-chunks? If yes, we get the first kind of expression, if not we get the second kind (note in particular rows 3 and 4).

Table 4.23. Examples of 2b + 1 for Level 3/16.

0000	1000	1000	1000	$X'_{4a}X'_{2a}(X_{1a}+X_{3b})$
0001	0001	0001	0000	$X_{4a}X_{2a}(X'_{1a}+X'_{3b})$
0001	0000	0001	0100	$X_{2a}X'_{3b}(X_{4a}+X_{1a})$
0001	0000	0100	0001	$X_{1a}X_{2a}X_{3b}+X'_{1a}X_{4a}X_{2a}X'_{3b}$
1000	0000	0010	0010	$X_{1a}X_{4a}X'_{2a}+X'_{1a}X'_{4a}X'_{2a}X'_{3b}$
0001	0100	0000	0010	$X'_{1a}X_{2a}X'_{3b}+X_{1a}X_{4a}X'_{2a}X'_{3b}$
0100	0000	0010	0010	$X_{1a}X_{4a}X'_{2a}+X'_{1a}X'_{4a}X_{2a}X_{3b}$
0100	0000	0001	0010	$X_{1a}X_{4a}X'_{3b}+X'_{1a}X'_{4a}X_{2a}X_{3b}$
0000	1000	0001	0001	$X_{1a}X_{4a}X_{2a}+X_{1a}X'_{4a}X'_{2a}X_{3b}$

All of the expressions of Level 3/16 are generated through the sum of two expressions of Level 2/16, such as:

$$X_{1a}X_{4a}(X_{2a}+X_{3b})=(X_{1a}X_{4a}X_{2a})+(X_{1a}X_{4a}X_{3b}),$$

$$X_{1a}X_{4a}X'_{3b}+X'_{1a}X'_{4a}X_{2a}X_{3b}=X_{1a}X_{4a}X'_{3b}+(X_{1a}\sim X_{4a}\sim X'_{2a}\sim X'_{3b}).$$

In turn, the product of two expressions of Level 3/16 can generate expressions of Level 2/16:

$$X_{1a}X_{4a}X_{2a}$$

$$=(X_{1a}X_{4a}X_{2a}+X_{1a}X'_{4a}X_{2a}X_{3b})(X_{1a}X_{4a}X_{2a}+X'_{1a}X'_{4a}X_{2a}X_{3b}),$$

$$X_{1a}X_{4a}X_{2a}X_{3b}+X'_{1a}X'_{4a}X'_{2a}X'_{3b}$$

$$=(X_{1a}X_{4a}X_{2a}+X'_{1a}X'_{4a}X'_{2a}X'_{3b})(X'_{1a}X'_{4a}X'_{2a}+X_{1a}X_{4a}X_{2a}X_{3b}).$$

In summary, we have two logical forms here: (i) a binary product times a sum of two terms and (ii) a sum of a ternary and a quaternary product. Of course, the basic form is the latter (showing no superposition of the columns). Moreover, we have

$$X_{1a}X_{4a}(X_{2a}+X_{3b})=X_{1a}X_{4a}X_{2a}+X_{1a}X_{4a}X'_{2a}X_{3b}. \qquad (4.21)$$

For the expression having the form $X_{1a}X_{4a}X'_{2a}+X'_{1a}X'_{4a}X'_{2a}X'_{3b}$, we have $2/16+1/16=3/16$, while for the expression of the form $X_{1a}X_{2a}(X_{4a}+X_{3b})$, we have $8/16\times 8/16\times 12/16=3/16$.

Note that the first one (and for some cases) could also be rewritten as $X'_{2a}(X_{1a}X_{4a} + X'_{1a}X'_{4a}X'_{3b})$. Also in this form, we have $8/16 \times (4/16+2/16) = 3/16$. At the opposite, the second expression cannot be rewritten as a sum of two ternary products. This situation suggests that we have essentially three kinds of a expressions that can occur at each level:

- Reduced expressions that cannot be written in expanded form, like the latter one.
- Expressions that in some cases can be written in more compact form, as for the first case.
- Basic or common expressions to which all other forms can be reduced.

4.3.3. *Levels 4/16 and 5/16*

For Level 4/16 we have

$$\binom{16}{4} = \binom{4}{1}\binom{4}{4} + 2\binom{4}{2}\binom{4}{3}\binom{4}{1} + \binom{4}{2}\binom{4}{2}\binom{4}{2}$$
$$+ 2\binom{4}{2}\binom{4}{2}\binom{4}{1}\binom{4}{1} + \binom{4}{1}\binom{4}{1}\binom{4}{1}\binom{4}{1}$$
$$= 4 + 192 + 216 + 1152 + 256 = 1820 \tag{4.22}$$

nodes. Let us compute this level in detail:

- We have 4 cases with a chunk of four 1s.
- When we have three 1s in a chunk and one 1 in another, there are 4 permutations in both kinds of chunks and $6 \times 2 = 12$ possible combinations of chunks, which gives $4 \times 4 \times 12 = 192$ nodes.
- For the case of two 1s in a chunk and two 1s in another, we 6 permutations for both chunks and 6 alternative configurations or combinations of permutations, which makes $6 \times 6 \times 6 = 216$ nodes.
- Then, we have the case with two 1s in a chunk but with the other two 1s distributed in two different chunks. This situation makes 6 permutations for the first chunk and 4 for the other two.

We have again $6 \times 2 = 12$ possible configurations of the chunks, which makes $96 \times 12 = 1152$.

- Finally, we have $4 \times 4 \times 4 \times 4 = 256$ nodes with one 1 in each different chunk.

Such level corresponds to Level 2/8. This means that, from the point of view of IDs, we expect to have either endogenous 2-chunks combined with endogenous 2-chunks or exogenous 2-chunks combined with exogenous 2-chunks. Regarding the logical expressions, we expect to find binary products and ternary equivalences. However, the combinatorics follow in part that of Level 4/8.

Let us now consider the patterns of the IDs. The five combinatorial groups give rise to following patterns:

1st case The 4c (CA1.1) pattern gives rise to 4 cases (different combinations of $X_{1.1}$ or $X'_{1.1}$ and 0000 0000).

2nd case We distinguish two patterns: (i) the 32 4b forms (endogenous even CAs) in combination with 0000 0000 produce 64 IDs (note that those pertaining to CA2.8 constitute 3D SD variables and their complements); (ii) the 8 3b forms (exogenous CAs) combined with the 1 patterns produce $64 \times 2 = 128$ IDs $(3/8 + 1/8)$.

3rd case Also here we distinguish two patterns: (i) the 28 cases with 4a give rise to two different patterns: 12 of them (all in CA1.1, representing binary equivalences or contravalences, as well as the 3D SD variables $X_{1.4}, X_{2.2}, X_{2.3}$ and their complements) combined with 0000 0000 give rise to 24 combinations, while the other 24 cases (endogenous odd CAs representing different combinations of chunks of the previous 3D SD variables), always together with 0000 0000, produce 48 combinations (72 nodes as a whole); (ii) the 12 2a chunks (endogenous odd CAs) with 12 2a chunks give rise to 144 IDs $(2/8 + 2/8)$. Of these, 48 sum 2-chunks in the same CA while 96 2-chunks in different CAs.

4th case We have two subcases: (i) the 48 3a forms (exogenous CAs) need to be combined with the 1 pattern $(3/8 + 1/8)$ and produce $48 \times 8 \times 2 = 768$ IDs; (ii) when we sum the 2b chunks (even CAs) and the 2a chunks (odd CAs) we get the other $12 \times 16 \times 2 = 384$ cases $(2/8 + 2/8)$.

5th case The 16 2b chunks (endogenous even CAs) combined with other 2b chunks give rise to $16 \times 16 = 256$ IDs (again $2/8 + 2/8$). Of these, 64 combine two 2-chunks pertaining to the same CA, while the other 192 combine with some 2-chunks of different CAs.

The 4 4c cases plus the first 24 4a cases, the first 48 of 64 2b + 2b cases, and the first 36 of 48 2a + 2a cases give rise to 112 binary products. All of them are the combination of two endogenous 2-chunks both pertaining to the same CA. In order to understand the combinatorics of binary products, let us consider a particular product of $\mathcal{B}_2$ and follow its ramifications in $\mathcal{B}_3$ and $\mathcal{B}_4$, as displayed in Table 4.24 (similar considerations are true for the other 2D binary products). Such a ramification follows distinctive patterns. We need to distinguish here self–products, like $X_{1a}X_{1b}$, from the other cases (reference Tables 3.12–3.13). Some interesting cases of self–product are shown in Table 4.25. In all other cases listed in Table 4.24, note that we have all intermediate positions (the two central chunks) of the relative distinctive pattern (or their complements).

Similarly for the new forms of Level 3/8, i.e. ternary equivalences, as displayed in Table 4.26. Also in this case, we have the combination of two 2-chunks both pertaining to the same CA, but only to endogenous even CAs. Note that some alternative codifications give the same result, for example

$$X_{1b} \sim X_{4a} \sim X_{2a} = X_{1a} \sim X_{4b} \sim X_{2b} : 1000 \ 0001 \ 1000 \ 0001,$$

$$(4.23\text{a})$$

$$X_{1b} \sim X_{4a} \sim X_{2b} = X_{1a} \sim X_{4b} \sim X_{2a} : 1000 \ 0010 \ 0100 \ 0001.$$

$$(4.23\text{b})$$

It is then evident that we have $4 \times 4 = 16$ ternary equivalences representing the rest of the first 64 2b + 2b cases.

However, we need also to consider the remaining 12 cases out of the first 48 2a + 2a cases, as displayed in Table 4.27. They are combinations of two 2-chunks pertaining to the same endogenous odd

Table 4.24. Generation of binary products in different dimensional algebras. In bold are the patterns shared by the products generated by a single previous product. In the first group the pattern is 0011 or 1100 to occupy diverse positions, in the second group 0101 and 1010, in the third group 0110 and 1001, in the fourth group several combinations of 01 and 10, in the fifth and sixth groups the same but with antiparallel patterns.

$\mathscr{B}_2$	$\mathscr{B}_3$			$\mathscr{B}_4$				
$X_1 X_2$ 0001	$X_{1.1}X_{1.4}$	0000	0011	$X_{1a}X_{1b}$	**0000**	**0000**	**0000**	**1111**
				$X_{4a}X_{4b}$	**0011**	**0000**	**0000**	**0011**
				$X_{1a}X_{4a}$	**0000**	**0000**	**0011**	**0011**
				$X_{1a}X_{4b}$	**0000**	**0000**	**1100**	**0011**
				$X_{1b}X_{4a}$	**0000**	**0011**	**0000**	**0011**
				$X_{1b}X_{4b}$	**0000**	**1100**	**0000**	**0011**
	$X_{1.1}X_{2.2}$	0000	0101	$X_{1a}X_{1b}$	**0000**	**0000**	**0000**	**1111**
				$X_{2a}X_{2b}$	**0101**	**0000**	**0000**	**0101**
				$X_{1a}X_{2a}$	**0000**	**0000**	**0101**	**0101**
				$X_{1a}X_{2b}$	**0000**	**0000**	**1010**	**0101**
				$X_{1b}X_{2a}$	**0000**	**0101**	**0000**	**0101**
				$X_{1b}X_{2b}$	**0000**	**1010**	**0000**	**0101**
	$X_{1.1}X_{2.3}$	0000	1001	$X_{1a}X_{1b}$	**0000**	**0000**	**0000**	**1111**
				$X_{3a}X_{3b}$	**0110**	**0000**	**0000**	**1001**
				$X_{1a}X_{3a}$	**0000**	**0000**	**1001**	**1001**
				$X_{1a}X_{3b}$	**0000**	**0000**	**0110**	**1001**
				$X_{1b}X_{3a}$	**0000**	**0110**	**0000**	**1001**
				$X_{1b}X_{3b}$	**0000**	**1001**	**0000**	**1001**
	$X_{1.4}X_{2.2}$	0001	0001	$X_{4a}X_{4b}$	**0011**	**0000**	**0000**	**0011**
				$X_{2a}X_{2b}$	**0101**	**0000**	**0000**	**0101**
				$X_{4a}X_{2a}$	**0001**	0001	0001	**0001**
				$X_{4a}X_{2b}$	**0001**	0010	0010	**0001**
				$X_{4b}X_{2a}$	**0001**	0100	0100	**0001**
				$X_{4b}X_{2b}$	**0001**	1000	1000	**0001**
	$X_{1.4}X_{2.3}$	0010	0001	$X_{4a}X_{4b}$	**0011**	**0000**	**0000**	**0011**
				$X_{3a}X_{3b}$	**0110**	**0000**	**0000**	**1001**
				$X_{4a}X_{3a}$	**0010**	0010	0001	**0001**
				$X_{4a}X_{3b}$	**0010**	0001	0010	**0001**
				$X_{4b}X_{3a}$	**0010**	0100	1000	**0001**
				$X_{4b}X_{3b}$	**0010**	1000	0100	**0001**
	$X_{2.2}X_{2.3}$	0100	0001	$X_{2a}X_{2b}$	**0101**	**0000**	**0000**	**0101**
				$X_{3a}X_{3b}$	**0110**	**0000**	**0000**	**1001**
				$X_{2a}X_{3a}$	**0100**	0100	0001	**0001**
				$X_{2a}X_{3b}$	**0100**	0001	0100	**0001**
				$X_{2b}X_{3a}$	**0100**	0010	1000	**0001**
				$X_{2b}X_{3b}$	**0100**	1000	0010	**0001**

Table 4.25.   A group of self–products.

$X_{1a}X_{4a}$	0000	0000	0011	0011
$X_{1a}X'_{4a}$	0000	0000	1100	1100
$X'_{1a}X_{4a}$	0011	0011	0000	0000
$X'_{1a}X'_{4a}$	1100	1100	0000	0000

Table 4.26.   Generation of ternary equivalences in different dimensional algebras. Note that the internal part of the 4D IDs (8 numbers) repeats the 3D IDs. This pattern confirms the strict relation between material equivalence and logical self–identity, as gone over in Subsections 3.2.3 and 3.2.7.

$\mathcal{B}_3$	$\mathcal{B}_4$			
	$X_{1a} \sim X_{4a} \sim X_{2a}$ **1000** 1000 0001 **0001**			
	$X_{1a} \sim X_{4a} \sim X_{2b}$ **1000** 0100 0010 **0001**			
$X_{1.1} \sim X_{1.4} \sim X_{2.2}$ 1000 0001	$X_{1a} \sim X_{4b} \sim X_{2a}$ **1000** 0010 0100 **0001**			
	$X_{1a} \sim X_{4b} \sim X_{2b}$ **1000** 0001 1000 **0001**			
	$X_{1a} \sim X_{4a} \sim X'_{2a}$ **0100** 0100 0010 **0010**			
	$X_{1a} \sim X_{4a} \sim X'_{2b}$ **0100** 1000 0001 **0010**			
$X_{1.1} \sim X_{1.4} \sim X'_{2.2}$ 0100 0010	$X_{1a} \sim X_{4b} \sim X'_{2a}$ **0100** 0001 1000 **0010**			
	$X_{1a} \sim X_{4b} \sim X'_{2b}$ **0100** 0010 0100 **0010**			
	$X_{1a} \sim X'_{4a} \sim X_{2a}$ **0010** 0010 0100 **0100**			
	$X_{1a} \sim X'_{4a} \sim X_{2b}$ **0010** 0001 1000 **0100**			
$X_{1.1} \sim X'_{1.4} \sim X_{2.2}$ 0010 0100	$X_{1a} \sim X'_{4b} \sim X_{2a}$ **0010** 1000 0001 **0100**			
	$X_{1a} \sim X'_{4b} \sim X_{2b}$ **0010** 0100 0010 **0100**			
	$X'_{1a} \sim X_{4a} \sim X_{2a}$ **0001** 0001 1000 **1000**			
	$X'_{1a} \sim X_{4a} \sim X_{2b}$ **0001** 0010 0100 **1000**			
$X'_{1.1} \sim X_{1.4} \sim X_{2.2}$ 0001 1000	$X'_{1a} \sim X_{4b} \sim X_{2a}$ **0001** 0100 0010 **1000**			
	$X'_{1a} \sim X_{4b} \sim X_{2b}$ **0001** 1000 0001 **1000**			

Table 4.27.   Equivalences with two 1s in two different chunks (2a + 2a).

1100	0000	0000	0011	$X_{1a} \sim X_{1b} \sim X_{4a}$
0011	0000	0000	1100	$X_{1a} \sim X_{1b} \sim X'_{4a}$
0000	1100	0011	0000	$X_{1a} \sim X'_{1b} \sim X_{4a}$
0000	0011	1100	0000	$X'_{1a} \sim X_{1b} \sim X_{4a}$
1010	0000	0000	0101	$X_{1a} \sim X_{1b} \sim X_{2a}$
0101	0000	0000	1010	$X_{1a} \sim X_{1b} \sim X'_{2a}$
0000	1010	0101	0000	$X_{1a} \sim X'_{1b} \sim X_{2a}$
0000	0101	1010	0000	$X'_{1a} \sim X_{1b} \sim X_{2a}$
1001	0000	0000	1001	$X_{1b} \sim X_{3a} \sim X_{3b}$
0110	0000	0000	0110	$X'_{1b} \sim X_{3a} \sim X_{3b}$
0000	1001	1001	0000	$X_{1b} \sim X'_{3a} \sim X_{3b}$
0000	0110	0110	0000	$X_{1b} \sim X_{3a} \sim X'_{3b}$

Table 4.28. Other combinations of 2a + 2a chunks. Note
that the reduction in the penultimate row is possible because
it involves columns a, d, j, k, which form couples (a-d and
j-k, a-j and d-k, a-k and d-j) and therefore superpositions in
all the SCAs in which there are X_{1b} and X_{3a}.

1100	0000	0000	0101	$X'_{1a}X'_{1b}X'_{4a} + X_{1a}X_{1b}X_{2a}$
1100	0000	0101	0000	$X'_{1a}X'_{1b}X'_{4a} + X_{1a}X'_{1b}X_{2a}$
0000	1010	0101	0000	$X'_{1a}X_{1b}X'_{2a} + X_{1a}X'_{1b}X_{2a}$
1100	0000	1001	0000	$X'_{1b}X'_{4b}X'_{2b} + X'_{1b}X_{2b}X_{3a}$
1001	0000	0110	0000	$X'_{1b}X'_{3a}$
1001	0000	0000	0110	$X'_{3a}X'_{3b}$

(and not even) CA. Also here are always a couple of terms having
the same root in a 3D SD variable, like X_{1a} and X_{1b}.

The 96 2a + 2a cases (with two 2-chunks pertaining to different endogenous odd CAs) give rise to the first 4 IDs displayed in
Table 4.28, which are represented by sums of two ternary products
(note characteristic complementations in the logical expressions)
and the last two binary products. There are some correspondences
between the first 4 expressions and some expressions of Level 4/8,
in the order: $X'_{1.1}X'_{1.4} + X_{1.1}X_{2.2}$, $X'_{1.1}X'_{1.4} + X_{1.1}X_{2.2}$, $X_{1.1}X_{2.2} + X'_{1.1}X'_{2.2}$, $X'_{1.1}X'_{1.4} + X_{1.1}X_{2.3}$. Another way to consider the problem
is to sum the left and right 2-chunks as independent parts. In such a
case, the above IDs could also be considered as sums of "expanded"
binary products. In such a case, we have, in the order (which partly
reflects the above sequence), $X'_{1.1}X'_{1.4}$ and $X_{1.1}X_{2.2}$, $X'_{1.1}X'_{1.4}$ and
$X_{1.1}X_{2.2}$, $X_{1.1}X'_{2.2}$ and $X'_{1.1}X_{2.2}$, $X'_{1.1}X'_{1.4}$ and $X'_{1.1}X'_{2.3}$. Note that
in this case whether the 1s occur in the first or second 1-chunk in
each 2-chunk does not matter.

Some of the 64 4b + **0** nodes are displayed in Table 4.29. Note
the correspondence with $X_{1.4}X_{2.2} + X_{1.1}X'_{2.2}$, $X_{1.4}X'_{2.2} + X_{1.1}X'_{1.4}$,
$X'_{1.4}X'_{2.2} + X_{1.1}X_{2.2}$, $X'_{1.4}X'_{2.2} + X_{1.1}X_{1.4}$. Being the 1s are confined
in a single 2-chunk, the expressions are closed to those of Level
4/8. Some examples of the 72 4a + **0** combinations are displayed in
Table 4.30 where we have the correspondences $X'_{1.4}X'_{2.3} + X_{1.4}X_{2.2}$,
$X'_{2.2}X_{2.3} + X_{1.4}X_{2.2}$, $X'_{1.4}X_{2.2} + X_{1.4}X'_{2.3}$. Some of the 128 3b + 1
cases are presented in Table 4.31. In this case, the two chunks that

Table 4.29.   Combinations of 4b + **0** chunks. Note that, differently from the previous cases that showed two 2-chunks coming both from either odd or even endogenous CAs, we have here one 2-chunk coming from an even endogenous CA and another from a odd endogenous CA.

0000	0000	0001	1011	$X_{1a}X_{4a}X_{2a} + X_{1a}X_{1b}X'_{2a}$
0000	0000	0001	1101	$X_{1a}X_{4a}X_{2a} + X_{1a}X_{1b}X'_{4a}$
0000	0000	1000	1101	$X_{1a}X'_{4a}X'_{2a} + X_{1a}X_{1b}X_{2a}$
1000	1011	0000	0000	$X'_{1a}X'_{4a}X'_{2a} + X_{1a}X_{1b}X_{4a}$

Table 4.30.   Examples of the second group of 4a + **0** combinations.

1001	0101	0000	0000	$X'_{1a}X'_{4a}X'_{3b} + X'_{1a}X_{4a}X_{2a}$
0011	1001	0000	0000	$X'_{1a}X'_{2a}X_{3b} + X'_{1a}X_{4a}X_{2a}$
0101	0110	0000	0000	$X'_{1a}X'_{4a}X_{2a} + X'_{1a}X_{4a}X'_{3b}$

Table 4.31.   Combinations of 3b + 1 chunks. Note that both of the 2-chunks come from exogenous CAs. In particular, we deal here with those cases that sum two components pertaining to complementary exogenous CAs (thus, one odd and one even).

1000	0000	0000	0111	$X'_{4a}X'_{4b}X'_{3a} + X_{1b}X_{4a}X_{4b}$
0000	1011	0001	0000	$X_{1b}X'_{3a}X_{3b} + X_{4a}X_{3a}X'_{3b}$
0100	0000	0111	0000	$X'_{1b}X'_{4a}X'_{2a} + X_{1a}X'_{1b}X_{4a}$
0111	0000	0100	0000	$X'_{1b}X'_{4a}X_{2a} + X'_{1a}X'_{1b}X'_{4a}$
0111	0000	0000	1000	$X'_{1a}X_{2a}X_{2b} + X'_{2a}X'_{2b}X_{3b}$

Table 4.32.   Examples of 3a + 1 and 2a + 2b combinations.

0000	0010	0010	0011	$X_{1a}X_{4a}X_{3b} + X_{4a}X'_{2a}X'_{3b}$
1001	0100	0001	0000	$X'_{1a}X'_{4a}X'_{3b} + X_{4a}X_{2a}X'_{3b}$
1000	0110	0000	0010	$X'_{2a}X'_{3a}X'_{3b} + X_{1b}X_{3a}X'_{3b}$
0000	0100	0011	0001	$X_{1b}X_{4b}X_{3a} + X'_{1b}X_{4a}X'_{4b}$
0001	0100	0000	0011	$X_{4a}X_{4b}X_{3a} + X_{1b}X_{4b}X_{3a}$
1000	0010	0000	0011	$X'_{1a}X'_{4b}X'_{2a} + X_{1a}X_{4a}X_{4b}$

present the four 1s no longer constitute a single 2-chunk. Note also that the issue in which a 1-chunk is located, the 1s are irrelevant.

Table 4.32 displays some examples of the 768 3a + 1 and the 384 2a + 2b combinations. Here, due to the distribution of 1s over

Table 4.33. Examples of 2b + 2b combinations.

1000	0100	0001	0010	$X'_{1a}X'_{4a}X'_{3b} + X_{1a}X_{4a}X'_{3b}$
0100	1000	0010	0001	$X'_{1a}X'_{4a}X_{3b} + X_{1a}X_{4a}X_{3b}$
0010	0001	0100	1000	$X'_{1a}X_{4a}X_{3b} + X_{1a}X'_{4a}X_{3b}$

3-chunks, there are not direct correspondences with 3D expressions. In fact, they can be found only for the first three rows by summing an expression of Level 1/8 and another of Level 3/8 in the same order: $X_{1.1}X_{1.4}+X_{1.4}X_{2.2}$, $X'_{2.2}X'_{2.3}+X_{1.1}X'_{2.2}$. $X'_{1.4}X'_{2.3}+X'_{1.1}X'_{2.3}$. Finally, Table 4.33 displays some examples of the 2b + 2b combinations. Here, no insightful correspondence can be found.

In summary, we have three logical forms (one more than Level 2/8): binary products, ternary equivalences, and sums of two ternary products. Note that sometimes the latter case can also be written as $X'_{1a}X'_{1b}X'_{4a} + X_{1a}X'_{1b}X_{2a} = X'_{1b}(X'_{1a}X'_{4a} + X_{1a}X_{2a})$. The common form is represented by a sum of two ternary products. In fact, any of the 1,820 nodes of Level 4/16 can be represented in this way, as it is evident for the other two forms. Indeed, $2/16+2/16 = 4/16$. Sums of two ternary products are cleraly a generalisation of the logical forms of Level 2/8, which present only some of the forms that are either reducible or invariant under reversal (ternary equivalences). In this case, $8/16 \times 8/16 = 4/16$. At the opposite, forms like $1/16 + 3/16$ are not allowed, since Level 4/16 is even. When this form is the case (e.g. expressions of Level 2/16 as sum of two of Level 1/16), it is because no other choice is possible. The same is true for all even levels: no combination of odd levels is allowed. This case is due to the fact that even levels duplicate levels that already occur in less dimensional algebras, while odd levels always represent new levels (and new logical forms).

Expressions of Level 4/16 can be generated through sums of two expressions of Level 3/16, for example:

$$X_{1a}X_{4a} = (X_{1a}X_{4a}X_{2a} + X_{1a}X_{4a}X'_{2a}X'_{3b})$$
$$+ (X_{1a}X_{4a}X'_{2a} + X_{1a}X_{4a}X_{2a}X'_{3b}),$$

$$X_{1a} \sim X_{4a} \sim X_{2a} = (X_{1a}X_{4a}X_{2a} + X'_{1a}X'_{4a}X'_{2a}X'_{3b})$$
$$+ (X'_{1a}X'_{4a}X'_{2a} + X_{1a}X_{4a}X_{2a}X_{3b}),$$
$$X'_{1a}X'_{2a}X_{3b} + X'_{1a}X_{4a}X_{2a} = (X'_{1a}X'_{2a}X_{3b} + X'_{1a}X_{4a}X_{2a}X'_{3b})$$
$$+ (X'_{1a}X_{4a}X_{2a} + X'_{1a}X'_{4a}X'_{2a}X'_{3b}).$$

In turn, expressions of Level 3/16 can be generated through products of two expressions of Level 4/16:

$$X_{1a}X_{4a}(X_{2a} + X_{3b})$$
$$= (X_{1a}X_{4a}X_{2a} + X_{1a}X_{4a}X'_{2a})(X_{1a}X_{2a}X'_{3b} + X_{1a}X_{4a}X_{3b}),$$
$$X_{1a}X_{4a}X'_{3b} + X'_{1a}X'_{4a}X_{2a}X_{3b}$$
$$= (X_{1a}X_{4a}X'_{3b} + X'_{1a}X'_{4a}X_{3b})(X'_{1a}X'_{4a}X_{2a} + X_{1a}X_{4a}X'_{3b}).$$

For Level 5/16, we have the combinatorics displaying

$$\binom{16}{5} = 2\binom{4}{2}\binom{4}{4}\binom{4}{1} + 2\binom{4}{2}\binom{4}{3}\binom{4}{2}$$
$$+ 2\binom{4}{2}\binom{4}{3}\binom{4}{1}\binom{4}{1} + 2\binom{4}{2}\binom{4}{2}\binom{4}{2}\binom{4}{1}$$
$$+ \binom{4}{1}\binom{4}{2}\binom{4}{1}\binom{4}{1}\binom{4}{1}$$
$$= 48 + 288 + 768 + 1728 + 1536 = 4368 \qquad (4.24)$$

nodes. In other words, we have the following combinations:

- Four 1s in a chunk and one 1 in another: 4 permutations in the latter kind of chunk and 12 different combinations of the chunks: 48 cases.
- Three 1s in a chunk and two in another: 4 permutations in the first case and 6 in the second. The possible combinations of the chunks are again 12, so that $24 \times 12 = 288$ nodes.
- Three 1s in a chunk and one 1 in two different chunks: in all three cases, 4 permutations for a chunk. Again there are 12 arrangements of the chunks, so that $64 \times 12 = 768$ nodes.

- Two 1s in two different chunks and one 1 in another: 6 permutations for the first two kinds of chunks and 4 for the latter, making 144 combinations. Since there are 12 arrangements of the chunks, $144 \times 12 = 1728$ nodes.
- Finally, two 1s in a chunk and one 1 in the other ones: $384 \times 4 = 1536$ nodes.

In respect to the IDs, following Level 5/8, we have a combination of an endogenous and an exogenous 2-chunk. In particular, we distinguish among the following cases:

1st case We have two subcases: (i) 4c (pertaining to CA1.1) + 1 (exogenous CA), making 32 IDs ($X_{1.1}$ + 1/8), and (ii) 5b (pertaining to exogenous CAs and representing in $\mathscr{B}_3$ sums of a 3D SD variable and a binary product) + 0000 0000, making 16 IDs (5/8 + 0/8).

2nd case 288 elements split into two groups: (i) 3b (pertaining to exogenous CAs and representing the first logical form of Level 3/8) + 2a (pertaining to endogenous odd CAs), making 192 IDs, and (ii) 5a (exogenous CAs, representing both logical forms of Level 5/8) + 0000 0000, making 96 IDs.

3rd case 768 elements split into two groups: (i) 3b (exogenous CAs) + 2b (endogenous even CAs), making 256 IDs, and (ii) 4b (endogenous even CAs) + 1 (exogenous CA), making 512 IDs. Note that 4b presents (in CA2.8) the complete 3D CA2 variables.

4th case The penultimate 1728 elements split into two subcases: (i) 4a (endogenous odd CAs) + 1 (exogenous CA), making 576 IDs, and (ii) 2a (endogenous odd CAs) + 3a (exogenous CA), making 1152 IDs. The form 3a presents both logical forms of Level 3/8.

5th case 1536 IDs are: 3a (exogenous CA) + 2b (endogenous even CAs). Note that 4a presents (in CA1.1) the complete 3D variables of CA1.

Let us group these different forms according to this general typology by starting with expressions like 5a + **0** and 5b + **0**, as displayed in Table 4.34. Referencing Table 3.20 is very helpful. The first

Table 4.34.   Expressions 5a + **0** and 5b + **0** of Level 5/16.

0000	0000	0011	0111	$X_{1a}(X_{4a} + X_{2a}X'_{3b})$
0000	0000	0011	1011	$X_{1a}(X_{4a} + X'_{2a}X_{3b})$
0000	0000	1100	1110	$X_{1a}(X'_{4a} + X'_{2a}X'_{3b})$
0000	0000	0011	1110	$X_{1a}(X'_{4b} + X_{4a}X'_{2a})$
1100	0111	0000	0000	$X'_{1a}(X'_{4b} + X'_{4a}X_{2a})$
0000	0000	1111	0001	$X_{1a}(X'_{1b} + X_{4a}X_{2a})$
0000	0000	0001	1111	$X_{1a}(X_{1b} + X_{4a}X_{2a})$
0000	0000	0010	1111	$X_{1a}(X_{1b} + X_{4a}X'_{2a})$
1111	1000	0000	0000	$X'_{1a}(X'_{1b} + X'_{4a}X'_{2a})$

five rows of Table 4.34 display the first case. Here, the first three rows have the following correspondences: $X_{1.4} + X_{1.1}X_{2.2}$, $X_{1.4} + X_{1.1}X'_{2.2}$, $X'_{1.4} + X_{1.1}X'_{2.2}$. The second 2-chunk in the first three rows can be found among the first 32 expressions in Table 3.20, while rows four and five among the last 24 expressions of the same table. Thus, for the subsequent two rows, no direct correspondence can be found. At most, the first of these two rows can be put in correspondence with $X'_{1.1}X_{1.4} + X_{1.1}X'_{2.2} + X'_{1.4}X'_{2.2}$. The last four rows of Table 4.34 display the 5b case: all these 2-chunks can be found in the first 32 rows of Table 3.20. Here, we have the correspondences $X'_{1.1} + X_{1.4}X_{2.2}$, $X_{1.1} + X_{1.4}X_{2.2}$, $X_{1.1} + X_{1.4}X'_{2.2}$, $X'_{1.1} + X'_{1.4}X'_{2.2}$.

Then, let us group the 4a + 1, 4b + 1, and 4c + 1 expressions, as in Table 4.35. The three groups in the table correspond to these three cases in such order. The first five rows of the first group (4a + 1) display a combination of a 3D SD variable ($X_{1.4}$) and different 2-chunks with a single 1. It appears that different locations of the 1 can affect the general form of the logical expression. In particular, we can establish some kind of correspondence by summing $X_{1.4}$ and a ternary product so that we get for the first two rows $X_{1.4}$ plus a ternary product that already includes $X_{1.4}$, and for the third row $X_{1.1}X_{1.4} + X'_{1.1}X'_{1.4}$. For the last three rows of the first group there is no clear correspondence, although the fourth and fifth rows represent again combinations of $X_{1.4}$ and a ternary product (but with a 1 displaced relative to the $X_{1.4}$'s ID), while the sixth row is such that the right 2-chunk does not represent a 3D SD variable. This case is also

Table 4.35. Expressions for 4a + 1, 4b + 1, and 4c + 1 of Level 5/16.

0011	0011	0000	0001	$X_{4a}(X_{1a}' + X_{2a}X_{3b})$
0011	0011	0010	0000	$X_{4a}(X_{1a}' + X_{2a}'X_{3b})$
0011	1100	0000	0001	$X_{4b}(X_{1a}' + X_{4a}X_{2a})$
0011	0011	1000	0000	$X_{1a}'X_{4a} + X_{1a}X_{4a}'X_{2a}'X_{3b}'$
1000	0000	0011	0011	$X_{1a}X_{4a} + X_{1a}'X_{4a}'X_{2a}'X_{3b}'$
0001	0000	0101	1001	$X_{1b}'X_{4b}X_{3a}' + X_{4b}'X_{2b}'X_{3a} + X_{1b}X_{4b}X_{2b}X_{3a}$
0001	0111	0000	0001	$X_{1a}'X_{2b}X_{3b}' + X_{1a}'X_{2a}X_{2b}' + X_{1a}X_{2a}X_{2b}X_{3b}$
0001	0111	0001	0000	$X_{1a}'X_{2b}X_{3b}' + X_{1a}'X_{2a}X_{2b}' + X_{1a}X_{2a}X_{2b}'X_{3b}'$
0010	1101	0000	0010	$X_{1a}'X_{1b}X_{4a}' + X_{1b}X_{4a}X_{2a} + X_{1a}X_{1b}'X_{4a}X_{2a}'$
0000	0001	0000	1111	$X_{1b}(X_{1a} + X_{4a}X_{2a})$
0000	0010	0000	1111	$X_{1b}(X_{1a} + X_{4a}X_{2a}')$
0000	1111	0000	0001	$X_{1b}(X_{1a}' + X_{4a}X_{2a})$
1111	0000	1000	0000	$X_{1b}'(X_{1a} + X_{4a}'X_{2a}')$
0000	1111	0000	1000	$X_{1b}(X_{1a}' + X_{4a}'X_{2a}')$
1000	0000	0000	1111	$X_{1a}X_{1b} + X_{1a}'X_{1b}'X_{4a}'X_{2a}'$

true for the second group. Here, the problem is represented by the fact that the first two rows represent a sum of a 2-chunk representing $X_{1.2.8.A}$ and $X_{1.1.9.B}'$ and $X_{2.4.16.B}$, respectively, while the third row shows a combination of $X_{1.3.14.A}$ with $X_{2.1.1.B}$, all of the relative SD variables not pertaining to CA1.1, used for codification. For the first five expressions of the third group, we can find the correspondences $X_{1.1} + X_{1.4}X_{2.2}$, $X_{1.1} + X_{1.4}X_{2.2}'$, $X_{1.1}' + X_{1.4}X_{2.2}$, $X_{1.1}' + X_{1.4}'X_{2.2}'$, and again $X_{1.1}' + X_{1.4}'X_{2.2}'$. Note that the different positions of the four 1s and the single 1 do matter, as it is evident in the last row.

Table 4.36 displays all combinations of 3s and 2s. The first group represents 3a + 2a combinations; the second group 3a + 2b combinations; the third group 3b + 2a combinations; the fourth and last group 3b + 2b combinations. For the first group, we find the correspondences $X_{1.1}'X_{1.4} + X_{1.4}X_{2.2}$ and $X_{1.1}'X_{1.4} + X_{1.1}X_{1.4} + X_{1.1}X_{1.4}'X_{2.2}$ (note that for the first ID, the 0011 occupies the first 1-chunk in both the 2-chunks). For the second group, we have in the same order: $X_{1.1}X_{1.4} + X_{1.4}X_{2.2}$, $X_{1.4}'X_{2.2} + X_{1.1}X_{1.4}'$, $X_{1.4}'X_{2.2}' + X_{1.1}'X_{1.4}X_{2.2}$, $X_{2.2}X_{2.3}' + X_{1.1}'X_{2.2}'X_{2.3}$. The third row of this group contains an equivalence and therefore it behaves differently from the first two rows, while in the last row we have mismatch of two 1s relative to

Table 4.36. Combinations of 3a-3b with 2a-2b of Level 5/16.

0011	0001	0011	0000	$X_{4a}(X_{1b}' + X_{1a}'X_{2a})$
0011	0100	0000	0011	$X_{1a}'X_{1b}'X_{4a} + X_{1a}X_{1b}X_{4a} + X_{1a}'X_{1b}X_{4a}'X_{2a}$
0001	0001	0001	0011	$X_{4a}(X_{2a} + X_{1a}X_{3b}')$
0100	0100	0100	1100	$X_{4a}'(X_{2a} + X_{1a}X_{1b})$
0010	0010	0010	0110	$X_{4a}X_{2a}' + X_{1a}X_{4a}'X_{2a}X_{3b}'$
0001	0100	0011	0100	$X_{2a}X_{3b}' + X_{1a}X_{4a}X_{2a}'X_{3b}$
0000	0111	0000	0011	$X_{1b}(X_{4a} + X_{1a}'X_{2a})$
0000	0111	0000	0101	$X_{1b}(X_{2a} + X_{1a}'X_{4a})$
0000	0011	0000	0111	$X_{1b}(X_{4a} + X_{1a}X_{2a})$
0001	1000	0111	0000	$X_{1a}X_{2a}X_{2b}' + X_{2a}'X_{2b}X_{3b} + X_{1a}'X_{2a}X_{2b}X_{3b}'$
1110	0000	1000	0001	$X_{1a}'X_{2a}'X_{2b}' + X_{2a}X_{2b}X_{3b} + X_{1a}X_{2a}'X_{2b}X_{3b}'$
0000	0111	1000	0001	$X_{1a}'X_{2a}X_{2b}' + X_{2a}'X_{2b}X_{3b}' + X_{1a}X_{2a}X_{2b}X_{3b}$

the couple of 1s. All of the three expressions of the third group can lead to $X_{1.1}X_{1.4} + X_{1.1}X_{2.2}$. For the fourth group there is no insightful correspondence.

In summary, there are three new logical forms:

$$X_{1a}X_{4a}'X_{2a} + X_{1a}X_{4a}X_{3b} + X_{1a}'X_{4a}'X_{2a}'X_{3b},$$

$$X_{1a}'X_{4a} + X_{1a}X_{4a}'X_{2a}'X_{3b}', \quad X_{4a}(X_{1a}' + X_{2a}X_{3b}).$$

The first two, at the opposite of the last one, are characterised by a ID displaying no superpositions, and the first is clearly the basic logical form. Thus, the last one reduces to the first one. For instance,

$$X_{1a}'X_{4a} + X_{4a}X_{2a}X_{3b} = X_{1a}'X_{4a}X_{3b} + X_{1a}'X_{4a}X_{3b}'$$
$$+ X_{1a}X_{4a}X_{2a}X_{3b}. \qquad (4.25)$$

It is easy to see that the logical expressions of Level 5/16 can be generated through either $1/16 + 2/16 + 2/16$ or $1/16 + 4/16$ or through $8/16 \times 10/16$. The first two cases correspond to the first two logical forms, while the last case to the third logical form.

The three expressions above can be generated through sums of logical forms of Level 4/16:

$$X_{1a}X_{4a}'X_{2a} + X_{1a}X_{4a}X_{3b} + X_{1a}'X_{4a}'X_{2a}'X_{3b}$$
$$= (X_{1a}X_{4a}'X_{2a} + X_{1a}X_{4a}X_{3b}) + (X_{1a}'X_{4a}X_{2a}'X_{3b}$$

$$+ X'_{1a}X'_{4a}X_{2a}X_{3b} + X_{1a}X'_{4a}X'_{2a}X_{3b} + X_{1a}X'_{4a}X_{2a}X_{3b})$$
$$= (X_{1a}X'_{4a}X_{2a} + X_{1a}X_{4a}X_{3b}) + (X'_{4a}X_{3b});$$
$$X'_{1a}X_{4a} + X_{1a}X'_{4a}X'_{2a}X'_{3b}$$
$$= [X'_{1a}X_{4a}(X_{3b} + X'_{3b})] + (X_{1a}X'_{4a}X'_{2a}X'_{3b}$$
$$\qquad + X'_{1a}X_{4a}X'_{2a}X'_{3b} + X'_{1a}X'_{4a}X'_{2a}X'_{3b} + X_{1a}X_{4a}X'_{2a}X'_{3b})$$
$$= (X'_{1a}X_{4a}) + (X'_{2a}X'_{3b});$$
$$X_{4a}(X'_{1a} + X_{2a}X_{3b})$$
$$= (X'_{1a}X_{4a}) + (X_{4a}X_{2a}X_{3b} + X'_{1a}X_{4a}X'_{3b}).$$

Reciprocally, we can derive expressions of Level 4/16 through products of two logical forms of Level 5/16:

$$X'_{1a}X_{4a}$$
$$= (X'_{1a}X_{4a} + X_{1a}X'_{4a}X'_{2a}X'_{3b})(X'_{1a}X_{4a} + X'_{1a}X_{4a}X'_{2a}X'_{3b});$$
$$X_{1a} \sim X_{4a} \sim X_{2a}$$
$$= (X_{1a}X_{4a}X_{2a} + X'_{1a}X'_{4a}X_{3b} + X'_{1a}X'_{4a}X'_{2a}X'_{3b})$$
$$\qquad \times (X'_{1a}X'_{4a}X'_{2a} + X_{1a}X_{4a}X_{3b} + X_{1a}X_{4a}X_{2a}X'_{3b});$$
$$X_{1a}X'_{4a}X_{3b} + X'_{1a}X_{4a}X_{2a}$$
$$= (X_{1a}X'_{4a}X_{3b} + X'_{1a}X_{4a}X_{2a} + X'_{1a}X'_{4a}X'_{2a}X'_{3b})$$
$$\qquad \times (X_{1a}X'_{4a}X_{3b} + X'_{1a}X_{4a}X_{2a} + X_{1a}X_{4a}X'_{2a}X'_{3b}).$$

4.3.4. Levels 6/16 and 7/16

For Level 6/16 we have

$$\binom{16}{6} = 2\binom{4}{2}\binom{4}{4}\binom{4}{2} + 2\binom{4}{2}\binom{4}{4}\binom{4}{1}\binom{4}{1}$$
$$+ \binom{4}{2}\binom{4}{3}\binom{4}{3} + 4\binom{4}{2}\binom{4}{3}\binom{4}{2}\binom{4}{1}$$

$$+ \binom{4}{1} \binom{4}{3} \binom{4}{1} \binom{4}{1} \binom{4}{1} + \binom{4}{3} \binom{4}{2} \binom{4}{2} \binom{4}{2}$$

$$+ \binom{4}{2} \binom{4}{2} \binom{4}{2} \binom{4}{1} \binom{4}{1}$$

$$= 72 + 192 + 96 + 2304 + 1024 + 864 + 3456 = 8008 \qquad (4.26)$$

nodes. Then, we have the following combinatorics:

- A chunk with four 1s and another with two 1s. In the last case we have 6 permutations, and, since there are 12 different arrangements of the chunks, we have $6 \times 12 = 72$ nodes.
- A chunk with four 1s and two chunks with a single 1 each. In the latter two chunks we have 4 permutations. The different permutations of the chunks give rise to 12 arrangements, so that we have $4 \times 4 \times 12 = 192$ nodes.
- Three 1s in each of two chunks, with 4 permutations each. The two chunks give rise to 6 different arrangements, so that we have $4 \times 4 \times 6 = 96$ nodes.
- Three 1s in a chunk, two 1s in a second chunk, and one 1 in a third chunk. These elements make $4 \times 6 \times 4 = 96$ permutations. Since the chunks are different, and thus the different arrangements of the chunks are $4 \times 6 = 24$, this combination makes $96 \times 24 = 2304$ nodes.
- Three 1s in a chunk and three chunks with a single 1 each. These elements make $4 \times 4 \times 4 \times 4 = 256$ permutations, and, since there are 4 permutations of the chunks, we have $256 \times 4 = 1024$ nodes.
- Three chunks with two 1s each. These elements make $6 \times 6 \times 6 = 216$ permutations. The permutations of the chunks are 4, so that there are $216 \times 4 = 864$ nodes.
- Finally, two chunks with two 1s and two chunks with one 1. These elements make $6 \times 6 \times 4 \times 4 = 576$ permutations. Having 6 different arrangements of the chunks, we have $576 \times 6 = 3456$ nodes.

Table 4.37. Ramifications of the first group of logical expressions of Level 3/8.

$\mathscr{B}_3$	$\mathscr{B}_4$			
$X_{1.1}(X'_{1.4} + X'_{2.2})$ 0000 1110	$X_{1a}(X'_{4a} + X'_{2a})$	0000	0000	1110	**1110**
	$X_{1a}(X'_{4a} + X'_{2b})$	0000	0000	1101	**1110**
	$X_{1a}(X'_{4b} + X'_{2a})$	0000	0000	1011	**1110**
	$X_{1a}(X'_{4b} + X'_{2b})$	0000	0000	0111	**1110**
	$X_{1b}(X'_{4a} + X'_{2a})$	0000	1110	0000	**1110**
	$X_{1b}(X'_{4a} + X'_{2b})$	0000	1101	0000	**1110**
	$X_{1b}(X'_{4b} + X'_{2a})$	0000	1011	0000	**1110**
	$X_{1b}(X'_{4b} + X'_{2b})$	0000	0111	0000	**1110**
$X_{1.1}(X'_{1.4} + X_{2.2})$ 0000 1101	$X_{1a}(X'_{4a} + X_{2a})$	0000	0000	1101	**1101**
	$X_{1a}(X'_{4a} + X_{2b})$	0000	0000	1110	**1101**
	$X_{1a}(X'_{4b} + X_{2a})$	0000	0000	0111	**1101**
	$X_{1a}(X'_{4b} + X_{2b})$	0000	0000	1011	**1101**
	$X_{1b}(X'_{4a} + X_{2a})$	0000	1101	0000	**1101**
	$X_{1b}(X'_{4a} + X_{2b})$	0000	1110	0000	**1101**
	$X_{1b}(X'_{4b} + X_{2a})$	0000	0111	0000	**1101**
	$X_{1b}(X'_{4b} + X_{2b})$	0000	1011	0000	**1101**
$X_{1.1}(X_{1.4} + X'_{2.2})$ 0000 1011	$X_{1a}(X_{4a} + X'_{2a})$	0000	0000	1011	**1011**
	$X_{1a}(X_{4a} + X'_{2b})$	0000	0000	0111	**1011**
	$X_{1a}(X_{4b} + X'_{2a})$	0000	0000	1110	**1011**
	$X_{1a}(X_{4b} + X'_{2b})$	0000	0000	1101	**1011**
	$X_{1b}(X_{4a} + X'_{2a})$	0000	1011	0000	**1011**
	$X_{1b}(X_{4a} + X'_{2b})$	0000	0111	0000	**1011**
	$X_{1b}(X_{4b} + X'_{2a})$	0000	1110	0000	**1011**
	$X_{1b}(X_{4b} + X'_{2b})$	0000	1101	0000	**1011**
$X_{1.1}(X_{1.4} + X_{2.2})$ 0000 0111	$X_{1a}(X_{4a} + X_{2a})$	0000	0000	0111	**0111**
	$X_{1a}(X_{4a} + X_{2b})$	0000	0000	1011	**0111**
	$X_{1a}(X_{4b} + X_{2a})$	0000	0000	1101	**0111**
	$X_{1a}(X_{4b} + X_{2b})$	0000	0000	1110	**0111**
	$X_{1b}(X_{4a} + X_{2a})$	0000	0111	0000	**0111**
	$X_{1b}(X_{4a} + X_{2b})$	0000	1011	0000	**0111**
	$X_{1b}(X_{4b} + X_{2a})$	0000	1101	0000	**0111**
	$X_{1b}(X_{4b} + X_{2b})$	0000	1110	0000	**0111**

We reproduce at this level logical structures that we already met in Level 3/8. Before providing a complete analysis of the IDs, let us consider this connection. In particular, by taking into account the second 4 nodes of Table 3.15, we get the ramifications displayed in Table 4.37, while some ramifications of the second group of expressions is shown in Table 4.38. Of course, there are many other possibilities not shown here, like self–products, which are considered below.

Table 4.38. Ramifications of the second group of logical expressions of Level 3/8. Note the mismatch in the position of the 1s.

$\mathscr{B}_3$	$\mathscr{B}_4$
$X'_{1.1}X'_{1.4} + X_{1.1}X_{1.4}X_{2.2}$ 1100 0001	$X'_{1a}X'_{4a} + X_{1a}X_{4a}X_{2a}$ 1100 1100 0001 0001
	$X'_{1a}X'_{4a} + X_{1a}X_{4a}X_{2b}$ 1100 1100 0010 0001
	$X'_{1a}X'_{4b} + X_{1a}X_{4b}X_{2a}$ 1100 0011 0100 0001
	$X'_{1a}X'_{4b} + X_{1a}X_{4b}X_{2b}$ 1100 0011 1000 0001
	$X'_{1b}X'_{4a} + X_{1b}X_{4a}X_{2a}$ 1100 0001 1100 0001
	$X'_{1b}X'_{4a} + X_{1b}X_{4a}X_{2b}$ 1100 0010 1100 0001
	$X'_{1b}X'_{4b} + X_{1b}X_{4b}X_{2a}$ 1100 0100 0011 0001
	$X'_{1b}X'_{4b} + X_{1b}X_{4b}X_{2b}$ 1100 1000 0011 0001
$X'_{1.1}X_{1.4} + X_{1.1}X'_{1.4}X'_{2.2}$ 0011 1000	$X'_{1a}X_{4a} + X_{1a}X'_{4a}X'_{2a}$ 0011 0011 1000 1000
	$X'_{1a}X_{4a} + X_{1a}X'_{4a}X'_{2b}$ 0011 0011 0100 1000
	$X'_{1a}X_{4b} + X_{1a}X'_{4b}X'_{2a}$ 0011 1100 0010 1000
	$X'_{1a}X_{4b} + X_{1a}X'_{4b}X'_{2b}$ 0011 1100 0001 1000
	$X'_{1b}X_{4a} + X_{1b}X'_{4a}X'_{2a}$ 0011 1000 0011 1000
	$X'_{1b}X_{4a} + X_{1b}X'_{4a}X'_{2b}$ 0011 0100 0011 1000
	$X'_{1b}X_{4b} + X_{1b}X'_{4b}X'_{2a}$ 0011 0010 1100 1000
	$X'_{1b}X_{4b} + X_{1b}X'_{4b}X'_{2b}$ 0011 0001 1100 1000
$X_{1.1}X'_{1.4} + X'_{1.1}X_{1.4}X_{2.2}$ 0001 1100	$X_{1a}X'_{4a} + X'_{1a}X_{4a}X_{2a}$ 0001 0001 1100 1100
	$X_{1a}X'_{4a} + X'_{1a}X_{4a}X_{2b}$ 0001 0010 1100 1100
	$X_{1a}X'_{4b} + X'_{1a}X_{4b}X_{2a}$ 0001 0100 0011 1100
	$X_{1a}X'_{4b} + X'_{1a}X_{4b}X_{2b}$ 0001 1000 0011 1100
	$X_{1b}X'_{4a} + X'_{1b}X_{4a}X_{2a}$ 0001 1100 0001 1100
	$X_{1b}X'_{4a} + X'_{1b}X_{4a}X_{2b}$ 0001 1100 0010 1100
	$X_{1b}X'_{4b} + X'_{1b}X_{4b}X_{2a}$ 0001 0011 0100 1100
	$X_{1b}X'_{4b} + X'_{1b}X_{4b}X_{2b}$ 0001 0011 1000 1100
$X_{1.1}X_{1.4} + X'_{1.1}X'_{1.4}X'_{2.2}$ 1000 0011	$X_{1a}X_{4a} + X'_{1a}X'_{4a}X'_{2a}$ 1000 1000 0011 0011
	$X_{1a}X_{4a} + X'_{1a}X'_{4a}X'_{2b}$ 1000 0100 0011 0011
	$X_{1a}X_{4b} + X'_{1a}X'_{4b}X'_{2a}$ 1000 0010 1100 0011
	$X_{1a}X_{4b} + X'_{1a}X'_{4b}X'_{2b}$ 1000 0001 1100 0011
	$X_{1b}X_{4a} + X'_{1b}X'_{4a}X'_{2a}$ 1000 0011 1000 0011
	$X_{1b}X_{4a} + X'_{1b}X'_{4a}X'_{2b}$ 1000 0011 0100 0011
	$X_{1b}X_{4b} + X'_{1b}X'_{4b}X'_{2a}$ 1000 1100 0010 0011
	$X_{1b}X_{4b} + X'_{1b}X'_{4b}X'_{2b}$ 1000 1100 0001 0011

Let us now analyse the possible structures of the IDs, where we have either endogenous–endogenous or exogenous–exogenous combinations of the two 2–chunks:

1st case Four 1s and two 1s. We have here two subcases: (i) the two 2-chunks 4c combined with the twelve 2-chunks 2a, which makes 48 IDs; (ii) the twelve 2-chunks 6a and left or right position, which makes 24 IDs. These two subcases are shown in Table 4.39. Note that all of these expressions but the last two show self–products.

2nd case Four 1s and two 1s in two different chunks. Also here we distinguish between two subcases: (i) the two 4c combined with the

Table 4.39. Combinations of 4c and 2a as well as of 6a and **0**. Both 2-chunks come from odd endogenous CAs.

1111	0000	0011	0000	$X'_{1b}(X'_{1a} + X_{4a})$
1111	0011	0000	0000	$X'_{1a}(X'_{1b} + X_{4a})$
1111	0000	0000	0011	$X'_{1a}X'_{1b} + X_{1a}X_{1b}X_{4a}$
1111	0000	0000	1100	$X'_{1a}X'_{1b} + X_{1a}X_{1b}X'_{4a}$
0000	0000	1100	1111	$X_{1a}(X_{1b} + X'_{4a})$
0000	0000	0011	1111	$X_{1a}(X'_{1b} + X'_{4a})$
0000	0000	0101	1111	$X_{1a}(X_{2a} + X'_{2b})$
0000	0000	1001	1111	$X_{1b}(X'_{2a} + X_{3b})$

Table 4.40. Combinations of 4c and 2b as well as of 5b and 1.

0001	0001	0000	1111	$X_{4a}(X_{1a} + X_{4b})$
0001	0010	0000	1111	$X_{1a}X_{2a}X_{2b} + X_{1a}X'_{2a}X'_{2b} + X'_{1a}X_{2b}X_{3b}$
1111	0001	0000	0001	$X'_{1a}X'_{1b} + X_{1b}X_{4a}X_{2a}$
1111	0100	0000	1000	$X'_{1b}X'_{4a}X'_{4b} + X'_{1b}X_{4a}X_{4b} + X_{1b}X'_{4a}X_{3a}$

Table 4.41. Combinations of 6b and **0** as well as of 3b and 3b.

1110	1110	0000	0000	$X'_{1a}(X'_{4a} + X'_{2a})$
1101	1101	0000	0000	$X'_{1a}(X'_{4a} + X_{2a})$
0000	0000	1011	1011	$X_{1a}(X_{4a} + X_{3a})$
0000	0000	1101	1110	$X_{1a}(X_{4a} + X'_{3b})$
0000	0111	0000	1011	$X_{1b}(X_{4a} + X_{3a})$
0000	1110	0000	0111	$X_{1a}X_{1b}X_{4a} + X'_{1a}X_{1b}X'_{4a} + X_{1b}X_{4a}X'_{2a}$

sixteen 2-chunks 2b, which makes 64 IDs; (ii) the eight 2-chunks 5b combined with the eight 2-chunks 1, which makes 128 IDs. These two subcases are displayed in Table 4.40. Note the mismatch between the two isolated 1s in the second and fourth rows. All of these expressions display self–products.

3rd case Three 1s and three 1s. Again two subcases: (i) The sixteen 2-chunks 6b combined with 0000 0000, which makes 32 IDs; (ii) the eight 2-chunks 3b combined with themselves, which makes 64 IDs. The two subcases are displayed in Table 4.41. Note the two mismatches in the last row relative to the single mismatch in the 3rd row. Some rows show self–products.

4th case Three, two, and one 1s. We have here three cases: (i) the forty-eight 2-chunks 3a combined with the eight 3b, which makes 768 IDs; (ii) the forty-eight 2-chunks 5a combined with the eight 2-chunks 1, which makes another 768 IDs; (iii) the thirty-two 2-chunks 4b combined with the twelve 2a, which again makes 768 IDs. These three subcases are shown in Table 4.42. Note that the first two rows combine two 2-chunks that are represented by the first 32 logical expressions of Table 3.15, although these are obviously different. For the second group, the first two rows have on the right a 2-chunk represented again by the first 32 logical expressions of Table 3.15, while the third row by the other 24 expressions. The last row of this group has a 2-chunk of the same form as the first two, but the isolated 1 is connected with a 3D ternary product with all terms complemented relative to the first expression.

5th case The thirty-two 4b combined with the sixteen 2b, which makes 1024 IDs. This case is shown in Table 4.43.

6th case The thirty-six 2-chunks 4a combined with the twelve 2a, which makes 864 IDs. This case is shown in Table 4.44. Note the differences in the disposition of the 1s.

Table 4.42.   Combinations of 3a and 3b, of 5a and 1, and of 4b and 2a.

0111	0000	0001	0011	$X_{4a}X_{4b} + X'_{1b}X'_{4b}X_{3a}$
0011	0010	0000	0111	$X_{4a}X_{3a}X_{3b} + X'_{1a}X_{4a}X'_{3b} + X_{1a}X'_{3a}X'_{3b}$
0000	1000	0011	1011	$X_{1a}X_{4a} + X'_{4a}X'_{2a}X_{3b}$
0100	0000	1010	1011	$X_{1a}X'_{2a} + X_{2a}X_{2b}X_{3b}$
0100	0000	1100	1011	$X_{1a}X'_{4a}X'_{2a} + X_{1a}X_{1b}X_{4a} + X'_{1b}X'_{4a}X_{2a}$
0100	0000	0011	1011	$X_{4a}X_{1b}X_{4b} + X_{4a}X'_{1b}X'_{4b} + X'_{4a}X'_{4b}X_{3a}$
0000	0011	0001	0111	$X_{4a}X_{2a}X_{3b} + X_{4a}X'_{2a}X'_{3b} + X_{1a}X_{2a}X'_{3b}$
1001	0000	0001	0111	$X'_{1a}X'_{1b}X'_{3b} + X_{1a}X'_{2b}X'_{3b} + X_{1a}X_{1b}X_{3b}$
0110	0000	1000	0111	$X'_{1b}X'_{2a}X_{3a} + X_{1b}X'_{3a}X'_{3b} + X_{2a}X_{3a}X_{3b}$

Table 4.43.   Combinations of 4b and 2b. Note the self–products.

1110	0100	1000	1000	$X'_{1a}X'_{1b}X'_{2a} + X'_{1a}X'_{4a}X_{2a} + X_{1a}X'_{4a}X'_{2a}$
0010	0111	0010	0010	$X'_{1a}X'_{1b}X_{2a} + X'_{1a}X'_{4a}X'_{2a} + X_{1a}X_{4a}X'_{2a}$
1101	0001	1000	1000	$X'_{1a}X'_{1b}X'_{4a} + X'_{1a}X_{4a}X_{2a} + X_{1a}X'_{4a}X'_{2a}$

Table 4.44.   Combinations of 4a and 2a. Note the self–products.

0101	0000	0101	0101	$X_{2a}(X_{1a} + X_{2b})$
0000	0101	0101	0101	$X_{2a}(X_{1a} + X'_{2b})$
0000	0101	1010	1010	$X_{1a}X'_{2a} + X'_{1a}X_{2a}X'_{2b}$
0000	0101	0110	0110	$X'_{1a}X_{1b}X_{2a} + X_{1a}X'_{4a}X_{2a} + X_{1a}X_{4a}X'_{2a}$

Table 4.45.   Combinations of 4a and 2b as well as of 3a and 3a.

1000	1000	1010	1001	$X'_{4a}X'_{2a} + X_{1a}X_{4a}X_{3b}$
0100	1000	0110	1010	$X'_{4a}X_{3b} + X_{1a}X_{4a}X_{2a}$
0101	0001	0101	0001	$X_{2a}(X_{4a} + X_{3b})$
0110	0001	0110	0001	$X_{3b}(X_{4a} + X_{2a})$

7th case Two, two, one, one 1s. Two subcases: (i) the thirty-six 2-chunks 4a combined with the sixteen 2b, which makes 1152 IDs; (ii) the forty-eight 3a combined with themselves, which makes 2304 IDs. Both such subcases are displayed in Table 4.45.

In conclusion, we have three logical forms:

$$X_{1a}(X_{4a} + X_{2a}), \quad X'_{1a}X_{4a} + X_{1a}X'_{4a}X'_{2a},$$
$$X'_{1a}X'_{4a}X'_{2a} + X'_{1a}X_{2a}X_{3b} + X_{1a}X'_{2a}X_{3b}.$$

The first two are inherited from Level 3/8, where I have shown that the former (whose ID displays superpositions) can be reduced to the latter. It is easy to verify that this second form can be transformed into the third one, so that it is the common logical form. The first form is $8/16 \times 12/16$, the second $4/16 + 2/16$, and the third $4/16 + 2/16$. In some cases, the new expressions can be rewritten as

$$X_{1b}X_{1a}X_{4a} + X_{1b}X'_{1a}X'_{4a} + X_{1b}X_{4a}X'_{2a}$$
$$= X_{1b}(X_{1a}X_{4a} + X'_{1a}X'_{4a} + X_{4a}X'_{2a}).$$

The form on the second row is allowed since $8/16 \times (4/16 + 8/16) = 6/16$. Note that $1/16 + 5/16$ and $3/16 + 3/16$ do not work since they are odd. In other words, we have two sums and a product (we are, in fact, in the first half of the algebra, starting from **0**).

Each of the three logical forms of Level 6/16 can be derived through sums of two logical expressions of Level 5/16:

$$X_{1a}(X_{4a} + X_{2a})$$
$$= [X_{1a}(X_{4a} + X_{2a}X_{3b})] + [X_{1a}(X_{4a} + X_{2a}X'_{3b})];$$
$$X'_{1a}X_{4a} + X_{1a}X'_{4a}X'_{2a}$$
$$= (X'_{1a}X_{4a} + X_{1a}X'_{4a}X'_{2a}X'_{3b}) + (X'_{1a}X_{4a} + X_{1a}X'_{4a}X'_{2a}X_{3b});$$
$$X'_{1a}X'_{4a}X'_{2a} + X'_{1a}X_{2a}X_{3b} + X_{1a}X'_{2a}X_{3b}$$
$$= (X'_{1a}X'_{4a}X'_{2a} + X'_{1a}X_{2a}X_{3b} + X_{1a}X_{4a}X'_{2a}X_{3b})$$
$$+ (X'_{1a}X'_{4a}X'_{2a} + X_{1a}X'_{2a}X_{3b} + X'_{1a}X_{4a}X_{2a}X_{3b}).$$

Reciprocally, we can derive logical forms of Level 5/16 from those of Level 6/16:

$$X_{1a}(X_{4a} + X_{1b}X_{2a})$$
$$= [X'_{1a}(X_{4a} + X_{1b})](X_{1b}X_{2a} + X_{1a}X'_{1b}X'_{2a});$$
$$X_{4a}X'_{2a} + X_{1a}X'_{4a}X_{2a}X'_{3b}$$
$$= (X'_{1a}X_{4a}X'_{2a} + X_{1a}X'_{4a}X_{2a} + X_{1a}X_{4a}X'_{2a})$$
$$\times (X'_{1a}X_{4a}X'_{2a} + X'_{4a}X_{2a}X'_{3b} + X_{1a}X_{4a}X'_{2a});$$
$$X'_{1a}X'_{4a}X'_{2a} + X_{1a}X_{1b}X_{4a} + X_{1a}X'_{1b}X'_{4a}X_{2a}$$
$$= (X'_{1a}X'_{4a}X'_{2a} + X_{1a}X_{1b}X_{4a} + X_{1a}X'_{1b}X_{2a})$$
$$\times (X'_{1a}X'_{4a}X'_{2a} + X_{1a}X_{1b}X_{4a} + X'_{1b}X'_{4a}X'_{4a}X_{2a}).$$

Level 7/16 presents

$$\binom{16}{7} = 2\binom{4}{2}\binom{4}{4}\binom{4}{3} + 4\binom{4}{2}\binom{4}{4}\binom{4}{2}\binom{4}{1}$$
$$+ \binom{4}{1}\binom{4}{4}\binom{4}{1}\binom{4}{1}\binom{4}{1} + 4\binom{4}{2}\binom{4}{3}\binom{4}{3}\binom{4}{1}$$
$$+ 4\binom{4}{2}\binom{4}{3}\binom{4}{2}\binom{4}{2} + 2\binom{4}{2}\binom{4}{3}\binom{4}{2}\binom{4}{1}\binom{4}{1}$$

$$+ \binom{4}{1} \binom{4}{2} \binom{4}{2} \binom{4}{2} \binom{4}{1}$$
$$= 48 + 576 + 256 + 768 + 1728 + 4608 + 3456 = 11440$$

$$(4.27)$$

nodes. The combinatorics is then the following:

- Four 1s in a chunk and three in another. There are 4 permutations in the latter chunk and, since the two chunks are different, 12 permutations of the chunks, which gives $4 \times 12 = 48$ nodes.
- Four 1s in a chunk, two in a second one, and 1 in a third one. The latter two show 6 and 4 permutations, respectively. Being the three chunks different, their possible permutations are 24, so that we have $24 \times 24 = 576$ nodes.
- Four 1s in a chunk and one 1 in each of three chunks. There are 64 permutations and 4 arrangements of the chunks, which makes $64 \times 4 = 256$ nodes.
- Three 1s in each of two chunks and one 1 in another. There are 64 permutations and 12 different combinations of the chunks, which makes $64 \times 12 = 768$ nodes.
- Three 1s in a chunk and two 1s in each of two chunks. There are 144 permutations and 12 combinations of the chunks, which makes $144 \times 12 = 1728$ nodes.
- Three 1s in a chunk, two in a second one, and one 1 in each of two chunks. There are $4 \times 6 \times 4 \times 4 = 384$ permutations that need to be multiplied by 12 arrangements of the chunks, giving rise to $384 \times 12 = 4608$ nodes.
- Two 1s in each of three chunks and a single 1 in another chunk. There are $6 \times 6 \times 6 \times 4 = 864$ permutations and 4 combinations of chunks, which makes $864 \times 4 = 3456$ nodes.

For the IDs, we have following cases:

(1) The first case is subdivided into two subcases: (i) the combination $7 + \mathbf{0}$ gives rise to 16 IDs, while (ii) the combination $4c + 3b$ to 32 IDs (Table 4.46). Having here four 1s in a chunk and three

Table 4.46. $7 + \mathbf{0}$ and $4c + 3b$.

1111	1110	0000	0000	$X'_{1a}(X'_{4a} + X'_{2a} + X'_{3b})$
0000	0000	0111	1111	$X_{1a}(X_{4a} + X_{2a} + X_{3b})$
1111	0000	1110	0000	$X'_{1b}(X'_{1a} + X_{4a} + X'_{2a})$
1111	0000	0111	0000	$X'_{1b}(X'_{1a} + X_{4a} + X_{2a})$

Table 4.47. Combinations of $6a + 1$, $5b + 2a$, and $4c + 3a$.

0000	0001	0011	1111	$X_{1a}(X_{1b} + X_{4a}) + X'_{1a}X_{1b}X_{4a}X_{2a}$
0001	0000	1100	1111	$X_{1a}(X'_{1b} + X'_{4a}) + X'_{1a}X'_{1b}X_{4a}X_{2a}$
0000	0011	0001	1111	$X_{1b}(X_{1a} + X_{4a}) + X_{1a}X'_{1b}X_{4a}X_{2a}$
1000	1111	0011	0000	$X'_{1a}X_{1b} + X_{1a}X'_{1b}X_{4a} + X'_{1a}X'_{1b}X'_{4a}X'_{2a}$
0001	0011	0000	1111	$X_{1b}(X_{1a} + X_{4a}) + X'_{1a}X'_{1b}X_{4a}X_{2a}$
0100	0011	1111	0000	$X_{1a}X'_{1b} + X'_{1a}X_{1b}X_{4a} + X'_{1a}X'_{1b}X'_{4a}X_{2a}$

Table 4.48. Combinations of $5b + 2b$.

0001	0001	1000	1111	$X_{1a}X_{1b} + X'_{1a}X_{4a}X_{2a} + X_{1a}X'_{1b}X'_{4a}X'_{2a}$
0001	0001	1111	1000	$X_{1a}X'_{1b} + X'_{1a}X_{4a}X_{2a} + X_{1a}X_{1b}X'_{4a}X'_{2a}$

in another, there are correspondences with expressions of Level 7/8, in particular, in the same order $X'_{1.1} + X'_{1.4} + X'_{2.2}$, $X_{1.4} + X_{2.2} + X_{2.3}$, $X'_{1.1} + X'_{1.4} + X'_{2.2}$, and $X'_{1.1} + X_{1.4} + X_{2.2}$.

(2) The second kind of combinatorics is subdivided into three sub-cases: (i) $6a + 1$ gives rise to 192 IDs, (ii) $5b + 2a$ also to 192 IDs, and (iii) the same for the combination $4c + 3a$ (Table 4.47). The first group and the first row of both the second and third groups show a combination of an expression of Level 6/16 and a quaternary product. For instance, for the first row such expression is instantiated in the last two chunks, and the same for the second row, while for the latter two cases in the second and last chunk. The other two expressions (fourth and sixth rows of the table) have no pattern of this kind. The considerations expressed up to now also apply to subsequent cases.

(3) The single case $5b + 2b$ gives rise to 256 IDs (Table 4.48). Also here, there is no pattern of the kind previously described.

(4) The two cases here are (i) $6b + 1$ to 256 IDs, and (ii) $4b + 3b$ to 512 IDs (Table 4.49).

Table 4.49.   Combinations of 6b + 1 and 4b + 3b.

1011	1110	0000	0001	$X'_{1a}(X'_{2a} + X'_{3b}) + X_{1a}X_{4a}X_{2a}X_{3b}$
0001	0000	0111	0111	$X_{1a}(X_{4a} + X_{2a}) + X'_{1a}X_{4a}X_{2a}X'_{3b}$
0000	0001	0111	0111	$X_{1a}(X_{4a} + X_{2a}) + X'_{1a}X_{4a}X_{2a}X_{3b}$
1000	0000	0111	0111	$X_{1a}(X_{4a} + X_{2a}) + X'_{1a}X'_{4a}X'_{2a}X'_{3b}$
0010	0000	0111	0111	$X_{1a}(X_{4a} + X_{2a}) + X'_{1a}X'_{4a}X'_{2a}X_{3b}$
0111	0000	0001	0111	$X_{2a}X_{2b} + X_{1a}X'_{2b}X'_{3b} + X'_{1a}X'_{2a}X'_{2b}X_{3b}$
0111	0000	1000	0111	$X_{2a}X_{2b} + X_{1a}X_{2a}X'_{3b} + X'_{1a}X'_{2a}X'_{2b}X_{3b}$

Table 4.50.   Combinations of 5a + 2a and 4a + 3b.

0011	0000	1011	0011	$X'_{1b}X_{4a} + X_{1a}X_{1b}X_{4a} + X_{1a}X'_{1b}X'_{4a}X'_{2a}$
0000	0011	0011	0111	$X_{1a}X_{4a} + X'_{1a}X_{1b}X_{4a} + X_{1a}X_{1b}X'_{4a}X_{2a}$
1100	0000	0011	0111	$X_{1a}X_{4a} + X'_{1a}X'_{1b}X'_{4a} + X_{1a}X_{1b}X'_{4a}X_{2a}$
0011	0011	0000	0111	$X_{1b}X_{4a} + X'_{1a}X'_{1b}X_{4a} + X_{1a}X_{1b}X'_{4a}X_{2a}$
0000	0111	0011	0011	$X_{1b}X_{4a} + X_{1a}X'_{1b}X_{4a} + X'_{1a}X_{1b}X'_{4a}X_{2a}$

Table 4.51.   Combinations of 5a + 2b and 4b + 3a.

0001	0001	0011	0111	$X_{4a}(X_{1a} + X_{2a}) + X_{1a}X'_{4a}X_{2a}X'_{3b}$
0100	0001	1100	1101	$X_{1a}X'_{4a} + X_{1b}X_{4a}X_{2a} + X'_{1a}X'_{1b}X'_{4a}X_{2a}$
0001	0001	1100	0111	$X_{1a}X_{4b} + X'_{1a}X_{4a}X_{2a} + X_{1a}X'_{4a}X_{4b}X_{2a}$
0001	0011	0001	0111	$X_{4a}(X_{2a} + X'_{3b}) + X_{1a}X'_{4a}X_{2a}X'_{3b}$
0001	0011	0001	0111	$X_{4a}(X_{1b} + X_{2a}) + X_{1a}X_{1b}X'_{4a}X_{2a}$

Table 4.52.   Combinations of 4a + 3a. Note the mismatch in the second row.

0011	0011	0001	0011	$X_{4a}(X'_{1a} + X_{2a} + X'_{3b})$
0001	0011	1100	0011	$X_{1b}X_{4a} + X_{1a}X'_{1b}X'_{4a} + X'_{1a}X'_{1b}X_{4a}X_{2a}$

(5) Two cases: (i) 5a + 2a determines 1152 IDs, while (ii) 4a + 3b makes 576 (Table 4.50).

(6) Two cases: (i) 5a + 2b gives rise to 1536 IDs, and (ii) 4b + 3a to 3072 IDs (Table 4.51).

(7) The combination 4a + 3a determines 3456 IDs (Table 4.52).

Note that the expression $X_{1a}X'_{1b} + X'_{1a}X_{4a}X_{2a} + X_{1a}X_{1b}X_{4a}X_{2a}$ (ID 0001 0001 1111 0001) could be reduced to $X_{1a}X'_{1b} + X_{4a}X_{2a}$, which

Table 4.53. Particular combinations of 5a + 2b.

0100 0001 1100 1101	$X_{1a}'X_{2a}X_{3b} + X_{1a}X_{4a}'X_{2a}' + X_{1a}X_{4a}'X_{2a} + X_{1a}X_{4a}X_{2a}X_{3b}$
0111 0011 0100 0001	$X_{4a}'X_{2a}X_{3b} + X_{1a}'X_{4a}X_{2a}' + X_{1a}'X_{4a}X_{2a} + X_{1a}X_{4a}X_{2a}X_{3b}$
0010 0001 1010 1011	$X_{1a}'X_{4a}X_{3b} + X_{1a}X_{2a}'X_{3b}' + X_{1a}X_{2a}'X_{3b} + X_{1a}X_{4a}X_{2a}X_{3b}$
0010 0010 0110 1011	$X_{1a}'X_{4a}X_{2a}' + X_{1a}X_{4a}'X_{3b} + X_{1a}X_{4a}X_{2a}' + X_{1a}X_{4a}X_{2a}X_{3b}$
0100 0001 0111 0011	$X_{1a}'X_{2a}X_{3b} + X_{1a}X_{2a}X_{3b} + X_{1a}X_{4a}X_{2a}' + X_{1a}X_{4a}X_{2a}X_{3b}'$

appears bizarre, since a sum of two binary products should pertain to Level 8/16 and similar ones. We have often met situations in which dropping or adding some negation could lead to inappropriate expressions (compare the previous expression with the second row of Table 4.48). The peculiarity here is that the two products show two couples of terms that are fully unrelated. Nevertheless, such a behaviour is not isolated. In fact, there are many sums of products with such a pattern occurring at this level: have a look at Table 4.53 and note a characteristic pattern. For instance, consider the first row, $X_{1a}'X_{2a}X_{3b} + X_{1a}X_{4a}'X_{2a}' + X_{1a}X_{4a}'X_{2a} + X_{1a}X_{4a}X_{2a}X_{3b}$. All terms but one (X_{3b}) appear in their affirmative and negative form. All of these expressions can be reduced to sums of two binary products. In the same order are $X_{1a}X_{4a}' + X_{2a}X_{3b}$, $X_{1a}'X_{4a} + X_{2a}X_{3b}$, $X_{1a}X_{2a}' + X_{4a}X_{3b}$, $X_{1a}X_{3b} + X_{4a}X_{2a}'$, $X_{1a}X_{4a} + X_{2a}X_{3b}$. The reason for this behaviour is that the two binary products share a 1 (that is, they superpose columns), and we have already observed such a peculiarity for other cases (Subsection 3.2.7). For instance, the two products $X_{1a}X_{4a}'$ and $X_{2a}X_{3b}$ share the 1 in tenth position, the two products $X_{1a}'X_{4a}$ and $X_{2a}X_{3b}$ share the 1 in seventh position, and so on.

On the other hand, the sum $X_{1a}X_{4a} + X_{1a}'X_{2a}$ displays the ID 0101 0101 0011 0011, and therefore pertains to Level 8/16. Thus, there is a difference between a sum of two binary products of the latter kind and the other ones. This shows that we have four different kinds of sums of two binary products:

- Sums of the form $XY + XY'$ give rise to SD variables.
- Sums of the form $XY + X'Z$ are characteristic of all Levels 4/8 (and 8/16).

- Sums of the form $XY + XZ$ need to be rewritten as $X(Y + Z)$ and are characteristic of Levels 3/8 (and 6/16).
- Sums of the form $XY + ZW$ could be defined as *apparent* sums since there is no common term between the two products and in $\mathscr{B}_4$ are typical of Level 7/16. Therefore, we need to find here alternative logical expressions for the relative nodes.

The latter situation is a recurrent behaviour of all algebras as far as the number of dimensions is sufficiently large to allow it. We have already met it for $\mathscr{B}_3$, although not explicitly discussed it. In fact, consider the sum $X_{1.1}X_{1.4} + X_{2.2}X_{2.3}$. This sum gives rise to the ID 0100 0011, which represents a node of Level 3/8 and not of Level 4/8, which is correctly expressed by $X_{1.1}X_{1.4} + X'_{1.1}X'_{1.4}X_{2.2}$. The reason appears simple; the incorrect sum represents a sum of products pertaining to different SCAs: 1.a or 1.b on one hand and 1.c or 1.d on the other. Moreover, for $\mathscr{B}_5$ the sum of the binary product $X_{1.1.1.1}X_{1.1.16.241}$ and of the ternary product $X_{1.4.4.205}X_{1.4.13.196}X_{2.2.6.171}$ gives an ID that in $\mathscr{B}_3$ is logically expressed as $X_{1.1}X_{1.4} + X'_{1.1}X'_{1.4}X_{2.2}$:

$$0001 \quad 0000 \quad 1000 \quad 0000 \quad 1111 \quad 1000 \quad 0000 \quad 1111,$$

which represents a node pertaining to Level 11/32 instead of 12/32, as we could expect given Levels 3/8 and 6/16.

At the opposite we have a kind of *saturation principle* for the new logical forms:

1. In $\mathscr{B}_1$ we have a single expression with a single SD variable;
2. All of the new forms in $\mathscr{B}_2$ (Levels 1/4 and 3/4) involve the two SD variables;
3. All of the new forms of $\mathscr{B}_3$ (Levels 1/8, 3/8, 5/8, and 7/8) involve three SD variables;
4. All of the new forms in $\mathscr{B}_4$ involve four SD variables.

For Level 7/16 we have essentially three basic forms, which can be cast as follows:

$$X_{1a}(X_{4a} + X_{2a} + X'_{3b}), \tag{4.28}$$

$$X_{4a}(X_{1a} + X_{2a}) + X_{1a}X'_{4a}X_{2a}X'_{3b}, \qquad (4.29)$$

$$X_{1b}X_{4a} + X_{1a}X'_{1b}X'_{4a} + X'_{1a}X'_{1b}X_{4a}X_{2a}. \qquad (4.30)$$

It can be seen that the first expression is $8/16 \times 14/16$, the second $6/16 + 1/16$, and the third $4/16 + 2/16 + 1/16$. The second form can be easily reduced to the third one, while for the first form we have

$$X_{4a}(X'_{1a} + X_{2a} + X'_{3b}) = X'_{1a}X_{4a} + X_{1a}X_{4a}X_{2a} + X_{1a}X_{4a}X'_{2a}X'_{3b}. \qquad (4.31)$$

The logical forms of Level $7/16$ can be derived through the sum of two expressions of Level $6/16$:

$$X_{1a}(X_{4a} + X_{2a} + X'_{3b})$$
$$= (X_{1a}X_{4a} + X_{1a}X_{2a}) + (X_{1a}X'_{3b} + X_{1a}X_{4a}X_{3b});$$
$$X_{4a}(X_{1a} + X_{2a}) + X_{1a}X'_{4a}X_{2a}X'_{3b}$$
$$= [X_{4a}(X_{1a} + X_{2a})] + (X_{1a}X_{4a}$$
$$+ X_{1a}X'_{4a}X_{2a}X'_{3b} + X_{1a}X_{4a}X_{2a}X'_{3b});$$
$$X_{1b}X_{4a} + X_{1a}X'_{1b}X'_{4a} + X'_{1a}X'_{1b}X_{4a}X_{2a}$$
$$= (X_{1b}X_{4a} + X'_{1a}X'_{1b}X_{4a}X_{2a}$$
$$+ X_{1a}X'_{1b}X_{4a}X_{2a}) + (X_{1b}X_{4a} + X_{1a}X'_{1b}X'_{4a}).$$

Reciprocally, logical forms of Level $6/16$ can be derived from those of Level $7/16$:

$$X_{1a}(X_{4a} + X_{2a})$$
$$= [X_{1a}(X_{4a} + X_{2a}) + X'_{1a}X'_{4a}X_{2a}X_{3b}]$$
$$\times [X_{1a}(X_{4a} + X_{2a}) + X'_{1a}X'_{4a}X'_{2a}X'_{3b}];$$
$$X'_{1a}X_{4a} + X_{1a}X'_{4a}X'_{2a}$$
$$= (X'_{1a}X_{4a} + X_{1a}X'_{4a}X'_{2a} + X'_{1a}X'_{4a}X_{2a}X'_{3b})$$
$$+ (X'_{1a}X_{4a} + X_{1a}X'_{4a}X'_{2a} + X'_{1a}X'_{4a}X_{2a}X_{3b});$$

$$X'_{1a}X_{4a}X'_{2a} + X'_{1a}X_{2a}X_{3b} + X_{1a}X'_{2a}X_{3b}$$
$$= (X'_{1a}X_{4a} + X_{1a}X'_{2a}X_{3b} + X'_{1a}X'_{4a}X_{2a}X_{3b})$$
$$\times (X'_{2a}X_{3b} + X'_{1a}X_{2a}X_{3b} + X'_{1a}X_{4a}X'_{2a}X'_{3b}).$$

4.3.5. Level 8/16

For Level 8/16, the intermediate one, we have

$$\binom{16}{8} = \binom{4}{2}\binom{4}{4}\binom{4}{4} + 4\binom{4}{2}\binom{4}{4}\binom{4}{3}\binom{4}{1}$$

$$+ 2\binom{4}{2}\binom{4}{4}\binom{4}{2}\binom{4}{2} + 2\binom{4}{2}\binom{4}{4}\binom{4}{2}\binom{4}{1}\binom{4}{1}$$

$$+ 2\binom{4}{2}\binom{4}{3}\binom{4}{3}\binom{4}{2} + \binom{4}{2}\binom{4}{3}\binom{4}{3}\binom{4}{1}\binom{4}{1}$$

$$+ 2\binom{4}{2}\binom{4}{3}\binom{4}{2}\binom{4}{2}\binom{4}{1} + \binom{4}{2}\binom{4}{2}\binom{4}{2}\binom{4}{2}$$

$$= 6 + 384 + 432 + 1152 + 1152 + 1536 + 6912 + 1296 = 12870$$

$$(4.32)$$

nodes. In other words,

- Four 1s in each of two chunks: 6 nodes.
- Four 1s in a chunk, three in a second one, and one 1 in a third. These chunks make 16 permutations that multiplied by 24 permutations of the chunks give rise to 384 nodes.
- Four 1s in a chunk and two in each of two chunks. These chunks make 36 permutations, so that we have $36 \times 12 = 432$ nodes.
- Four 1s in a chunk, two in a second one, and one 1 in each of two chunks: $96 \times 12 = 1152$ nodes.
- Three 1s in each of two chunks and two 1s in another. These chunks make $96 \times 12 = 1152$ nodes.
- Three 1s in each of two chunks and one 1 in each of other two. These chunks make $256 \times 6 = 1536$ nodes.

Table 4.54.   Ramifications of binary equivalences of Level 4/8.

$\mathcal{B}_3$			$\mathcal{B}_4$				
$X_{1.1} \sim X_{1.4}$	1100	0011	$X_{1a} \sim X_{4a} = X_{1b} \sim X_{4b}$	**1100**	1100	0011	**0011**
			$X_{1a} \sim X_{4b} = X_{1b} \sim X_{4a}$	**1100**	0011	1100	**0011**
$X_{1.1} \sim X_{2.2}$	1010	0101	$X_{1a} \sim X_{2a} = X_{1b} \sim X_{2b}$	**1010**	1010	0101	**0101**
			$X_{1a} \sim X_{2b} = X_{1b} \sim X_{2a}$	**1010**	0101	1010	**0101**
$X_{1.1} \sim X_{2.3}$	1001	1001	$X_{1a} \sim X_{3a} = X_{1b} \sim X_{3b}$	**1001**	1001	1001	**1001**
			$X_{1a} \sim X_{3b} = X_{1b} \sim X_{3a}$	**1001**	0110	0110	**1001**

- Three 1s in a chunk, two in each of two chunks, and one 1 in a fourth chunk. These chunks make $576 \times 12 = 6912$ nodes.
- Four chunks with two 1s in each. These chunks make 1296 nodes.

Regarding the IDs, first let us follow the ramifications starting from expressions of Level 4/8, as displayed in Tables 4.54–4.55 (generation of SD variables has been already shown at the beginning of the chapter).

We have

(1) In the first case (i) 8 + **0**, with 2 IDs, and (ii) 4c + 4c, with 4 IDs.

(2) (i) 7 +1, (ii) 5b + 3b, and (iii) 4c + 4b: each of the three subcases give rise to 128 IDs (Table 4.56). Apart from the 3 SD variables, the last expressions are easily found by considering the relative pieces of their ID.

As already anticipated, all logical expressions of Levels 1/2 are either the neg–reversal of themselves (SD variables) or are the neg–reversal of each other (where material equivalences are the neg–reversal of their relative contravalences, due to the fact that both are invariant under reversal and are the negation of each other). For instance, the neg–reversal of the fourth row of the previous table is

0000 0111 1111 0100	$X_{1a}X_{1b}' + X_{1a}'X_{1b}X_{4a} + X_{1b}X_{4a}'X_{2a}$

(see also Equation (2.8)).

Note in particular the inversion of the complementation of terms X_{1b} and X_{4a}.

Table 4.55. Ramifications of sums of binary products of Level 4/8.

$\mathscr{B}_3$	$\mathscr{B}_4$			
	$X'_{1a}X'_{4a} + X_{1a}X'_{2a}$ **1100**	**1100**	**1010**	**1010**
	$X'_{1a}X'_{4a} + X_{1a}X'_{2b}$ **1100**	**1100**	**0101**	**1010**
	$X'_{1a}X'_{4b} + X_{1a}X'_{2a}$ **1100**	**0011**	**1010**	**1010**
$X'_{1.1}X'_{1.4} + X_{1.1}X'_{2.2}$ 1100 1010	$X'_{1a}X'_{4b} + X_{1a}X'_{2b}$ **1100**	**0011**	**0101**	**1010**
	$X'_{1b}X'_{4a} + X_{1b}X'_{2a}$ **1100**	**1010**	**1100**	**1010**
	$X'_{1b}X'_{4a} + X_{1b}X'_{2b}$ **1100**	**0101**	**1100**	**1010**
	$X'_{1b}X'_{4b} + X_{1b}X'_{2a}$ **1100**	**1010**	**0011**	**1010**
	$X'_{1b}X'_{4b} + X_{1b}X'_{2b}$ **1100**	**0101**	**0011**	**1010**
	$X_{1a}X'_{4a} + X'_{1a}X'_{2a}$ **1010**	**1010**	**1100**	**1100**
	$X_{1a}X'_{4a} + X'_{1a}X'_{2b}$ **1010**	**0101**	**1100**	**1100**
	$X_{1a}X'_{4b} + X'_{1a}X'_{2a}$ **1010**	**1010**	**0011**	**1100**
$X_{1.1}X'_{1.4} + X'_{1.1}X'_{2.2}$ 1010 1100	$X_{1a}X'_{4b} + X'_{1a}X'_{2b}$ **1010**	**0101**	**0011**	**1100**
	$X_{1b}X'_{4a} + X'_{1b}X'_{2a}$ **1010**	**1100**	**1010**	**1100**
	$X_{1b}X'_{4a} + X'_{1b}X'_{2b}$ **1010**	**1100**	**0101**	**1100**
	$X_{1b}X'_{4b} + X'_{1b}X'_{2a}$ **1010**	**0011**	**1010**	**1100**
	$X_{1b}X'_{4b} + X'_{1b}X'_{2b}$ **1010**	**0011**	**0101**	**1100**
	$X_{1a}X_{4a} + X'_{1a}X'_{2a}$ **1010**	**1010**	**0011**	**0011**
	$X_{1a}X_{4a} + X'_{1a}X'_{2b}$ **1010**	**0101**	**0011**	**0011**
	$X_{1a}X_{4b} + X'_{1a}X'_{2a}$ **1010**	**1010**	**1100**	**0011**
$X_{1.1}X_{1.4} + X'_{1.1}X'_{2.2}$ 1010 0011	$X_{1a}X_{4b} + X'_{1a}X'_{2b}$ **1010**	**0101**	**1100**	**0011**
	$X_{1b}X_{4a} + X'_{1a}X'_{2a}$ **1010**	**0011**	**1010**	**0011**
	$X_{1b}X_{4a} + X'_{1b}X'_{2b}$ **1010**	**0011**	**0101**	**0011**
	$X_{1b}X_{4b} + X'_{1b}X'_{2a}$ **1010**	**1100**	**1010**	**0011**
	$X_{1b}X_{4b} + X'_{1b}X'_{2b}$ **1010**	**1100**	**0101**	**0011**
	$X'_{1a}X'_{4a} + X_{1a}X_{2a}$ **1100**	**1100**	**0101**	**0101**
	$X'_{1a}X'_{4a} + X_{1a}X_{2b}$ **1100**	**1100**	**1010**	**0101**
	$X'_{1a}X'_{4b} + X_{1a}X_{2a}$ **1100**	**0011**	**0101**	**0101**
$X'_{1.1}X'_{1.4} + X_{1.1}X_{2.2}$ 1100 0101	$X'_{1a}X'_{4b} + X_{1a}X_{2b}$ **1100**	**0011**	**1010**	**0101**
	$X'_{1b}X'_{4a} + X_{1b}X_{2a}$ **1100**	**0101**	**1100**	**0101**
	$X'_{1b}X'_{4a} + X_{1b}X_{2b}$ **1100**	**1010**	**1100**	**0101**
	$X'_{1b}X'_{4b} + X_{1b}X_{2a}$ **1100**	**0101**	**0011**	**0101**
	$X'_{1b}X'_{4b} + X_{1b}X_{2a}$ **1100**	**1010**	**0011**	**0101**

(3) (i) 6a + 2a: 288 IDs, and (ii) 4c + 4a: 144 IDs (Table 4.57).

(4) (i) 6a + 2b: 384 IDs, and (ii) 5b + 3a: 768 IDs (Table 4.58).

(5) (i) 6b + 2a: 384 IDs, and (ii) 5a + 3b: 768 IDs (Table 4.59).

(6) (i) 6b + 2b: 512 IDs, and (ii) 4b + 4b: 1024 IDs (Table 4.60). Note the different forms of the first two rows. This difference is due to the fact that we find in Level 4/8 a sum of two binary products with the ID 1011 1000 (expressed by $X'_{1.1}X_{1.4} + X'_{1.4}X_{2.2}$), and

Table 4.56. Combinations 7 + 1, 5b + 3b, and 4c + 4b of Level 8/16.

0000	0001	0111	1111	$X_{1.1.2}$
0001	0000	1111	0111	$X_{1.2.1}$
0000	0111	0001	1111	$X_{1.1.8}$
1101	0000	0001	1111	$X_{1a}X_{1b} + X'_{1a}X'_{1b}X'_{4a} + X'_{1b}X_{4a}X_{2a}$
0111	0000	0001	1111	$X_{4a}X_{4b} + X'_{1b}X'_{4b}X_{3a} + X_{1b}X'_{4a}X'_{4b}$
1000	0111	1111	0000	$X_{2a}X_{2b} + X'_{1a}X_{2a}X_{3b} + X_{1a}X_{2a}X_{2b}$
0001	0111	0000	1111	$X_{1b}X'_{4b} + X_{1b}X_{4b}X_{3a} + X_{4a}X_{4b}X_{3a}$

Table 4.57. Combinations 6a + 2a and 4c + 4a of Level 8/16.

0000	1010	1111	1010	$X'_{2a}X_{2b} + X_{1a}X'_{2b}$
0101	0000	1010	1111	$X_{2a}X_{2b} + X_{1a}X'_{2a}$
1100	1100	0000	1111	$X'_{1a}X'_{4a} + X_{1a}X_{1b}$
1001	0110	1111	0000	$X'_{1b}X'_{3a} + X_{3a}X'_{3b}$

Table 4.58. Combinations 6a + 2b and 5b + 3a of Level 8/16.

0001	0001	0011	1111	$X_{1a}X_{1b} + X'_{1a}X_{4a}X_{2a} + X_{1a}X'_{1b}X_{4a}$
1000	0010	1111	0101	$X_{1a}X_{2a} + X'_{1a}X'_{2a}X_{3b} + X_{1a}X'_{2a}X_{2b}$
0001	1001	0100	1111	$X_{1b}X_{3b} + X'_{1b}X_{2a}X_{3a} + X_{1b}X'_{3a}X_{3b}$
0001	0001	1111	0100	$X_{2a}X'_{2b} + X_{1a}X'_{2a}X_{2b} + X_{2a}X_{2b}X'_{3b}$

Table 4.59. Combinations 6b + 2a and 5a + 3b of Level 8/16.

0000	0011	0111	0111	$X_{1a}X_{4a} + X'_{1a}X_{1b}X_{4a} + X_{4a}X'_{4a}X_{2a}$
1010	0000	0111	1011	$X'_{2a}X'_{2b} + X_{1a}X_{2a}X_{2b} + X_{1a}X_{2b}X_{3b}$
0011	0111	0000	0111	$X_{1b}X_{4a} + X'_{1a}X'_{1b}X_{4a} + X_{1b}X'_{4a}X_{2a}$
1001	1011	0000	1011	$X_{1b}X_{3b} + X'_{1b}X'_{3a}X'_{3b} + X_{1b}X'_{2a}X'_{3b}$

we find also the pattern 0001 0111, but this node is expressed by $X_{1.2}$.

(7) (i) 5a + 3a: 4608 IDs, and (ii) 4a + 4b: 2304 IDs (Table 4.61).

(8) 4a + 4a: 1296 IDs (Table 4.62).

The two IDs of the case 8 + **0** are X_{1a} and its complement. Similarly, the 4 IDs of the case 4c + 4c give rise to X_{1b}, its complement, and to the two binary equivalences and contravalences between X_{1a} and X_{1b}.

Table 4.60. Combinations 6b + 2b and 4b + 4b of Level 8/16.

0001	0001	0111	0111	$X_{1a}X_{4a} + X'_{1a}X_{4a}X_{2a} + X_{1a}X'_{4a}X_{2a}$
1011	1011	1000	1000	$X'_{1a}X_{4a} + X'_{4a}X'_{2a}$
0001	0111	0001	0111	$X_{1b}X_{4a} + X'_{1b}X_{4a}X_{2a} + X_{1b}X'_{4a}X_{2a}$
0100	1011	0001	0111	$X_{1b}X_{4a} + X'_{1b}X'_{4b}X_{3a} + X_{1b}X'_{4a}X_{3a}$

Table 4.61. Combinations 5a + 3a and 4a + 4b.

1000	0101	0101	0111	$X_{2a}X'_{2b} + X_{1a}X_{2a}X_{2b} + X'_{2a}X'_{2b}X_{3b}$
1001	0010	0110	1011	$X'_{1b}X'_{3a} + X_{1b}X'_{2a}X'_{3b} + X_{1b}X_{3a}X_{3b}$
0011	0011	1000	1011	$X'_{1a}X_{4a} + X_{1a}X_{1b}X_{4a} + X_{1a}X'_{4a}X'_{2a}$
0101	1010	0001	0111	$X'_{1a}X_{2b} + X_{1a}X_{2a}X_{2b} + X_{1a}X'_{2b}X'_{3b}$

Table 4.62. Combinations 4a + 4a.

| 1010 | 1010 | 1001 | 1001 | $X'_{1a}X'_{2a} + X_{1a}X_{4a}X_{2a} + X_{1a}X'_{4a}X'_{2a}$ |
| 0110 | 0110 | 0011 | 0011 | $X_{4a}X'_{2a} + X_{1a}X_{4a}X_{2a} + X'_{1a}X'_{4a}X_{2a}$ |

Of course, this level shows all the expressions already present at Level 4/8: SD variables with their complements (256 as whole), binary equivalences (6) and contravalences (6), and sums of three binary products with their complements. To this we need to add sums of both a binary product and two ternary products. The latter is the common form, being a generalisation of all previous logical forms. We can also express the half of the sums of two binary products in negative form, as we already did for Level 4/8. For the negative of the last form, we have

$$(X'_{1a}X'_{2a} + X_{1a}X_{4a}X_{2a} + X_{1a}X'_{4a}X'_{2a})'$$
$$= (X_{1a} + X_{2a})(X'_{1a} + X'_{4a} + X'_{2a})(X'_{1a} + X_{4a} + X_{2a}). \quad (4.33)$$

Then, we have the following four forms:

$$X_{1a}, \ X_{1a} \sim X_{4a}, \ X'_{1a}X_{4a} + X'_{4a}X'_{2a},$$

and

$$X_{1a}X_{4a} + X'_{1a}X_{4a}X_{2a} + X_{1a}X'_{4a}X_{2a}.$$

These forms can be easily derived from forms of the previous level:

$$X_{1a} = [X_{1a}(X_{4a} + X_{2a} + X_{3a})] + [X_{1a}(X'_{4a} + X'_{2a} + X'_{3a})],$$

$$X_{1a} \sim X_{4a}$$
$$= (X_{1a}X_{4a} + X'_{1a}X'_{4a}X_{3b} + X'_{1a}X'_{4a}X_{2a}X_{3b})$$
$$+ (X'_{1a}X'_{4a} + X_{1a}X_{4a}X'_{3b} + X_{1a}X_{4a}X_{2a}X_{3b}),$$

$$X'_{1a}X_{4a} + X'_{4a}X_{2a}$$
$$= (X'_{1a}X_{4a} + X'_{4a}X_{2a}X_{3b} + X_{1a}X'_{4a}X_{2a}X'_{3b})$$
$$+ (X'_{4a}X_{2a} + X'_{1a}X_{4a}X_{3b} + X'_{1a}X_{4a}X_{2a}X'_{3b}),$$

$$X'_{1a}X_{4a} + X_{1a}X'_{4a}X_{2a} + X_{1a}X'_{2a}X'_{3b}$$
$$= (X'_{1a}X_{4a} + X_{1a}X'_{4a}X_{2a} + X_{1a}X'_{4a}X'_{2a}X'_{3b})$$
$$+ (X'_{1a}X_{4a} + X_{1a}X'_{2a}X'_{3b} + X_{1a}X'_{4a}X_{2a}X_{3b}).$$

Reciprocally, we can derive the logical forms of Level 7/16 from those of Level 8/16:

$$X_{1a}(X_{4a} + X_{2a} + X_{3b})$$
$$= (X_{1a}X_{3b} + X_{1a}X_{2a}X'_{3b} + X_{4a}X'_{2a}X'_{3b})$$
$$\times (X_{1a}X_{4a} + X_{1a}X'_{4a}X_{3b} + X'_{4a}X_{2a}X'_{3b}),$$

$$X_{1a}(X_{4a} + X_{2a}) + X'_{1a}X_{4a}X_{2a}X'_{3b}$$
$$= (X_{1a}X_{4a} + X'_{1a}X_{2a}X'_{3b} + X_{1a}X'_{4a}X_{2a})$$
$$\times (X_{1a}X_{4a} + X'_{1a}X_{4a}X'_{3b} + X_{1a}X'_{4a}X_{2a}),$$

$$X_{1a}X_{4a} + X_{1a}X'_{4a}X_{2a} + X'_{1a}X_{4a}X'_{2a}X_{3b}$$
$$= (X_{1a}X_{4a} + X_{1a}X'_{4a}X_{2a} + X'_{1a}X'_{2a}X_{3b})$$
$$\times (X_{1a}X_{4a} + X_{1a}X'_{4a}X_{2a} + X'_{1a}X_{4a}X'_{2a}).$$

4.3.6. *Summary of the following levels*

Having explained the general mechanism, in the following I only summarise the combinatorics and show some examples. Level 9/16 presents

$$\binom{16}{9} = 11440 \tag{4.34}$$

nodes and is characterised by the following combinatorics:

- Four 1s in each of two chunks plus a 1 in one chunk: $4 \times 12 = 48$ nodes.
- Four 1s, three 1s, and two 1s: $24 \times 24 = 576$ nodes.
- Four 1s, three 1s, and two times one 1: $64 \times 12 = 768$ nodes.
- Four 1s, two times two 1s, and one 1: $144 \times 12 = 1728$ nodes.
- Three times three 1s: $64 \times 4 = 256$ nodes.
- Two times three 1s, two 1s, and one 1: $384 \times 12 = 4608$ nodes.
- Three 1s and three times two 1s: $864 \times 4 = 3456$ nodes.

For some examples of the logical expressions see Table 4.63.

Level 10/16 has

$$\binom{16}{10} = 8008 \tag{4.35}$$

nodes and the following combinatorics:

- Two times four 1s and two 1s: $6 \times 12 = 72$ nodes.
- Two times four 1s and two times one 1: $16 \times 6 = 96$ nodes.
- Four 1s and two times three 1s: $16 \times 12 = 192$ nodes.
- Four 1s, three 1s, two 1s, and one 1: $96 \times 24 = 2304$ nodes.
- Four 1s and three times two 1s: $216 \times 4 = 864$ nodes.
- Three times three 1s and one 1: $256 \times 4 = 1024$ nodes.
- Two times three 1s and two times two 1s: $576 \times 6 = 3456$ nodes.

Note the analogies of the first two rows of Table 4.64 with the 3D expressions $X'_{1.1} + X'_{1.4}X'_{2.2}$ and $X'_{1.1} + X'_{1.4}X'_{2.3}$, respectively.

Table 4.63.   Sample of logical expression for Level 9/16.

0000 0001 1111 1111	$X_{1a} + X_{4a}X_{2a}X_{3b}$
1110 1111 1000 1000	$(X'_{1a} + X'_{4a}X_{2a})(X_{1a} + X'_{4a} + X'_{2a} + X_{3b})$
1000 1111 1110 1000	$(X'_{2a} + X'_{2b})(X'_{1a} + X_{2b} + X'_{3b})(X_{1a} + X_{2a} + X_{2b} + X'_{3b})$

Table 4.64. Sample of logical expression for Level 10/16.

1111 1111 1000 1000	$X'_{1a} + X'_{4a}X'_{2a}$
1111 1111 1000 0100	$X'_{1a} + X'_{4a}X'_{3a}$
1111 0001 1111 1000	$(X'_{1a} + X'_{1b} + X'_{4a})(X_{1a} + X'_{1b} + X'_{4a})(X'_{1b} + X'_{4a} + X_{2a})$
1111 0111 1100 0100	$(X'_{1a} + X'_{4a})(X_{4a} + X_{2a} + X'_{3b})$

Table 4.65. Sample of logical expression for Level 11/16.

1111 1111 1100 1000	$X'_{1a} + X'_{4a}(X'_{2a} + X_{3b})$
1111 1111 0011 0001	$X'_{1a} + X_{4a}(X_{2a} + X_{3b})$
0111 1111 1100 1100	$(X'_{1a} + X'_{4a})(X_{1a} + X_{4a} + X_{2a} + X_{3b})$
1110 1000 1111 1110	$(X_{1a} + X'_{2b} + X_{3b})(X_{1a} + X'_{2a} + X_{2b})(X'_{1a} + X'_{2a} + X'_{2b} + X'_{3b})$

Level 11/6 presents

$$\binom{16}{11} = 4368 \tag{4.36}$$

nodes and the following combinatorics:

- Two times four 1s and three 1s: $4 \times 12 = 48$ nodes.
- Two times four 1s, two 1s, and one 1: $24 \times 12 = 288$ nodes.
- Four 1s, two times three 1s, and one 1: $64 \times 12 = 768$ nodes.
- Four 1s, three 1s, and two times two 1s: $144 \times 12 = 1728$ nodes.
- Three times three 1s and two 1s: $384 \times 4 = 1536$ nodes.

Some examples of logical expressions are displayed in Table 4.65.
    Level 12/16 displays

$$\binom{16}{12} = 1820 \tag{4.37}$$

nodes and the following combinatorics:

- Three times four 1s: 4 nodes.
- Two times four 1s, three 1s, and one 1: $16 \times 12 = 192$ nodes.
- Two times four 1s and two times two 1s: $36 \times 6 = 216$ nodes.
- Four 1s, two times three 1s, and two 1s: $96 \times 12 = 1152$ nodes.
- Four times three 1s: 256 nodes.

Table 4.66. Sample of logical expression for Level 12/16.

1111	1111	1100	1100	$X'_{1a} + X'_{4a}$
1111	1111	1001	0110	$X'_{1a} + X'_{3b}$
0111	0111	1110	1110	$X_{1a} \sim X_{4a} \sim X_{2a}$
0110	1010	1111	1111	$(X_{1a} + X_{4a} + X'_{3b})(X_{1a} + X'_{4a} + X'_{2a})$

Table 4.67. Sample of logical expression for Level 13/16.

1111	1111	1110	1100	$X'_{1a} + X'_{4a} + X'_{2a}X_{3b}$
0011	1011	1111	1111	$X_{1a} + X_{4a} + X'_{2a}X_{3b}$
0111	1111	1101	1101	$(X'_{1a} + X'_{4a} + X_{2a})(X_{1a} + X_{4a} + X_{2a} + X_{3b})$
1011	1111	1110	1101	$(X'_{1a} + X'_{4a} + X_{3b})(X_{1a} + X_{4a} + X'_{2a} + X'_{3b})$

Note the analogies between the first two rows of Table 4.66 with $X'_{1.1} + X'_{1.4}$ and $X'_{1.1} + X'_{2.3}$, respectively.

Level 13/16 presents

$$\binom{16}{13} = 560 \tag{4.38}$$

nodes and the following combinatorics:

- Three times four 1s and one 1: $4 \times 4 = 16$ nodes.
- Two times four 1s, three 1s, and two 1s: $24 \times 12 = 288$ nodes.
- Four 1s and three times three 1s: $64 \times 4 = 256$ nodes.

A sample of the logical expressions for this level are shown in Table 4.67.

Level 14/16 displays

$$\binom{16}{14} = 120 \tag{4.39}$$

nodes and the following combinatorics:

- Three times four 1s and two 1s: $6 \times 4 = 24$ nodes.
- Two times four 1s and two times three 1s: $16 \times 6 = 96$ nodes.

Of course, Level 15/16 displays $4 \times 4 = 16$ nodes. A sample of expressions of the last two levels is shown in Table 4.68.

Table 4.68.   Sample of logical expression for Levels 14/16 and 15/16.

1111	1111	1110	1110	$X'_{1a} + X'_{4a} + X'_{2a}$
1011	1011	1111	1111	$X_{1a} + X_{4a} + X'_{2a}$
1111	1101	1011	1111	$X_{1a} \sim X'_{4a} \sim X_{2a} \sim X_{3b}$
1111	1011	1111	1111	$X_{1a} + X_{4a} + X'_{2a} + X_{3b}$
1111	1111	1011	1111	$X'_{1a} + X_{4a} + X'_{2a} + X'_{3b}$

4.4.   Generalisations

4.4.1.   *Combinatorics of nodes*

The use of a single chunk of 4 digits (instead of using different chunks with $2, 4, 8, \ldots$ digits) has several advantages. First, it is very easy to find out which is the distribution of 1s. Second, it allows a single formalism across the different dimensional algebras. Finally, it considerably simplifies the notation. The combinatorial function $C(x; y; z)$ means that we have x 1s in a chunk (always of 4 digits) and y chunks with x 1s over a total number z of chunks. In other words, $C(x; y; z)$ is composed of two parts: the combinatorics in a single chunk and the combinatorics of the chunks themselves. Instead of single numbers, we can have series of numbers. For instance, $C(1, 2; 2, 1; 4)$ means that we have two chunks with a single 1 and one chunk with two 1s over 4 chunks. Of course, we could have also taken two-digit combinatorics as basic, but this would make calculations more complicated.

Let us first list the basic combinatorics for $\mathscr{B}_2$, as displayed in Table 4.69, which turn out to be the permutations inside a single chunk. With such a combinatorics it is easy to compute the combinations for $\mathscr{B}_3$, as in Table 4.70. There are some interesting regularities to note. As it is obvious, the sum of the first number always gives the number α of Level α/m, as in $C(3, 4; 1, 1; 2) = C(3; 1; 2)C(4; 1; 1)$ we have that $3 + 4 = 7$. Moreover, apart from the forms like C_{aa} ($a = 1, 2, 3, 4$), all combinatorial functions have 2 as coefficient.

This is evident by labelling the combinatorics of Table 4.69 C_1, C_2, C_3, C_4, in the same order, allowing us to rewrite the combinations of Table 4.70 as in Table 4.71. Note that the products

Table 4.69.   Permutations inside a single chunk.

$C(1;1;1) = 4$	$C(2;1;1) = 6$	$C(3;1;1) = 4$	$C(4;1;1) = 1$

Table 4.70.   Combinatorics for $\mathscr{B}_3$. The table should be read from the bottom to the top.

8/8	$C(4;2;2) = C(4;1;1)C(4;1;1)$
7/8	$C(3,4;1,1;2) = C(3;1;2)C(4;1;1)$
6/8	$C(2,4;1,1;2) = C(2;1;2)C(4;1;1),\ C(3;2;2) = C(3;1;1)C(3;1;1)$
5/8	$C(1,4;1,1;2) = C(1;1;2)C(4;1;1),\ C(2,3;1,1;2) = C(2;1;2)C(3;1;1)$
4/8	$C(4;1;2) = 2C(4;1;1),\ C(2;2;2) = C(2;1;1)C(2;1;1),$
	$C(1,3;1,1;2) = C(1;1;2)C(3;1;1)$
3/8	$C(3;1;2) = 2C(3;1;1),\ C(1,2;1,1;2) = C(1;1;2)C(2;1;1)$
2/8	$C(2;1;2) = 2C(2;1;1),\ C(1;2;2) = C(1;1;1)C(1;1;1)$
1/8	$C(1;1;2) = 2C(1;1;1)$

Table 4.71.   Reformulation of the combinatorics for $\mathscr{B}_3$.

8/8	$C_{44} = C_{44}$
7/8	$C_{34} = 2C_3C_4$
6/8	$C_{24} = 2C_2C_4, C_{33} = C_3C_3$
5/8	$C_{14} = 2C_1C_4, C_{23} = 2C_2C_3$
4/8	$2C_4, C_{22} = C_2C_2, C_{13} = 2C_1C_3$
3/8	$2C_3, C_{12} = 2C_1C_2$
2/8	$2C_2, C_{11} = C_1C_1$
1/8	$2C_1$

combine either elements that in $\mathscr{B}_2$ are on the same level or on levels that are opposite relative to the cobination in $\mathscr{B}_3$.

The step further is to compute on this basis the combinatorics of $\mathscr{B}_4$, as displayed in Table 4.72. Finally, we can compare the three kinds of combinatorics as in Table 4.73. Note that the combinations in $\mathscr{B}_4$ follow distinctive patterns ($\forall a, b, c = 1, 2, 3, 4$):

- $2C_a$ transforms into $2 \cdot 2C_a$;
- C_{aa} transforms into $6C_{aa}$;
- C_{aa} with C_{bb} transforms into $C_{aa}C_{bb}$;
- $2C_a$ with $2C_b$ or C_{ab} (C_{aa}) transforms into $6C_{ab}$ ($6C_{aa}$);

Table 4.72.   Combinatorics of $\mathcal{B}_4$. These combinatorics follow the order of the presentation in the previous section.

16/16	$C(4;4;4) = C_{44}C_{44}$
15/16	$C(3,4;1,3;4) = 2C_{34}C_{44}$
14/16	$C(2,4;1,3;4) = 2C_{24}C_{44}, C(3,4;2,2;4) = 6C_{33}C_{44}$
13/16	$C(1,4;1,3;4) = 2C_{14}C_{44}, C(2,3,4;1,1,2;4) = 6C_{23}C_{44}, C(3,4;3,1;4) = 2C_{33}C_{34}$
12/16	$C(4;3;4) = 4C_4C_{44}, C(1,3,4;1,1,2;4) = 6C_{13}C_{44},$ $C(2,4;2,2;4) = 6C_{22}C_{44}, C(2,3,4;1,2,1;4) = 6C_{24}C_{33}, C(3;4;4) = C_{33}C_{33}$
11/16	$C(3,4;1,2;4) = 12C_3C_{44}, C(1,2,4;1,1,2;4) = 6C_{12}C_{44}, C(1,3,4;1,2,1;4) = 3C_{14}C_{33},$ $C(2,3,4;2,1,1;4) = 3C_{23}C_{24}, C(2,3;1,3;4) = 2C_{23}C_{33}$
10/16	$C(2,4;1,2;4) = 6C_4C_{24}, C(1,4;2,2;4) = 6C_{11}C_{44}, C(3,4;2,1;4) = 6C_3C_{34},$ $C(1,2,3,4;1,1,1,1;4) = 6C_{13}C_{24}, C(2,4;3,1;4) = 2C_{22}C_{24},$ $C(1,3;1,3;4) = 2C_{13}C_{33}, C(2,3;2,2;4) = 6C_{22}C_{33}$
9/16	$C(1,4;1,2;4) = 6C_4C_{14}, C(2,3,4;1,1,1;4) = 12C_4C_{23},$ $C(1,3,4;2,1,1;4) = 6C_{11}C_{34}, C(1,2,4;1,2,1;4) = 6C_{14}C_{22},$ $C(3;3;4) = 4C_3C_{33}, C(1,2,3;1,1,2;4) = 6C_{12}C_{33}, C(2,3;3,1;4) = 2C_{22}C_{23}$
8/16	$C(4;2;4) = 6C_{44}, C(1,3,4;1,1,1;4) = 12C_4C_{13}, C(2,4;2,1;4) = 12C_4C_{22},$ $C(1,2,4;2,1,1;4) = 6C_{11}C_{24}, C(2,3;1,2;4) = 6C_3C_{23}, C(1,3;2,2;4) = 6C_{11}C_{33},$ $C(1,2,3;1,2,1;4) = 6C_{13}C_{22}, C(2;4;4) = C_{22}C_{22}$
7/16	$C(3,4;1,1;4) = 6C_{34}, C(1,2,4;1,1,1;4) = 12C_4C_{12}, C(1,4;3,1;4) = 2C_{11}C_{14},$ $C(1,3;1,2;4) = 6C_3C_{13}, C(2,3;2,1;4) = 12C_3C_{22}, C(1,2,3;2,1,1;4) = 3C_{12}C_{13}, C(1,2;1,3;4) = C_{12}C_{22}$
6/16	$C(2,4;1,1;4) = 6C_{24}, C(1,4;2,1;4) = 12C_4C_{11}, C(3;2;4) = 6C_{33},$ $C(1,2,3;1,1,1;4) = 12C_2C_{13}, C(1,3;3,1;4) = 2C_{11}C_{13}, C(2;3;4) = 4C_2C_{22}, C(1,2;2,2;4) = 6C_{11}C_{22}$
5/16	$C(1,4;1,1;4) = 6C_{14}, C(2,3;1,1;4) = 6C_{23}, C(1,3;2,1;4) = 12C_3C_{11},$ $C(1,2;1,2;4) = 12C_2C_{12}, C(1,2,3;1,1,1;4) = 12C_3C_{11},$ $C(1,2;2,1;4) = 2C_{11}C_{12}$
4/16	$C(4;1;4) = 4C_4, C(1,3;1,1;4) = 6C_{13}, C(2;2;4) = 6C_{22},$ $C(1,2;2,1;4) = 12C_2C_{11}, C(1;4;4) = C_{11}C_{11}$
3/16	$C(3;1;4) = 4C_3, C(1,2;1,1;4) = 6C_{12}, C(1;3;4) = 4C_1C_{11}$
2/16	$C(2;1;4) = 4C_2, C(1;2;4) = 6C_{11}$
1/16	$C(1;1;4) = 4C_1$

Table 4.73.   Comparison of combinatorics across different algebras.

$\mathscr{B}_2$	$\mathscr{B}_3$	$\mathscr{B}_4$
C_4	C_{44}	$C_{44}C_{44}$
		$2C_{34}C_{44}$
	C_{34}	$2C_{24}C_{44}, 6C_{33}C_{44}$
		$2C_{14}C_{44}, 6C_{23}C_{44}, 2C_{33}C_{34}$
C_3	C_{24}, C_{33}	$2 \cdot 2C_4C_{44}, 6C_{13}C_{44}, 6C_{22}C_{44},$ $6C_{24}C_{33}, C_{33}C_{33}$
		$6 \cdot 2C_3C_{44}, 6C_{12}C_{44}, 3C_{14}C_{33},$ $3C_{23}C_{24}, 2C_{23}C_{33}$
	C_{14}, C_{23}	$3 \cdot 2C_4C_{24}, 6C_{11}C_{44}, 3 \cdot 2C_3C_{34}, 6C_{13}C_{24},$ $2C_{22}C_{24}, 2C_{13}C_{33}, 6C_{22}C_{33}$
		$3 \cdot 2C_4C_{14}, 6 \cdot 2C_4C_{23}, 6C_{11}C_{34}, 6C_{14}C_{22},$ $2 \cdot 2C_3C_{33}, 6C_{12}C_{33}, 2C_{22}C_{23}$
C_2	$2C_4, C_{22}, C_{13}$	$6C_{44}, 6 \cdot 2C_4C_{13}, 6 \cdot 2C_4C_{22},$ $6C_{11}C_{24}, 3 \cdot 2C_3C_{23}, 6C_{11}C_{33}, 6C_{13}C_{22}, C_{22}C_{22}$
		$6C_{34}, 6 \cdot 2C_4C_{12}, 2C_{11}C_{14},$ $3 \cdot 2C_3C_{13}, 6 \cdot 2C_3C_{22}, 3C_{12}C_{13}, C_{12}C_{22}$
	$2C_3, C_{12}$	$6C_{24}, 6 \cdot 2C_4C_{11}, 6C_{33},$ $6 \cdot 2C_2C_{13}, 2C_{11}C_{13}, 2 \cdot 2C_2C_{22}, 6C_{11}C_{22}$
		$6C_{14}, 6C_{23}, 6 \cdot 2C_3C_{11}, 6 \cdot 2C_2C_{12}, 2C_{11}C_{12}$
C_1	$2C_2, C_{11}$	$2 \cdot 2C_4, 6C_{13}, 6C_{22}, 6 \cdot 2C_2C_{11}, C_{11}C_{11}$
		$2 \cdot 2C_3, 6C_{12}, 2 \cdot 2C_1C_{11}$
	$2C_1$	$2 \cdot 2C_2, 6C_{11}$
		$2 \cdot 2C_1$

- $2C_a$ with C_{ab} ($2C_b$ with C_{ab}) transforms into $3 \cdot 2C_aC_{ab}$ ($3 \cdot 2C_bC_{ab}$);
- $2C_a$ with C_{bc} or $2C_aC_{bb}$ transforms into $6 \cdot 2C_aC_{bc}$ ($6 \cdot 2C_aC_{bb}$).

4.4.2.   *Summary of the logical forms*

The number of contingent logical forms (excluding both tautologies and contradictions) for $\mathscr{B}_1 - \mathscr{B}_4$ is summarised in Figure 4.3. This result can be generalised as follows: let $f(m/2^n)$ denote a function giving the number of logical forms for Level $m/2^n$ of algebra $\mathscr{B}_n$. Then, for each algebra we have the series

$$f(2^{n-1}/2^n) = n, 2^{n-1}[f(2^{n-1}/2^n) - 1], 2^{n-2}[f(2^{n-1}/2^n) - 2],$$

$$2^{n-3}[f(2^{n-1}/2^n) - 3], \ldots$$

$$= n, 2^{n-1}(n - 1), 2^{n-2}(n - 2), 2^{n-3}(n - 3), \ldots \quad (4.40)$$

Levels	1/16	1/8	3/16	1/4	5/16	3/8	7/16	1/2	9/16	5/8	11/16	3/4	13/16	7/8	15/16
$\mathcal{B}_1$								1							
$\mathcal{B}_2$				1				2				1			
$\mathcal{B}_3$		1		2		2		3		2		2		1	
$\mathcal{B}_4$	1	2	2	3	3	3	3	4	3	3	3	3	2	2	1

Figure 4.3. Number of logical forms for $\mathcal{B}_1$–$\mathcal{B}_4$.

The first term is clearly the central level. For instance, for $n = 4$, we have four logical forms for such level, then we have eight levels with three logical forms, four levels with two logical forms, and two levels with a single logical form.

Of course, another issue is to guess the generation of the specific logical forms. Since we try to individuate the essential patterns, a good methodology is to reduce the variety of indexed SD variables and their complements to a unique basic form (X), according to Figure 4.4. We already know that in many cases the presence of complements is important for the final logical form. Nevertheless, such a simplification allows us to understand that we deal here with a kind of algebra.

A further step consists of considering the components that are repeated over and over in order to understand the rules of such a generation. We focus here only on the *new* forms relative to low–dimensional algebra (I do not consider equivalences since they are only particular forms of sums or products). In this way the patterns emerge more clearly. To this purpose, we start with the form a (X_0) of $\mathcal{B}_1$ and proceed recursively, according to Figure 4.5. Passing to higher–level algebras, the new forms are generated through three different kinds of operations:

- For getting some of the formulæ of the nD algebra for any odd level $\alpha/m(n)$, we multiply a formula of the lowest possible level of the $(n - 1)$D algebra by X (and proceed mounting the levels of both algebras), when $\alpha < 1/2$, or we add X to a formula of the lowest possible level of the $(n - 1)$D algebra when $\alpha > 1/2$ (and proceed again mounting the levels of both algebras).

	$\mathscr{B}_1$	$\mathscr{B}_2$	$\mathscr{B}_3$	$\mathscr{B}_4$
15/16				$[X]+[X+X+X]$
7/8			$[X]+[X+X]$	$[X+X+X],\ [X \dashv X \dashv X \dashv X]$
13/16				$[X]+[X+XX]$ $[X+X+X][[X]+[X+X+X]]$
3/4		$[X]+[X]$	$[X+X],\ [X \dashv X \dashv X]$	$X+X,\ X \dashv X \dashv X$ $[X+X+X][X+X+X]$
11/16				$[X]+[X(X+X)]$ $[X+X][[X]+[X+X+X]]$ $[X+X+X][[X+X+X][[X]+[X+X+X]]]$
5/8			$[X+X][[X]+[X+X]],$ $[X]+[XX]$	$[X+X][X+X+X],$ $[X+XX]$ $[X+X+X][X+X+X][X+X+X]$
9/16				$[X]+[XXX]$ $[X+XX][[X]+[X+X+X]]$ $[X+X][[X+X+X][[X]+[X+X+X]]]$
1/2	$[X]$	$X,\ [X \sim X]$	$X,\ X \sim X,\ [XX]+[XX]$	$X,\ X \sim X,\ [XX+XX]$ $[XX]+[[XXX]+[XXX]]$
7/16				$[X][(X+X+X)],$ $[X(X+X)]+[[X][XXX]],$ $[XX]+[[XXX]+[X][XXX]]$
3/8			$[XX]+[[X][XX]],\ \ [X][X+X]$	$[XX]+[XXX],\ [X(X+X)]$ $[[XXX]+[XXX]]+[XXX]$
5/16				$[X][X+XX]$ $[XX]+[[X][XXX]]$ $[XXX]+[[XXX]+[X][XXX]]$
1/4		$[X][X]$	$[XX],\ [X \sim X \sim X]$	$XX,\ X \sim X \sim X$ $[XXX]+[XXX]$
3/16				$[X][X(X+X)]$ $[XXX]+[[X][XXX]]$
1/8			$[X][XX]$	$[XXX],\ [X \sim X \sim X \sim X]$
1/16				$[X][XXX]$

Figure 4.4. The algebra of the generation of logical forms. The square brackets denote the different components and their use for the first time, so that we have the series $[X]$, $[X][X]$, $[XX]$. In other words, we fix the forms in a recursive way.

	$\mathcal{B}_1$	$\mathcal{B}_2$	$\mathcal{B}_3$	$\mathcal{B}_4$
15/16				a+j
7/8			j=a+c	
13/16				a+h jx(a+j)=a+i
3/4		c=a+a		jxj
11/16				a+f cx(a+j) jx(a+i)
5/8			h=a+b i=cxj	jx(jxj)
9/16				a+e hx(a+j) ix(a+j)
1/2	a		d=b+b=cxc	b+(e+e)=cx(jxj)
7/16				axj f+(axe) g+(axe)
3/8			f=axc g=b+e	e+(e+e)
5/16				axh b+(axe) e+(axg)
1/4		b=axa		e+e
3/16				axf e+(axe)=axg
1/8			e=axb	
1/16				axe

Figure 4.5. The generation of the new logical forms starting from an initial a = X. Note in particular the patterns of $\mathcal{B}_4$.

- We get other new formulæ of the nD algebra for odd Levels $\alpha/m(n)$ by adding the expression of the lowest level of such an algebra to the expression of the nearest lower levels of less–dimensional algebra when $\alpha < 1/2$ or we multiply the expression of the highest level of the nD algebra with the expression of the nearest higher levels of less–dimensional algebras when $\alpha > 1/2$.
- For even levels, we add (for $\alpha \leq 1/2$) or multiply (for $\alpha \geq 1/2$) identical expressions of the lowest or highest level, respectively.

When I say lowest or highest levels, I am excluding tautology and contradiction. These three kinds of operations are depicted in red, green, and blue, respectively. Note that for $\mathscr{B}_2$ these operations coincide and therefore are in black. In fact, b = axa is clearly a red operation, but it is also a green operation since a is located the nearest higher level of the less–dimensional algebra $\mathscr{B}_1$, and similarly for c = a + a. Note that the b and c are somehow inverted, due to the fact that in $\mathscr{B}_1$ we have only a. The blue series is instantiated here, with the sole difference that the two expressions (b and c) pertain to different levels.

The previous analysis shows that, although for n dimensions going to ∞ the number of nodes and relative logical expressions is also infinite, for any algebra with finite dimensions we can foresee their number and their form. Thus, we can definitively assert that there is nothing arbitrary in logic, but everything is fixed in advance in the structural properties of Boolean algebra.

Of course, we could also ask whether there are alternative generations with other logical forms. Such generations cannot be completely excluded. In fact, in Subsection 3.2.7 we alternatively use expressions containing material equivalences as a part. Nevertheless, as already mentioned, this appears to be less economic and thus violating Occam's razor. Further study will clarify the reasons for the uniqueness of the generation or the pattern ruling alternative generations of the same value.

Chapter 5

$$\mathscr{B}_5$$

5.1. SD Variables and CAs

$\mathscr{B}_5$ is characterised by the following numbers: $m(5) = 32, s'(5) = 32,$
$768, l(4) = 33,$

$$\begin{aligned}
k(5) = {} & 1 + 32 + 496 + 4960 + 35960 + 201376 + 906192 + 3365856 \\
& + 10518300 + 28048800 + 64512240 + 129024480 + 225792840 \\
& + 347373600 + 471435600 + 565722720 + 601080390 + 565722720 \\
& + 471435600 + 347373600 + 225792840 + 129024480 + 64512240 \\
& + 28048800 + 10518300 + 3365856 + 906192 + 201376 \\
& + 35960 + 4960 + 496 + 32 + 1 \\
= {} & 4,294,967,296,
\end{aligned} \tag{5.1}$$

where the third row represents levels $15/32$, $16/32$, and $17/32$. Moreover,

$$r(5) = 2^{32+4} = 2^{36}, \tag{5.2}$$

and

$$p(5) = 32!. \tag{5.3}$$

Let us look at an example of the way in which SD variables are generated. Taking $X_{1.1.1}$ as the root variable, we can generate all 256 relative daughter variables as in Table 5.1.

Table 5.1. Building of $\mathscr{B}_5$'s variables. Note that the third and fourth columns reproduce all the nodes of $\mathscr{B}_3$ according to the series of non–negative integers (the fifth and sixth ones also, but in almost reversed order).

$X_{1.1.1.1}$	0000	0000	0000	0000	1111	1111	1111	1111
$X_{1.1.1.2}$	0000	0000	0000	0001	0111	1111	1111	1111
$X_{1.1.1.3}$	0000	0000	0000	0010	1011	1111	1111	1111
$X_{1.1.1.4}$	0000	0000	0000	0011	0011	1111	1111	1111
$X_{1.1.1.5}$	0000	0000	0000	0100	1101	1111	1111	1111
$X_{1.1.1.6}$	0000	0000	0000	0101	0101	1111	1111	1111
$X_{1.1.1.7}$	0000	0000	0000	0110	1001	1111	1111	1111
$X_{1.1.1.8}$	0000	0000	0000	0111	0001	1111	1111	1111
$X_{1.1.1.9}$	0000	0000	0000	1000	1110	1111	1111	1111
$X_{1.1.1.10}$	0000	0000	0000	1001	0110	1111	1111	1111
$X_{1.1.1.11}$	0000	0000	0000	1010	1010	1111	1111	1111
$X_{1.1.1.12}$	0000	0000	0000	1011	0010	1111	1111	1111
$X_{1.1.1.13}$	0000	0000	0000	1100	1100	1111	1111	1111
$X_{1.1.1.14}$	0000	0000	0000	1101	0100	1111	1111	1111
$X_{1.1.1.15}$	0000	0000	0000	1110	1000	1111	1111	1111
$X_{1.1.1.16}$	0000	0000	0000	1111	0000	1111	1111	1111
$X_{1.1.1.17}$	0000	0000	0001	0000	1111	0111	1111	1111
$X_{1.1.1.18}$	0000	0000	0001	0001	0111	0111	1111	1111
$X_{1.1.1.19}$	0000	0000	0001	0010	1011	0111	1111	1111
$X_{1.1.1.20}$	0000	0000	0001	0011	0011	0111	1111	1111
$X_{1.1.1.21}$	0000	0000	0001	0100	1101	0111	1111	1111
$X_{1.1.1.22}$	0000	0000	0001	0101	0101	0111	1111	1111
$X_{1.1.1.23}$	0000	0000	0001	0110	1001	0111	1111	1111
$X_{1.1.1.24}$	0000	0000	0001	0111	0001	0111	1111	1111
$X_{1.1.1.25}$	0000	0000	0001	1000	1110	0111	1111	1111
$X_{1.1.1.26}$	0000	0000	0001	1001	0110	0111	1111	1111
$X_{1.1.1.27}$	0000	0000	0001	1010	1010	0111	1111	1111
$X_{1.1.1.28}$	0000	0000	0001	1011	0010	0111	1111	1111
$X_{1.1.1.29}$	0000	0000	0001	1100	1100	0111	1111	1111
$X_{1.1.1.30}$	0000	0000	0001	1101	0100	0111	1111	1111
$X_{1.1.1.31}$	0000	0000	0001	1110	1000	0111	1111	1111
$X_{1.1.1.32}$	0000	0000	0001	1111	0000	0111	1111	1111
$X_{1.1.1.33}$	0000	0000	0010	0000	1111	1011	1111	1111
$X_{1.1.1.34}$	0000	0000	0010	0001	0111	1011	1111	1111
$X_{1.1.1.35}$	0000	0000	0010	0010	1011	1011	1111	1111
$X_{1.1.1.36}$	0000	0000	0010	0011	0011	1011	1111	1111
$X_{1.1.1.37}$	0000	0000	0010	0100	1101	1011	1111	1111

Table 5.1. (*Continued*)

$X_{1.1.1.38}$	0000	0000	0010	0101	0101	1011	1111	1111
$X_{1.1.1.39}$	0000	0000	0010	0110	1001	1011	1111	1111
$X_{1.1.1.40}$	0000	0000	0010	0111	0001	1011	1111	1111
$X_{1.1.1.41}$	0000	0000	0010	1000	1110	1011	1111	1111
$X_{1.1.1.42}$	0000	0000	0010	1001	0110	1011	1111	1111
$X_{1.1.1.43}$	0000	0000	0010	1010	1010	1011	1111	1111
$X_{1.1.1.44}$	0000	0000	0010	1011	0010	1011	1111	1111
$X_{1.1.1.45}$	0000	0000	0010	1100	1100	1011	1111	1111
$X_{1.1.1.46}$	0000	0000	0010	1101	0100	1011	1111	1111
$X_{1.1.1.47}$	0000	0000	0010	1110	1000	1011	1111	1111
$X_{1.1.1.48}$	0000	0000	0010	1111	0000	1011	1111	1111
$X_{1.1.1.49}$	0000	0000	0011	0000	1111	0011	1111	1111
$X_{1.1.1.50}$	0000	0000	0011	0001	0111	0011	1111	1111
$X_{1.1.1.51}$	0000	0000	0011	0010	1011	0011	1111	1111
$X_{1.1.1.52}$	0000	0000	0011	0011	0011	0011	1111	1111
$X_{1.1.1.53}$	0000	0000	0011	0100	1101	0011	1111	1111
$X_{1.1.1.54}$	0000	0000	0011	0101	0101	0011	1111	1111
$X_{1.1.1.55}$	0000	0000	0011	0110	1001	0011	1111	1111
$X_{1.1.1.56}$	0000	0000	0011	0111	0001	0011	1111	1111
$X_{1.1.1.57}$	0000	0000	0011	1000	1110	0011	1111	1111
$X_{1.1.1.58}$	0000	0000	0011	1001	0110	0011	1111	1111
$X_{1.1.1.59}$	0000	0000	0011	1010	1010	0011	1111	1111
$X_{1.1.1.60}$	0000	0000	0011	1011	0010	0011	1111	1111
$X_{1.1.1.61}$	0000	0000	0011	1100	1100	0011	1111	1111
$X_{1.1.1.62}$	0000	0000	0011	1101	0100	0011	1111	1111
$X_{1.1.1.63}$	0000	0000	0011	1110	1000	0011	1111	1111
$X_{1.1.1.64}$	0000	0000	0011	1111	0000	0011	1111	1111
$X_{1.1.1.65}$	0000	0000	0100	0000	1111	1101	1111	1111
$X_{1.1.1.66}$	0000	0000	0100	0001	0111	1101	1111	1111
$X_{1.1.1.67}$	0000	0000	0100	0010	1011	1101	1111	1111
$X_{1.1.1.68}$	0000	0000	0100	0011	0011	1101	1111	1111
$X_{1.1.1.69}$	0000	0000	0100	0100	1101	1101	1111	1111
$X_{1.1.1.70}$	0000	0000	0100	0101	0101	1101	1111	1111
$X_{1.1.1.71}$	0000	0000	0100	0110	1001	1101	1111	1111
$X_{1.1.1.72}$	0000	0000	0100	0111	0001	1101	1111	1111
$X_{1.1.1.73}$	0000	0000	0100	1000	1110	1101	1111	1111
$X_{1.1.1.74}$	0000	0000	0100	1001	0110	1101	1111	1111
$X_{1.1.1.75}$	0000	0000	0100	1010	1010	1101	1111	1111
$X_{1.1.1.76}$	0000	0000	0100	1011	0010	1101	1111	1111

(*Continued*)

Boolean Structures

Table 5.1. (*Continued*)

$X_{1.1.1.77}$	0000	0000	0100	1100	1100	1101	1111	1111
$X_{1.1.1.78}$	0000	0000	0100	1101	0100	1101	1111	1111
$X_{1.1.1.79}$	0000	0000	0100	1110	1000	1101	1111	1111
$X_{1.1.1.80}$	0000	0000	0100	1111	0000	1101	1111	1111
$X_{1.1.1.81}$	0000	0000	0101	0000	1111	0101	1111	1111
$X_{1.1.1.82}$	0000	0000	0101	0001	0111	0101	1111	1111
$X_{1.1.1.83}$	0000	0000	0101	0010	1011	0101	1111	1111
$X_{1.1.1.84}$	0000	0000	0101	0011	0011	0101	1111	1111
$X_{1.1.1.85}$	0000	0000	0101	0100	1101	0101	1111	1111
$X_{1.1.1.86}$	0000	0000	0101	0101	0101	0101	1111	1111
$X_{1.1.1.87}$	0000	0000	0101	0110	1001	0101	1111	1111
$X_{1.1.1.88}$	0000	0000	0101	0111	0001	0101	1111	1111
$X_{1.1.1.89}$	0000	0000	0101	1000	1110	0101	1111	1111
$X_{1.1.1.90}$	0000	0000	0101	1001	0110	0101	1111	1111
$X_{1.1.1.91}$	0000	0000	0101	1010	1010	0101	1111	1111
$X_{1.1.1.92}$	0000	0000	0101	1011	0010	0101	1111	1111
$X_{1.1.1.93}$	0000	0000	0101	1100	1100	0101	1111	1111
$X_{1.1.1.94}$	0000	0000	0101	1101	0100	0101	1111	1111
$X_{1.1.1.95}$	0000	0000	0101	1110	1000	0101	1111	1111
$X_{1.1.1.96}$	0000	0000	0101	1111	0000	0101	1111	1111
$X_{1.1.1.97}$	0000	0000	0110	0000	1111	1001	1111	1111
$X_{1.1.1.98}$	0000	0000	0110	0001	0111	1001	1111	1111
$X_{1.1.1.99}$	0000	0000	0110	0010	1011	1001	1111	1111
$X_{1.1.1.100}$	0000	0000	0110	0011	0011	1001	1111	1111
$X_{1.1.1.101}$	0000	0000	0110	0100	1101	1001	1111	1111
$X_{1.1.1.102}$	0000	0000	0110	0101	0101	1001	1111	1111
$X_{1.1.1.103}$	0000	0000	0110	0110	1001	1001	1111	1111
$X_{1.1.1.104}$	0000	0000	0110	0111	0001	1001	1111	1111
$X_{1.1.1.105}$	0000	0000	0110	1000	1110	1001	1111	1111
$X_{1.1.1.106}$	0000	0000	0110	1001	0110	1001	1111	1111
$X_{1.1.1.107}$	0000	0000	0110	1010	1010	1001	1111	1111
$X_{1.1.1.108}$	0000	0000	0110	1011	0010	1001	1111	1111
$X_{1.1.1.109}$	0000	0000	0110	1100	1100	1001	1111	1111
$X_{1.1.1.110}$	0000	0000	0110	1101	0100	1001	1111	1111
$X_{1.1.1.111}$	0000	0000	0110	1110	1000	1001	1111	1111
$X_{1.1.1.112}$	0000	0000	0110	1111	0000	1001	1111	1111
$X_{1.1.1.113}$	0000	0000	0111	0000	1111	0001	1111	1111
$X_{1.1.1.114}$	0000	0000	0111	0001	0111	0001	1111	1111
$X_{1.1.1.115}$	0000	0000	0111	0010	1011	0001	1111	1111

Table 5.1. (*Continued*)

$X_{1.1.1.116}$	0000	0000	0111	0011	0011	0001	1111	1111
$X_{1.1.1.117}$	0000	0000	0111	0100	1101	0001	1111	1111
$X_{1.1.1.118}$	0000	0000	0111	0101	0101	0001	1111	1111
$X_{1.1.1.119}$	0000	0000	0111	0110	1001	0001	1111	1111
$X_{1.1.1.120}$	0000	0000	0111	0111	0001	0001	1111	1111
$X_{1.1.1.121}$	0000	0000	0111	1000	1110	0001	1111	1111
$X_{1.1.1.122}$	0000	0000	0111	1001	0110	0001	1111	1111
$X_{1.1.1.123}$	0000	0000	0111	1010	1010	0001	1111	1111
$X_{1.1.1.124}$	0000	0000	0111	1011	0010	0001	1111	1111
$X_{1.1.1.125}$	0000	0000	0111	1100	1100	0001	1111	1111
$X_{1.1.1.126}$	0000	0000	0111	1101	0100	0001	1111	1111
$X_{1.1.1.127}$	0000	0000	0111	1110	1000	0001	1111	1111
$X_{1.1.1.128}$	0000	0000	0111	1111	0000	0001	1111	1111
$X_{1.1.1.129}$	0000	0000	1000	0000	1111	1110	1111	1111
$X_{1.1.1.130}$	0000	0000	1000	0001	0111	1110	1111	1111
$X_{1.1.1.131}$	0000	0000	1000	0010	1011	1110	1111	1111
$X_{1.1.1.132}$	0000	0000	1000	0011	0011	1110	1111	1111
$X_{1.1.1.133}$	0000	0000	1000	0100	1101	1110	1111	1111
$X_{1.1.1.134}$	0000	0000	1000	0101	0101	1110	1111	1111
$X_{1.1.1.135}$	0000	0000	1000	0110	1001	1110	1111	1111
$X_{1.1.1.136}$	0000	0000	1000	0111	0001	1110	1111	1111
$X_{1.1.1.137}$	0000	0000	1000	1000	1110	1110	1111	1111
$X_{1.1.1.138}$	0000	0000	1000	1001	0110	1110	1111	1111
$X_{1.1.1.139}$	0000	0000	1000	1010	1010	1110	1111	1111
$X_{1.1.1.140}$	0000	0000	1000	1011	0010	1110	1111	1111
$X_{1.1.1.141}$	0000	0000	1000	1100	1100	1110	1111	1111
$X_{1.1.1.142}$	0000	0000	1000	1101	0100	1110	1111	1111
$X_{1.1.1.143}$	0000	0000	1000	1110	1000	1110	1111	1111
$X_{1.1.1.144}$	0000	0000	1000	1111	0000	1110	1111	1111
$X_{1.1.1.145}$	0000	0000	1001	0000	1111	0110	1111	1111
$X_{1.1.1.146}$	0000	0000	1001	0001	0111	0110	1111	1111
$X_{1.1.1.147}$	0000	0000	1001	0010	1011	0110	1111	1111
$X_{1.1.1.148}$	0000	0000	1001	0011	0011	0110	1111	1111
$X_{1.1.1.149}$	0000	0000	1001	0100	1101	0110	1111	1111
$X_{1.1.1.150}$	0000	0000	1001	0101	0101	0110	1111	1111
$X_{1.1.1.151}$	0000	0000	1001	0110	1001	0110	1111	1111
$X_{1.1.1.152}$	0000	0000	1001	0111	0001	0110	1111	1111
$X_{1.1.1.153}$	0000	0000	1001	1000	1110	0110	1111	1111
$X_{1.1.1.154}$	0000	0000	1001	1001	0110	0110	1111	1111

(*Continued*)

Boolean Structures

Table 5.1. (*Continued*)

$X_{1.1.1.155}$	0000	0000	1001	1010	1010	0110	1111	1111
$X_{1.1.1.156}$	0000	0000	1001	1011	0010	0110	1111	1111
$X_{1.1.1.157}$	0000	0000	1001	1100	1100	0110	1111	1111
$X_{1.1.1.158}$	0000	0000	1001	1101	0100	0110	1111	1111
$X_{1.1.1.159}$	0000	0000	1001	1110	1000	0110	1111	1111
$X_{1.1.1.160}$	0000	0000	1001	1111	0000	0110	1111	1111
$X_{1.1.1.161}$	0000	0000	1010	0000	1111	1010	1111	1111
$X_{1.1.1.162}$	0000	0000	1010	0001	0111	1010	1111	1111
$X_{1.1.1.163}$	0000	0000	1010	0010	1011	1010	1111	1111
$X_{1.1.1.164}$	0000	0000	1010	0011	0011	1010	1111	1111
$X_{1.1.1.165}$	0000	0000	1010	0100	1101	1010	1111	1111
$X_{1.1.1.166}$	0000	0000	1010	0101	0101	1010	1111	1111
$X_{1.1.1.167}$	0000	0000	1010	0110	1001	1010	1111	1111
$X_{1.1.1.168}$	0000	0000	1010	0111	0001	1010	1111	1111
$X_{1.1.1.169}$	0000	0000	1010	1000	1110	1010	1111	1111
$X_{1.1.1.170}$	0000	0000	1010	1001	0110	1010	1111	1111
$X_{1.1.1.171}$	0000	0000	1010	1010	1010	1010	1111	1111
$X_{1.1.1.172}$	0000	0000	1010	1011	0010	1010	1111	1111
$X_{1.1.1.173}$	0000	0000	1010	1100	1100	1010	1111	1111
$X_{1.1.1.174}$	0000	0000	1010	1101	0100	1010	1111	1111
$X_{1.1.1.175}$	0000	0000	1010	1110	1000	1010	1111	1111
$X_{1.1.1.176}$	0000	0000	1010	1111	0000	1010	1111	1111
$X_{1.1.1.177}$	0000	0000	1011	0000	1111	0010	1111	1111
$X_{1.1.1.178}$	0000	0000	1011	0001	0111	0010	1111	1111
$X_{1.1.1.179}$	0000	0000	1011	0010	1011	0010	1111	1111
$X_{1.1.1.180}$	0000	0000	1011	0011	0011	0010	1111	1111
$X_{1.1.1.181}$	0000	0000	1011	0100	1101	0010	1111	1111
$X_{1.1.1.182}$	0000	0000	1011	0101	0101	0010	1111	1111
$X_{1.1.1.183}$	0000	0000	1011	0110	1001	0010	1111	1111
$X_{1.1.1.184}$	0000	0000	1011	0111	0001	0010	1111	1111
$X_{1.1.1.185}$	0000	0000	1011	1000	1110	0010	1111	1111
$X_{1.1.1.186}$	0000	0000	1011	1001	0110	0010	1111	1111
$X_{1.1.1.187}$	0000	0000	1011	1010	1010	0010	1111	1111
$X_{1.1.1.188}$	0000	0000	1011	1011	0010	0010	1111	1111
$X_{1.1.1.189}$	0000	0000	1011	1100	1100	0010	1111	1111
$X_{1.1.1.190}$	0000	0000	1011	1101	0100	0010	1111	1111
$X_{1.1.1.191}$	0000	0000	1011	1110	1000	0010	1111	1111
$X_{1.1.1.192}$	0000	0000	1011	1111	0000	0010	1111	1111
$X_{1.1.1.193}$	0000	0000	1100	0000	1111	1100	1111	1111

Table 5.1. (*Continued*)

$X_{1.1.1.194}$	0000	0000	1100	0001	0111	1100	1111	1111
$X_{1.1.1.195}$	0000	0000	1100	0010	1011	1100	1111	1111
$X_{1.1.1.196}$	0000	0000	1100	0011	0011	1100	1111	1111
$X_{1.1.1.197}$	0000	0000	1100	0100	1101	1100	1111	1111
$X_{1.1.1.198}$	0000	0000	1100	0101	0101	1100	1111	1111
$X_{1.1.1.199}$	0000	0000	1100	0110	1001	1100	1111	1111
$X_{1.1.1.200}$	0000	0000	1100	0111	0001	1100	1111	1111
$X_{1.1.1.201}$	0000	0000	1100	1000	1110	1100	1111	1111
$X_{1.1.1.202}$	0000	0000	1100	1001	0110	1100	1111	1111
$X_{1.1.1.203}$	0000	0000	1100	1010	1010	1100	1111	1111
$X_{1.1.1.204}$	0000	0000	1100	1011	0010	1100	1111	1111
$X_{1.1.1.205}$	0000	0000	1100	1100	1100	1100	1111	1111
$X_{1.1.1.206}$	0000	0000	1100	1101	0100	1100	1111	1111
$X_{1.1.1.207}$	0000	0000	1100	1110	1000	1100	1111	1111
$X_{1.1.1.208}$	0000	0000	1100	1111	0000	1100	1111	1111
$X_{1.1.1.209}$	0000	0000	1101	0000	1111	0100	1111	1111
$X_{1.1.1.210}$	0000	0000	1101	0001	0111	0100	1111	1111
$X_{1.1.1.211}$	0000	0000	1101	0010	1011	0100	1111	1111
$X_{1.1.1.212}$	0000	0000	1101	0011	0011	0100	1111	1111
$X_{1.1.1.213}$	0000	0000	1101	0100	1101	0100	1111	1111
$X_{1.1.1.214}$	0000	0000	1101	0101	0101	0100	1111	1111
$X_{1.1.1.215}$	0000	0000	1101	0110	1001	0100	1111	1111
$X_{1.1.1.216}$	0000	0000	1101	0111	0001	0100	1111	1111
$X_{1.1.1.217}$	0000	0000	1101	1000	1110	0100	1111	1111
$X_{1.1.1.218}$	0000	0000	1101	1001	0110	0100	1111	1111
$X_{1.1.1.219}$	0000	0000	1101	1010	1010	0100	1111	1111
$X_{1.1.1.220}$	0000	0000	1101	1011	0010	0100	1111	1111
$X_{1.1.1.221}$	0000	0000	1101	1100	1100	0100	1111	1111
$X_{1.1.1.222}$	0000	0000	1101	1101	0100	0100	1111	1111
$X_{1.1.1.223}$	0000	0000	1101	1110	1000	0100	1111	1111
$X_{1.1.1.224}$	0000	0000	1101	1111	0000	0100	1111	1111
$X_{1.1.1.225}$	0000	0000	1110	0000	1111	1000	1111	1111
$X_{1.1.1.226}$	0000	0000	1110	0001	0111	1000	1111	1111
$X_{1.1.1.227}$	0000	0000	1110	0010	1011	1000	1111	1111
$X_{1.1.1.228}$	0000	0000	1110	0011	0011	1000	1111	1111
$X_{1.1.1.229}$	0000	0000	1110	0100	1101	1000	1111	1111
$X_{1.1.1.230}$	0000	0000	1110	0101	0101	1000	1111	1111
$X_{1.1.1.231}$	0000	0000	1110	0110	1001	1000	1111	1111
$X_{1.1.1.232}$	0000	0000	1110	0111	0001	1000	1111	1111

(*Continued*)

Table 5.1.   (*Continued*)

$X_{1.1.1.233}$	0000	0000	1110	1000	1110	1000	1111	1111
$X_{1.1.1.234}$	0000	0000	1110	1001	0110	1000	1111	1111
$X_{1.1.1.235}$	0000	0000	1110	1010	1010	1000	1111	1111
$X_{1.1.1.236}$	0000	0000	1110	1011	0010	1000	1111	1111
$X_{1.1.1.237}$	0000	0000	1110	1100	1100	1000	1111	1111
$X_{1.1.1.238}$	0000	0000	1110	1101	0100	1000	1111	1111
$X_{1.1.1.239}$	0000	0000	1110	1110	1000	1000	1111	1111
$X_{1.1.1.240}$	0000	0000	1110	1111	0000	1000	1111	1111
$X_{1.1.1.241}$	0000	0000	1111	0000	1111	0000	1111	1111
$X_{1.1.1.242}$	0000	0000	1111	0001	0111	0000	1111	1111
$X_{1.1.1.243}$	0000	0000	1111	0010	1011	0000	1111	1111
$X_{1.1.1.244}$	0000	0000	1111	0011	0011	0000	1111	1111
$X_{1.1.1.245}$	0000	0000	1111	0100	1101	0000	1111	1111
$X_{1.1.1.246}$	0000	0000	1111	0101	0101	0000	1111	1111
$X_{1.1.1.247}$	0000	0000	1111	0110	1001	0000	1111	1111
$X_{1.1.1.248}$	0000	0000	1111	0111	0001	0000	1111	1111
$X_{1.1.1.249}$	0000	0000	1111	1000	1110	0000	1111	1111
$X_{1.1.1.250}$	0000	0000	1111	1001	0110	0000	1111	1111
$X_{1.1.1.251}$	0000	0000	1111	1010	1010	0000	1111	1111
$X_{1.1.1.252}$	0000	0000	1111	1011	0010	0000	1111	1111
$X_{1.1.1.253}$	0000	0000	1111	1100	1100	0000	1111	1111
$X_{1.1.1.254}$	0000	0000	1111	1101	0100	0000	1111	1111
$X_{1.1.1.255}$	0000	0000	1111	1110	1000	0000	1111	1111
$X_{1.1.1.256}$	0000	0000	1111	1111	0000	0000	1111	1111

Note that the internal part of the ID for each row (16 numbers) in the table individuates one of the 128 4D SD variables (while all rows from 129 to 256 are their complements in inverted order). In particular, we can individuate 8 blocks:

- Rows 1–16: all these internal parts represent the 4D variables generated by $X_{1.1}$ (CA1) and correspond to the list in Table 4.2.
- Rows 17–32: are parts representing the 4D variables generated by $X_{1.2}$ (CA2).
- Rows 33–48: are parts representing the 4D variables generated by $X_{1.3}$ (CA2).
- Rows 49–64: are parts representing the 4D variables generated by $X_{1.4}$ (CA1).

- Rows 65–80: are parts representing the 4D variables generated by $X_{2.1}$ (CA2).
- Rows 81–96: are parts representing the 4D variables generated by $X_{2.2}$ (CA1).
- Rows 97–112: are parts representing the 4D variables generated by $X_{2.3}$ (CA1).
- Rows 113–128: are parts representing the 4D variables generated by $X_{2.4}$ (CA2).

For each of the 16 main CAs of $\mathscr{B}_4$, we have 128 main CAs of $\mathscr{B}_5$, that is, 2048 main CAs as a whole. Let us explore this combinatorics. For instance, the first of the 128 main CAs generated by CA1.1 (i.e. CA1.1.1) is displayed in Table 5.2.

Table 5.2.   Main CA1.1.1.

$X_{1.1.1.1}$	0000	0000	0000	0000	1111	1111	1111	1111
$X_{1.1.1.256}$	0000	0000	1111	1111	0000	0000	1111	1111
$X_{1.1.16.16}$	0000	1111	0000	1111	0000	1111	0000	1111
$X_{1.1.16.241}$	0000	1111	1111	0000	1111	0000	0000	1111
$X_{1.4.4.52}$	0011	0011	0011	0011	0011	0011	0011	0011
$X_{1.4.4.205}$	0011	0011	1100	1100	1100	1100	0011	0011
$X_{1.4.13.61}$	0011	1100	0011	1100	1100	0011	1100	0011
$X_{1.4.13.196}$	0011	1100	1100	0011	0011	1100	1100	0011
$X_{2.2.6.86}$	0101	0101	0101	0101	0101	0101	0101	0101
$X_{2.2.6.171}$	0101	0101	1010	1010	1010	1010	0101	0101
$X_{2.2.11.91}$	0101	1010	0101	1010	1010	0101	1010	0101
$X_{2.2.11.166}$	0101	1010	1010	0101	0101	1010	1010	0101
$X_{2.3.7.103}$	0110	0110	0110	0110	1001	1001	1001	1001
$X_{2.3.7.154}$	0110	0110	1001	1001	0110	0110	1001	1001
$X_{2.3.10.106}$	0110	1001	0110	1001	0110	1001	0110	1001
$X_{2.3.10.151}$	0110	1001	1001	0110	1001	0110	0110	1001

Of course, the sum of the last index (the fourth one) for each doublet makes $257 = 256+1$. Note that, on the outlines of Table 4.3, we have picked all variables whose internal part of the ID corresponds to the external part (rows 1–16 for $X_{1.1.x.y}$, rows 49–64 for $X_{1.4.x.y}$, rows 81–96 for $X_{2.2.x.y}$, rows 97–112 for $X_{2.3.x.y}$). For building the main CA1.1.2 we need to follow the directions (up or down) summarised in Table 4.4. So, we obtain the combinatorics displayed in Table 5.3.

Table 5.3. Main CA1.1.2.

$X_{1.1.1.2}$	0000	0000	0000	0001	0111	1111	1111	1111
$X_{1.1.1.255}$	0000	0000	1111	1110	1000	0000	1111	1111
$X_{1.1.16.15}$	0000	1111	0000	1110	1000	1111	0000	1111
$X_{1.1.16.242}$	0000	1111	1111	0001	0111	0000	0000	1111
$X_{1.4.4.51}$	0011	0011	0011	0010	1011	0011	0011	0011
$X_{1.4.4.206}$	0011	0011	1100	1101	0100	1100	0011	0011
$X_{1.4.13.62}$	0011	1100	0011	1101	0100	0011	1100	0011
$X_{1.4.13.195}$	0011	1100	1100	0010	1011	1100	1100	0011
$X_{2.2.6.85}$	0101	0101	0101	0100	1101	0101	0101	0101
$X_{2.2.6.172}$	0101	0101	1010	1011	0010	1010	0101	0101
$X_{2.2.11.92}$	0101	1010	0101	1011	0010	0101	1010	0101
$X_{2.2.11.165}$	0101	1010	1010	0100	1101	1010	1010	0101
$X_{2.3.7.104}$	0110	0110	0110	0111	0001	1001	1001	1001
$X_{2.3.7.153}$	0110	0110	1001	1000	1110	0110	1001	1001
$X_{2.3.10.105}$	0110	1001	0110	1000	1110	1001	0110	1001
$X_{2.3.10.152}$	0110	1001	1001	0111	0001	0110	0110	1001

As is evident, all internal parts of the ID again follow the same structure as before. For CA1.2.1, we need to shift the third index (what makes the external parts of the IDs as rooted in CA2) and go back to the fourth index of CA1.1.1, as displayed in Table 5.4.

Table 5.4. Main CA1.2.1.

$X_{1.1.2.1}$	0000	0001	0000	0001	0111	1111	0111	1111
$X_{1.1.2.256}$	1111	1110	1111	1110	1000	0000	1000	0000
$X_{1.1.15.16}$	0000	1110	0000	1110	1000	1111	1000	1111
$X_{1.1.15.241}$	1111	0001	1111	0001	0111	0000	0111	0000
$X_{1.4.3.52}$	0011	0010	0011	0010	1011	0011	1011	0011
$X_{1.4.3.205}$	1100	1101	1100	1101	0100	1100	0100	1100
$X_{1.4.14.61}$	0011	1101	0011	1101	0100	0011	0100	0011
$X_{1.4.14.196}$	1100	0010	1100	0010	1011	1100	1011	1100
$X_{2.2.5.86}$	0101	0100	0101	0100	1101	0101	1101	0101
$X_{2.2.5.171}$	1010	1011	1010	1011	0010	1010	0010	1010
$X_{2.2.12.91}$	0101	1011	0101	1011	0010	0101	0010	0101
$X_{2.2.12.166}$	1010	0100	1010	0100	1101	1010	1101	1010
$X_{2.3.8.103}$	0110	0111	0110	0111	0001	1001	0001	1001
$X_{2.3.8.154}$	1001	1000	1001	1000	1110	0110	1110	0110
$X_{2.3.9.106}$	0110	1000	0110	1000	1110	1001	1110	1001
$X_{2.3.9.151}$	1001	0111	1001	0111	0001	0110	0001	0110

Analogously for CA1.2.2, we have the index displayed in Table 5.5.

Table 5.5. Main CA1.2.2.

$X_{1.1.2.2}$	0000	0001	0000	0001	0111	1111	0111	1111
$X_{1.1.2.255}$	1111	1110	1111	1110	1000	0000	1000	0000
$X_{1.1.15.15}$	0000	1110	0000	1110	1000	1111	1000	1111
$X_{1.1.15.242}$	1111	0001	1111	0001	0111	0000	0111	0000
$X_{1.4.3.51}$	0011	0010	0011	0010	1011	0011	1011	0011
$X_{1.4.3.206}$	1100	1101	1100	1101	0100	1100	0100	1100
$X_{1.4.14.62}$	0011	1101	0011	1101	0100	0011	0100	0011
$X_{1.4.14.195}$	1100	0010	1100	0010	1011	1100	1011	1100
$X_{2.2.5.85}$	0101	0100	0101	0100	1101	0101	1101	0101
$X_{2.2.5.172}$	1010	1011	1010	1011	0010	1010	0010	1010
$X_{2.2.12.92}$	0101	1011	0101	1011	0010	0101	0010	0101
$X_{2.2.12.165}$	1010	0100	1010	0100	1101	1010	1101	1010
$X_{2.3.8.104}$	0110	0111	0110	0111	0001	1001	0001	1001
$X_{2.3.8.153}$	1001	1000	1001	1000	1110	0110	1110	0110
$X_{2.3.9.105}$	0110	1000	0110	1000	1110	1001	1110	1001
$X_{2.3.9.152}$	1001	0111	1001	0111	0001	0110	0001	0110

This allows the following generalisation. Let us consider the first eight main sets out of the 4D main sets. For the series of the variables generated by the root variable $X_{1.1}$ we have the series displayed in Table 5.6.

Table 5.6. 5D variables generated by $X_{1.1}$.

$X_{1.1.1.1} - X_{1.1.2.2}$	$X_{1.1.3.3} - X_{1.1.4.4}$	$X_{1.1.5.5} - X_{1.1.6.6}$	$X_{1.1.7.7} - X_{1.1.8.8}$
$X_{1.1.16.16} - X_{1.1.15.15}$	$X_{1.1.14.14} - X_{1.1.13.13}$	$X_{1.1.12.12} - X_{1.1.11.11}$	$X_{1.1.10.10} - X_{1.1.9.9}$

The final (fourth) index of the SD variables determines the last (third) index of the CA. We need to represent a column under (and above) each item covering all the 256 cases generated by a single 4D root variable, as shown in Table 5.1 for $X_{1.1.1}$. This scheme is also repeated for the other cases, but always following the cyclic character of the 4D variables displayed in Tables 4.3–4.4, so that for the root variables $X_{1.4.4}$ and $X_{1.4.13}$ we have the cycle shown in Table 5.7.

Table 5.7.   5D variables generated by $X_{1.4}$.

$X_{1.4.4.52} - X_{1.4.3.51}$	$X_{1.4.2.50} - X_{1.4.1.49}$	$X_{1.4.8.56} - X_{1.4.7.55}$	$X_{1.4.6.54} - X_{1.4.5.53}$
$X_{1.4.13.61} - X_{1.4.14.62}$	$X_{1.4.15.63} - X_{1.4.16.64}$	$X_{1.4.9.57} - X_{1.4.10.58}$	$X_{1.4.11.59} - X_{1.4.12.60}$

For the last two cases we have the arrangement shown in Table 5.8.

Table 5.8.   5D variables generated by $X_{2.2}$ and $X_{2.3}$.

$X_{2.2.6.86} - X_{2.2.5.85}$	$X_{2.2.8.88} - X_{2.2.7.87}$	$X_{2.2.2.82} - X_{2.2.1.81}$	$X_{2.2.4.84} - X_{2.2.3.83}$
$X_{2.2.11.91} - X_{2.2.12.92}$	$X_{2.2.9.89} - X_{2.2.10.90}$	$X_{2.2.15.95} - X_{2.2.16.96}$	$X_{2.2.13.93} - X_{2.2.14.94}$
$X_{2.3.7.103} - X_{2.3.8.104}$	$X_{2.3.5.101} - X_{2.3.6.102}$	$X_{2.3.3.99} - X_{2.3.4.100}$	$X_{2.3.1.97} - X_{2.3.2.98}$
$X_{2.3.10.106} - X_{2.3.9.105}$	$X_{2.3.12.108} - X_{2.3.11.107}$	$X_{2.3.14.110} - X_{2.3.13.109}$	$X_{2.3.16.112} - X_{2.3.15.111}$

Of course, after the first eight main CAs, we have an inversion
of each of the two doublets (of a quadruplet) rooted in a common
3D variable. This scheme is evident with CA1.1.9, as displayed in
Table 5.9.

Table 5.9.   Main CA1.1.9.

$X_{1.1.1.9}$	0000	0000	0000	1000	1110	1111	1111	1111
$X_{1.1.1.248}$	0000	0000	1111	0111	0001	0000	1111	1111
$X_{1.1.16.8}$	0000	1111	0000	0111	0001	1111	0000	1111
$X_{1.1.16.249}$	0000	1111	1111	1000	1110	0000	0000	1111
$X_{1.4.4.60}$	0011	0011	0011	1011	0010	0011	0011	0011
$X_{1.4.4.197}$	0011	0011	1100	0100	1101	1100	0011	0011
$\bar{X}_{1.4.13.53}$	0011	1100	0011	0100	1101	0011	1100	0011
$X_{1.4.13.204}$	0011	1100	1100	1011	0010	1100	1100	0011
$X_{2.2.6.94}$	0101	0101	0101	1101	0100	0101	0101	0101
$X_{2.2.6.163}$	0101	0101	1010	0010	1011	1010	0101	0101
$\bar{X}_{2.2.11.83}$	0101	1010	0101	0010	1011	0101	1010	0101
$X_{2.2.11.174}$	0101	1010	1010	1101	0100	1010	1010	0101
$\bar{X}_{2.3.7.111}$	0110	0110	0110	1110	1000	1001	1001	1001
$X_{2.3.7.146}$	0110	0110	1001	0001	0111	0110	1001	1001
$\bar{X}_{2.3.10.98}$	0110	1001	0110	0001	0111	1001	0110	1001
$X_{2.3.10.159}$	0110	1001	1001	1110	1000	0110	0110	1001

Then, always following Table 4.3, after the the first 16 CAs, we
have an inversion of blocks of rows. Thus, the variables $X_{1.1.x.y}$ range
now over rows 17–32 and $X_{1.4.x.y}$ over rows 33–48, while variables
$X_{2.2.x.y}$ and $X_{2.3.x.y}$ range over rows 65–80 and 113–128, respectively

(where $1 \leq x \leq 16, 1 \leq y \leq 256$). Thus, for CA1.1.17 we get the combinations shown in Table 5.10.

Table 5.10. Main CA1.1.17.

$X_{1.1.1.17}$	0000	0000	0001	0000	1111	0111	1111	1111
$X_{1.1.1.240}$	0000	0000	1110	1111	0000	1000	1111	1111
$X_{1.1.16.32}$	0000	1111	0001	1111	0000	0111	0000	1111
$X_{1.1.16.225}$	0000	1111	1110	0000	1111	1000	0000	1111
$X_{1.4.4.36}$	0011	0011	0010	0011	0011	1011	0011	0011
$X_{1.4.4.221}$	0011	0011	1101	1100	1100	0100	0011	0011
$X_{1.4.13.45}$	0011	1100	0010	1100	1100	1011	1100	0011
$X_{1.4.13.212}$	0011	1100	1101	0011	0011	0100	1100	0011
$X_{2.2.6.70}$	0101	0101	0100	0101	0101	1101	0101	0101
$X_{2.2.6.187}$	0101	0101	1011	1010	1010	0010	0101	0101
$X_{2.2.11.75}$	0101	1010	0100	1010	1010	1101	1010	0101
$X_{2.2.11.182}$	0101	1010	1011	0101	0101	0010	1010	0101
$X_{2.3.7.119}$	0110	0110	0111	0110	1001	0001	1001	1001
$X_{2.3.7.138}$	0110	0110	1000	1001	0110	1110	1001	1001
$X_{2.3.10.122}$	0110	1001	0111	1001	0110	0001	0110	1001
$X_{2.3.10.135}$	0110	1001	1000	0110	1001	1110	0110	1001

Note that each of those major blocks of 16 main CAs correspond to the general characters of each of the 4D main CAs. The third major block is always even (CA2) but permuting the blocks of rows. For instance, CA1.1.33 is shown in Table 5.11.

Table 5.11. Main CA1.1.33.

$X_{1.1.1.33}$	0000	0000	0010	0000	1111	1011	1111	1111
$X_{1.1.1.224}$	0000	0000	1101	1111	0000	0100	1111	1111
$X_{1.1.16.48}$	0000	1111	0010	1111	0000	1011	0000	1111
$X_{1.1.16.209}$	0000	1111	1101	0000	1111	0100	0000	1111
$X_{1.4.4.20}$	0011	0011	0001	0011	0011	0111	0011	0011
$X_{1.4.4.237}$	0011	0011	1110	1100	1100	1000	0011	0011
$X_{1.4.13.29}$	0011	1100	0001	1100	1100	0111	1100	0011
$X_{1.4.13.228}$	0011	1100	1110	0011	0011	1000	1100	0011
$X_{2.2.6.118}$	0101	0101	0111	0101	0101	0001	0101	0101
$X_{2.2.6.139}$	0101	0101	1000	1010	1010	1110	0101	0101
$X_{2.2.11.123}$	0101	1010	0111	1010	1010	0001	1010	0101
$X_{2.2.11.134}$	0101	1010	1000	0101	0101	1110	1010	0101
$X_{2.3.7.71}$	0110	0110	0100	0110	1001	1101	1001	1001
$X_{2.3.7.186}$	0110	0110	1011	1001	0110	0010	1001	1001
$X_{2.3.10.74}$	0110	1001	0100	1001	0110	1101	0110	1001
$X_{2.3.10.183}$	0110	1001	1011	0110	1001	0010	0110	1001

The fourth major block is again odd (CA1), with similar permutation of the rows' blocks, as is clear by considering CA1.1.49, as displayed in Table 5.12.

Table 5.12. Main CA1.1.49.

$X_{1.1.1.49}$	0000	0000	0011	0000	1111	0011	1111	1111
$X_{1.1.1.208}$	0000	0000	1100	1111	0000	1100	1111	1111
$X_{1.1.16.64}$	0000	1111	0011	1111	0000	0011	0000	1111
$X_{1.1.16.193}$	0000	1111	1100	0000	1111	1100	0000	1111
$X_{1.4.4.4}$	0011	0011	0000	0011	0011	1111	0011	0011
$X_{1.4.4.253}$	0011	0011	1111	1100	1100	0000	0011	0011
$X_{1.4.13.13}$	0011	1100	0000	1100	1100	1111	1100	0011
$X_{1.4.13.244}$	0011	1100	1111	0011	0011	0000	1100	0011
$X_{2.2.6.102}$	0101	0101	0110	0101	0101	1001	0101	0101
$X_{2.2.6.155}$	0101	0101	1001	1010	1010	0110	0101	0101
$X_{2.2.11.107}$	0101	1010	0110	1010	1010	1001	1010	0101
$X_{2.2.11.150}$	0101	1010	1001	0101	0101	0110	1010	0101
$X_{2.3.7.87}$	0110	0110	0101	0110	1001	0101	1001	1001
$X_{2.3.7.170}$	0110	0110	1010	1001	0110	1010	1001	1001
$X_{2.3.10.90}$	0110	1001	0101	1001	0110	0101	0110	1001
$X_{2.3.10.167}$	0110	1001	1010	0110	1001	1010	0110	1001

The fifth major block is even with exchange of the two halves. For instance, CA1.1.65 is shown in Table 5.13.

Table 5.13. Main CA1.1.65.

$X_{1.1.1.65}$	0000	0000	0100	0000	1111	1101	1111	1111
$X_{1.1.1.192}$	0000	0000	1011	1111	0000	0010	1111	1111
$X_{1.1.16.80}$	0000	1111	0100	1111	0000	1101	0000	1111
$X_{1.1.16.177}$	0000	1111	1011	0000	1111	0010	0000	1111
$X_{1.4.4.116}$	0011	0011	0111	0011	0011	0001	0011	0011
$X_{1.4.4.141}$	0011	0011	1000	1100	1100	1110	0011	0011
$X_{1.4.13.125}$	0011	1100	0111	1100	1100	0001	1100	0011
$X_{1.4.13.132}$	0011	1100	1000	0011	0011	1110	1100	0011
$X_{2.2.6.22}$	0101	0101	0001	0101	0101	0111	0101	0101
$X_{2.2.6.235}$	0101	0101	1110	1010	1010	1000	0101	0101
$X_{2.2.11.27}$	0101	1010	0001	1010	1010	0111	1010	0101
$X_{2.2.11.230}$	0101	1010	1110	0101	0101	1000	1010	0101
$X_{2.3.7.39}$	0110	0110	0010	0110	1001	1011	1001	1001
$X_{2.3.7.218}$	0110	0110	1101	1001	0110	0100	1001	1001
$X_{2.3.10.42}$	0110	1001	0010	1001	0110	1011	0110	1001
$X_{2.3.10.215}$	0110	1001	1101	0110	1001	0100	0110	1001

The sixth major block is odd and also with exchange of the two halves, as CA1.1.81 shows (Table 5.14).

Table 5.14. Main CA1.1.81.

$X_{1.1.1.81}$	0000	0000	0101	0000	1111	0101	1111	1111
$X_{1.1.1.176}$	0000	0000	1010	1111	0000	1010	1111	1111
$X_{1.1.16.96}$	0000	1111	0101	1111	0000	0101	0000	1111
$X_{1.1.16.161}$	0000	1111	1010	0000	1111	1010	0000	1111
$X_{1.4.4.100}$	0011	0011	0110	0011	0011	1001	0011	0011
$X_{1.4.4.157}$	0011	0011	1001	1100	1100	0110	0011	0011
$X_{1.4.13.109}$	0011	1100	0110	1100	1100	1001	1100	0011
$X_{1.4.13.148}$	0011	1100	1001	0011	0011	0110	1100	0011
$X_{2.2.6.6}$	0101	0101	0000	0101	0101	1111	0101	0101
$X_{2.2.6.251}$	0101	0101	1111	1010	1010	0000	0101	0101
$X_{2.2.11.11}$	0101	1010	0000	1010	1010	1111	1010	0101
$X_{2.2.11.246}$	0101	1010	1111	0101	0101	0000	1010	0101
$X_{2.3.7.55}$	0110	0110	0011	0110	1001	0011	1001	1001
$X_{2.3.7.202}$	0110	0110	1100	1001	0110	1100	1001	1001
$X_{2.3.10.58}$	0110	1001	0011	1001	0110	0011	0110	1001
$X_{2.3.10.199}$	0110	1001	1100	0110	1001	1100	0110	1001

The seventh major block is odd and displays the two inversions, as is evident for CA1.1.97 (Table 5.15).

Table 5.15. Main CA1.1.97.

$X_{1.1.1.97}$	0000	0000	0110	0000	1111	1001	1111	1111
$X_{1.1.1.160}$	0000	0000	1001	1111	0000	0110	1111	1111
$X_{1.1.16.112}$	0000	1111	0110	1111	0000	1001	0000	1111
$X_{1.1.16.145}$	0000	1111	1001	0000	1111	0110	0000	1111
$X_{1.4.4.84}$	0011	0011	0101	0011	0011	0101	0011	0011
$X_{1.4.4.173}$	0011	0011	1010	1100	1100	1010	0011	0011
$X_{1.4.13.93}$	0011	1100	0101	1100	1100	0101	1100	0011
$X_{1.4.13.164}$	0011	1100	1010	0011	0011	1010	1100	0011
$X_{2.2.6.54}$	0101	0101	0011	0101	0101	0011	0101	0101
$X_{2.2.6.203}$	0101	0101	1100	1010	1010	1100	0101	0101
$X_{2.2.11.59}$	0101	1010	0011	1010	1010	0011	1010	0101
$X_{2.2.11.198}$	0101	1010	1100	0101	0101	1100	1010	0101
$X_{2.3.7.7}$	0110	0110	0000	0110	1001	1111	1001	1001
$X_{2.3.7.250}$	0110	0110	1111	1001	0110	0000	1001	1001
$X_{2.3.10.10}$	0110	1001	0000	1001	0110	1111	0110	1001
$X_{2.3.10.247}$	0110	1001	1111	0110	1001	0000	0110	1001

The eighth major block displays the same behaviour as the previous one but is even, as is clear when considering main CA1.1.113, as displayed in Table 5.16.

Table 5.16.   Main CA1.1.113.

$X_{1.1.1.113}$	0000	0000	0111	0000	1111	0001	1111	1111
$X_{1.1.1.144}$	0000	0000	1000	1111	0000	1110	1111	1111
$X_{1.1.16.128}$	0000	1111	0111	1111	0000	0001	0000	1111
$X_{1.1.16.129}$	0000	1111	1000	0000	1111	1110	0000	1111
$X_{1.4.4.68}$	0011	0011	0100	0011	0011	1101	0011	0011
$X_{1.4.4.189}$	0011	0011	1011	1100	1100	0010	0011	0011
$X_{1.4.13.77}$	0011	1100	0100	1100	1100	1101	1100	0011
$X_{1.4.13.180}$	0011	1100	1011	0011	0011	0010	1100	0011
$X_{2.2.6.38}$	0101	0101	0010	0101	0101	1011	0101	0101
$X_{2.2.6.219}$	0101	0101	1101	1010	1010	0100	0101	0101
$X_{2.2.11.43}$	0101	1010	0010	1010	1010	1011	1010	0101
$X_{2.2.11.214}$	0101	1010	1101	0101	0101	0100	1010	0101
$X_{2.3.7.23}$	0110	0110	0001	0110	1001	0111	1001	1001
$X_{2.3.7.234}$	0110	0110	1110	1001	0110	1000	1001	1001
$X_{2.3.10.26}$	0110	1001	0001	1001	0110	0111	0110	1001
$X_{2.3.10.231}$	0110	1001	1110	0110	1001	1000	0110	1001

For completeness, I also give the table of the latter CA (Table 5.17).

Table 5.17.   Main CA1.1.128.

$X_{1.1.1.128}$	0000	0000	0111	1111	0000	0001	1111	1111
$X_{1.1.1.129}$	0000	0000	1000	0000	1111	1110	1111	1111
$X_{1.1.16.113}$	0000	1111	0111	0000	1111	0001	0000	1111
$X_{1.1.16.144}$	0000	1111	1000	1111	0000	1110	0000	1111
$X_{1.4.4.77}$	0011	0011	0100	1100	1100	1101	0011	0011
$X_{1.4.4.180}$	0011	0011	1011	0011	0011	0010	0011	0011
$X_{1.4.13.68}$	0011	1100	0100	0011	0011	1101	1100	0011
$X_{1.4.13.189}$	0011	1100	1011	1100	1100	0010	1100	0011
$X_{2.2.6.43}$	0101	0101	0010	1010	1010	1011	0101	0101
$X_{2.2.6.214}$	0101	0101	1101	0101	0101	0100	0101	0101
$X_{2.2.11.38}$	0101	1010	0010	0101	0101	1011	1010	0101
$X_{2.2.11.219}$	0101	1010	1101	1010	1010	0100	1010	0101
$X_{2.3.7.26}$	0110	0110	0001	1001	0110	0111	1001	1001
$X_{2.3.7.231}$	0110	0110	1110	0110	1001	1000	1001	1001
$X_{2.3.10.23}$	0110	1001	0001	0110	1001	0111	0110	1001
$X_{2.3.10.234}$	0110	1001	1110	1001	0110	1000	0110	1001

In summary, we have the combinatorial series for the 128 CA1.1's displayed in Table 5.18.

Table 5.18. Combinatorics of 5D variables. The numbers represent the range of the variable y as well as the labels of the 4D SD variables. The number x is determined by the relative doublet.

$X_{1.1.x.y}$	1-16	17-32	33-48	49-64	65-80	81-96	97-112	113-128
	256-241	240-225	224-209	208-193	192-177	176-161	160-145	144-129
$X_{1.4.x.y}$	49-64	33-48	17-32	1-16	113-128	97-112	81-96	65-80
	208-193	224-209	240-225	256-241	144-129	160-145	176-161	192-177
$X_{2.2.x.y}$	81-96	65-80	113-128	97-112	17-32	1-16	49-64	33-48
	176-161	192-177	144-129	160-145	240-225	256-241	208-193	224-209
$X_{2.3.x.y}$	97-112	113-128	65-80	81-96	33-48	49-64	1-16	17-32
	160-145	144-129	192-177	176-161	224-209	208-193	256-241	240-225

Thus, we get the 16 patterns of the fourth index shown in Table 5.19.

Table 5.19. The 16 patterns that cover the 5D CAs (each represents a CA). For instance, CA1.1.1 instantiate pattern A and CA1.1.2 pattern B. The patterns follow those of the 4D CAs with relative indices inversions and summation (Table 4.4). The numbers in bold start and end each of the blocks of Table 5.18 (on the same row half block, which together with the other two values in bold on the same columns complete the block). This is the only arrangement that satisfies both requirements that (i) the sum for each pair of numbers in complementary columns (in any row) is 257, and (ii) each column of any CA has an equal number of 0s and 1s apart from the first and last ones.

A	1	**16**	52	61	86	91	103	106	151	154	166	171	196	205	**241**	**256**
B	2	15	51	62	85	92	104	105	152	153	165	172	195	206	242	255
C	3	14	50	63	88	89	101	108	149	156	168	169	194	207	243	254
D	4	13	**49**	**64**	87	90	102	107	150	155	167	170	**193**	**208**	244	253
E	5	12	56	57	82	95	99	110	147	158	162	175	199	201	245	252
F	6	11	55	58	**81**	**96**	100	109	148	157	**161**	**176**	200	202	246	251
G	7	10	54	59	84	93	**97**	**112**	**145**	**160**	164	173	198	203	247	250
H	8	9	53	60	83	94	98	111	146	159	163	174	197	204	248	249
I	**17**	**32**	36	45	70	75	119	122	135	138	182	187	212	221	**225**	**240**
J	18	31	35	46	69	76	120	121	136	137	181	188	211	222	226	239
K	19	30	34	47	72	73	117	124	133	140	184	185	210	223	227	238
L	20	29	**33**	**48**	71	74	118	123	134	139	183	186	**209**	**224**	228	237
M	21	28	40	41	66	79	115	126	131	142	178	191	216	217	229	236
N	22	27	39	42	**65**	**80**	116	125	132	141	**177**	**192**	215	218	230	235
O	23	26	38	43	68	77	**113**	**128**	**129**	**144**	180	189	214	219	231	234
P	24	25	37	44	67	78	114	127	130	143	179	190	213	220	232	233

These patterns are repeated over and over, but with changes depending on which are the leading variables, such as the SD variables that start and end synchronously with a block (look at the first two double columns of Table 5.18). For instance, for the block 33-64 (whose leading variables are $X_{1.4.x.y}$) we have the sequence *LKJIPONM DCBAHGFE*, and for the block 65-96 ($X_{2.2.x.y}$) we have *NMPOJILK FEHGBADC*. Of course, all other seven groups of 128 main CAs are fully parallel to what is seen here, and the same is true for all other derivations of CA2 (other $8 \times 128 = 1024$ main CAs). Figure 5.1 resumes this whole structure of 2048 main CAs, in agreement with Equation (4.5). The number of main CAs makes a series starting with number 2 for $\mathscr{B}_3$, where the number x of them for a $\mathscr{B}_n$ algebra (with $n \geq 3$) is given by the product of the number of SD variables, and the number of CAs both of the $n - 1$-th algebra:

$$2 \times 1, 2^3 \times 2, 2^7 \times 2^4, 2^{15} \times 2^{11}, \ldots \tag{5.4}$$

5.2. SCAs

For a single CA we could obtain 4,368 combinations of five variables. However, the combinatorial law for getting sets of five variables is given by the combinatorial series determining the number of columns having a particular number of 1s:

$$\binom{5}{0}, \binom{5}{1}, \binom{5}{2}, \binom{5}{3}, \binom{5}{4}, \binom{5}{5}, \tag{5.5}$$

that is, we have the series 1, 5, 10, 10, 5, 1. Of course, all restrictions already considered for $\mathscr{B}_4$ in Section 4.2 are also valid here. Moreover, there are additional ones. First of all, let us partition the SCAs into two big groups: those with doublets and those without doublets.

Let us first consider SCAs with doublets. Here, we need to consider that for $\mathscr{B}_5$ we have characteristic recurrent patterns of the 1-chunks

	codes 1.1.*z*	codes 1.2.*z*	codes 1.3.*z*	codes 1.4.*z*	codes 1.5.*z*	codes 1.6.*z*	codes 1.7.*z*	codes 1.8.*z*	codes 2.1.*z*	codes 2.2.*z*	codes 2.3.*z*	codes 2.4.*z*	codes 2.5.*z*	codes 2.6.*z*	codes 2.7.*z*	codes 2.8.*z*
codes *x*.*y*.1-16																
codes *x*.*y*.17-32																
codes *x*.*y*.33-48																
codes *x*.*y*.49-64																
codes *x*.*y*.65-80																
codes *x*.*y*.81-96																
codes *x*.*y*.97-112																
codes *x*.*y*.113-128																

Figure 5.1. The 2048 5D main CAs covering all 32,768 5D SD variables. The variable x ranges over 1 and 2, the variable y over 1-128, and the variable z over 1-8.

composing the IDs. For instance, if we take CA1.1.1 (Table 5.2), we see that we have:

- $X_{1.1.x.y}$: 4-4, 2-2-2-2; 1-1-1-1-1-1-1-1, 1-2-1-1-2-1;
- $X_{1.4.x.y}$: 8, 2-4-2; 1-1-1-2-1-1-1, 1-2-2-2-1.
- $X_{2.2.x.y}$: 8, 2-4-2; 1-1-1-2-1-1-1, 1-2-2-2-1;
- $X_{2.3.x.y}$: 4-4, 2-2-2-2; 1-1-1-1-1-1-1-1, 1-2-1-1-2-1.

The numbers indicate the subsequent positions of either 1s or 0s. As is evident, each pattern is doubled and the last two rows are the mirror of the first two. Of course, each CA shows its distinctive patterns. For instance, CA1.1.2 (Table 5.3) displays following patterns:

- 3-1-1-3, 2-1-1-1-1-2; 1-1-1-1-1-1-1-1, 1-2-1-1-2-1;
- 3-1-1-3, 2-1-1-1-1-2; 1-1-1-1-1-1-1-1, 1-2-1-1-2-1.
- 3-1-1-3, 2-1-1-1-1-2; 1-1-1-1-1-1-1-1, 1-2-1-1-2-1;
- 3-1-1-3, 2-1-1-1-1-2; 1-1-1-1-1-1-1-1, 1-2-1-1-2-1.

Note that there are no such patterns for $\mathscr{B}_4$ as only CA1.1 shows recurrent patterns of 1-chunks, and when we use 2 digits (half chunk), almost everywhere we would have the form 1-1-1-1-1-1-1-1. Coming back to $\mathscr{B}_5$, when we have doublets we need to take into account that we cannot have two couples with identical patterns, although one occurs in the first set of eight SD variables and the other in the other. With such restrictions, we have (where a semicolon separates the first set of eight SD variables from the second one):

- A doublet with all bs: a-b,b,b; b. a-b,b; b,b. a-b; b,b,b. This makes $4 \times 3 \times 4 + 4 \times 3 \times 5 + 4 \times 4 = 124$ combinations, plus another 124 by exchanging the left and right sides, for a total number of 248 combinations.
- A doublet with all as: a-b,a,a; a. a-b,a; a,a. a-b; a,a,a. This makes $4 \times 3 \times 4 + 4 \times 3 \times 5 + 4 \times 4 = 124$, plus another 124 for a total number of 248 combinations.
- A doublet with a single a: a-b,a,b; b. a-b,a; b,b. a-b; a,b,b. a-b,b; a,b. a-b,b,b; a. This makes $4 \times 6 \times 4 + 4 \times 3 \times 6 + 4 \times 4 \times 3 + 4 \times 3 \times 4 \times 2 + 4 \times 3 \times 4 = 360$ combinations, plus another 360 for a total number 720 combinations.

- A doublet with two as: a-b,a,a; b. a-b,a; a,b. a-b; a,a,b. a-b,b; a,a. a-b,a,b; a. a-b; a,a,b. This makes $4 \times 3 \times 4 + 4 \times 3 \times 4 \times 2 + 4 \times 6 \times 2 + 4 \times 3 \times 6 + 4 \times 6 \times 4 + 4 \times 6 \times 2 = 432$, plus another 408, for a total number of 816 combinations.

The total number of combinations of this group is 2032.

Let us consider the second group (no doublets). Here, we cannot have two as and two bs occurring in either the first or the second set. Neither can we have two as and two bs occurring even in the two sets, but with the two as and the two bs pertaining to the same quadruplets, respectively (i.e. having a common 3D SD variable), or one a and one b pertaining to the same quadruplet and one a and one b, again pertaining to the same quadruplet. Thus, the allowed combinations are the following ones:

- A single a: a,b,b,b; b. a,b,b; b,b. a,b; b,b,b. This makes $4 \times 4 + 4 \times 3 \times 6 + 4 \times 3 \times 4 = 136$ combinations. By exchanging left and right we get another 136 combinations, for a total number of 272 combinations.
- Two as: a; a,b,b,b. a,b;a,b,b. This makes $4 \times 4 + 4 \times 3 \times 4 \times 2 = 112$ combinations, plus another 112 by exchanging left and right, for a total number of 224 combinations.
- Three as: a,a; a,b,b. a,b; a,a,b. a,a,a,b; b. This makes $6 \times 4 \times 3 + 4 \times 3 \times 6 \times 2 + 4 \times 4 = 232$ combinations, plus another 232, for a total number of 464 combinations.

The total number of all combinations for this group is 960. Thus, we get as a whole 2992 SCAs for every CA. The total number of the SCAs of $\mathscr{B}_5$ is then $2992 \times 2048 = 6,127,616$. It is interesting to note that the total number of SCAs for any algebra with dimension $n \geq 3$ is proportional to $2^{m(n-1)-1}$. Indeed, we have the series

$$2^3, 7 \cdot 2^7, 11 \cdot 17 \cdot 2^{15}, \ldots \tag{5.6}$$

5.3. Examples of Nodes

In this section I focus on a single codification and make use of the SCA displayed in Table 5.20. For the sake of simplicity, I label these

Table 5.20.   Example of 5D SCA.

$X_{1.1.1.1}$	0000	0000	0000	0000	1111	1111	1111	1111
$X_{1.1.16.241}$	0000	1111	1111	0000	1111	0000	0000	1111
$X_{1.4.4.205}$	0011	0011	1100	1100	1100	1100	0011	0011
$X_{1.4.13.196}$	0011	1100	1100	0011	0011	1100	1100	0011
$X_{2.2.6.171}$	0101	0101	1010	1010	1010	1010	0101	0101

Table 5.21.   Levels 1/32 and 2/32.

0000	0000	0000	0000	0000	0000	0000	0010	$X_1 X_{16} X_4 X_{13} X_6'$
0000	0001	0000	0000	0000	0000	0000	0000	$X_1' X_{16} X_4 X_{13}' X_6$
0000	0011	0000	0000	0000	0000	0000	0000	$X_1' X_{16} X_4 X_{13}'$
0000	0000	0000	0000	0100	0000	0000	0010	$X_1 X_{16} X_4 X_6'$
0000	0001	0000	1000	0000	0000	0000	0000	$X_1' X_4 X_{13}' X_6$
0100	0000	0000	0000	0000	0000	0000	0010	$X_1 \sim X_{16} \sim X_4 \sim X_{13} \sim X_6'$

variables using, as usual, the forelast index: $X_1, X_{16}, X_4, X_{13}, X_6$.
Moreover, I restrict the following summary to the nodes from Level
1/32 to Level 16/32.

Levels 1/32 and 2/32 have 32 and

$$\binom{32}{2} = 496 \tag{5.7}$$

nodes, respectively. In Table 5.21 are some examples of Levels 1/32
and 2/32, displaying the forms (i) a × (a × e), (ii) a × e, and a penta–
equivalence (see Figure 4.5).

Levels 3/32 and 4/32 have

$$\binom{32}{3} = 4,960 \quad \text{and} \quad \binom{32}{4} = 35,960 \tag{5.8}$$

nodes, respectively. In Table 5.22 I summarised these two levels,
respectively displaying the forms (i) a × (a × f), a × (a × g) =
(a × e) + (a × (a × e)), (ii) e, (a × e) + (a × e), and a quaternary
equivalence. Levels 5/32 and 6/32 have

$$\binom{32}{5} = 201,376 \quad \text{and} \quad \binom{32}{6} = 906,192 \tag{5.9}$$

Table 5.22. Levels 3/32 and 4/32.

0000	0011	0000	1000	0000	0000	0000	0000	$X_1'X_4X_{13}'(X_{16}+X_6)$
0000	0010	1100	0000	0000	0000	0000	0000	$X_1'X_{16}X_4(X_{13}+X_6')$
0000	0000	0000	0000	1100	0000	0000	0010	$X_1X_{16}X_4X_{13}'+X_1X_{16}X_4X_{13}X_6'$
0000	0000	0000	0000	1010	0000	0000	0101	$X_1X_{16}X_6$
0000	0011	1100	0000	0000	0000	0000	0000	$X_1'X_{16}X_4$
0000	0011	0000	0000	1100	0000	0000	0000	$X_{16}X_4X_{13}'$
0000	0001	1011	0000	0000	0000	0000	0000	$X_1'X_{16}X_4X_6+X_1'X_{16}X_4'X_{13}'$
0100	0000	0010	0000	0000	0100	0000	0010	$X_1\sim X_4\sim X_{13}\sim X_6'$

Table 5.23. Levels 5/32 and 6/32.

0000	0011	1100	0000	0000	0000	0000	0010	$X_{16}X_4(X_1'+X_{13}X_6')$
0000	0000	0000	0000	1100	0000	0000	0111	$X_1X_{16}(X_4+X_{13}'X_6)$
1000	0010	0001	0100	0000	0000	0001	0000	$X_1'X_{13}'X_6'+X_1X_{16}'X_4X_{13}'X_6$
0000	0011	0000	0000	0000	0000	0001	0011	$X_1X_{16}X_4X_{13}+X_1'X_{16}X_4X_{13}'$
								$+X_1X_{16}'X_4X_{13}'X_6$
0000	0000	0000	0000	0111	0000	0000	1011	$X_1X_{16}(X_{13}+X_6')$
0100	0001	0010	1000	0100	0000	0000	0010	$X_1'X_{13}'X_6+X_1X_{16}X_4X_6'$
0000	0001	0000	1000	1100	0000	0000	0011	$X_1'X_4X_{13}'X_6+X_1X_{16}X_4X_{13}$
								$+X_1X_{16}X_4X_{13}'$

Table 5.24. Levels 7/32 and 8/32.

0011	0000	1100	0000	0000	1000	0000	0011	$X_4X_{13}(X_1+X_{16}+X_6)$
0001	0000	0000	0000	0111	0000	0000	1011	$X_1X_{16}(X_{13}+X_6')$
								$+X_1'X_{16}'X_4X_{13}X_6$
1001	0010	0001	0100	0000	0000	0000	0011	$X_1'X_{13}'X_6'+X_1X_{16}X_4X_{13}$
								$+X_1'X_{16}'X_4X_{13}X_6$
0000	0000	0000	0000	1100	1100	0011	0011	X_1X_4
0100	0001	0010	1000	1000	0010	0001	0100	$X_{13}'X_6$
0100	0001	0010	1000	1100	0000	0000	0011	$X_1X_{16}X_4+X_1'X_{13}'X_6$
1000	0011	0001	1100	0000	0000	0000	0011	$X_1'X_{13}'X_6'+X_1X_{16}X_4X_{13}$
								$+X_1'X_4X_{13}'X_6$
1100	0000	0000	0011	1100	0000	0000	0011	$X_1\sim X_{16}\sim X_{4a}$

nodes, respectively. In Table 5.23 are shown Levels 5/32 and 6/32 with the forms (i) $a\times(a\times h)$, $a\times(b+(a\times e))=e+(a\times(a\times e))$, $(a\times e)+(a\times(a\times g))$, and (ii) $a\times f$, $a\times g$, $(a\times e)+(a\times e)+(a\times e)$. Levels 7/32 and 8/32 show

$$\binom{32}{7}=3,365,856 \quad \text{and} \quad \binom{32}{8}=10,519,300 \qquad (5.10)$$

Table 5.25.   Levels 9/32 and 10/32.

0000 0000 0100 0000 1111 0000 0000 1111	$X_{16}(X_1 + X_4 X_{13} X_6')$
0000 1110 1111 0000 0010 0000 0000 0100	$X_{16}(X_1' + X_4' X_6)(X_1 + X_4' + X_{13} + X_6')$
0000 1111 1011 0000 0001 0000 0000 1000	$X_{16}(X_1' + X_4')(X_1' + X_4 + X_6')$
	$\times (X_1 + X_4' + X_{13}' + X_6)$
0001 0000 0000 0000 1111 0000 0000 1111	$X_1 X_{16} + X_1' X_{16}' X_4 X_{13} X_6$
0000 0000 0000 0000 1100 1110 0011 0111	$X_1(X_4 + X_{13}' X_6)$
0100 0001 0010 1000 1100 0010 0001 0110	$X_{13}' X_6 + X_1 X_{16} X_4 X_6'$
0000 1000 0000 0001 0010 1110 0100 0111	$X_1 X_4' X_6 + X_1 X_4 X_{13}$
	$+ X_1' X_4' X_{13} X_6'$
0011 0001 0000 1000 1110 0000 0000 0111	$X_1 X_{16} X_6 + X_1' X_4 X_{13}' X_6$
	$+ X_1 X_{16} X_4 X_6' + X_1' X_{16}' X_4 X_{13}$

Table 5.26.   Levels 11/32 and 12/32.

0000 1111 1111 0000 0100 0000 0000 0011	$X_1' X_{16} + X_{16} X_4(X_{13} + X_6')$
1010 0000 0000 0001 0000 1111 1111 0000	$X_{16}'(X_1 + X_6')(X_1 + X_4' + X_{13} + X_6)$
0000 1110 1101 0000 0011 0000 0000 1011	$X_{16}(X_1' + X_4 + X_{13})(X_1 + X_{13} + X_6')$
	$\times (X_1' + X_4' + X_{13} + X_6)$
1100 0011 0011 1100 0100 0000 0001 0010	$X_1' X_{13}' + X_1 X_{16} X_4 X_6'$
	$+ X_1 X_{16}' X_4 X_{13}' X_6$
0100 0100 0010 0010 1010 1010 0101 0101	$X_6(X_1 + X_4')$
0000 1000 0000 0001 1101 1100 1011 0011	$X_1 X_4 + X_4' X_{13} X_6'$
0001 0010 1000 0100 1100 1010 0011 0101	$X_4 X_{13} X_6 + X_4 X_{13}' X_6' + X_1 X_{13}' X_6$
0000 1011 0000 0001 0010 1110 0100 0111	$X_1 X_4' X_6 + X_1 X_4 X_{13}$
	$+ X_1' X_4' X_{13} X_6' + X_1' X_{16} X_4 X_{13}'$

nodes, respectively. Table 5.24 displays Levels 7/32 and 8/32 with relative forms, namely, (i) a × (a × j), a × (f + (a × e)), a × (g + (a × e)) = e + (a × e) + (a × (a × e)), (ii) b, e + e. a × (b + (e + e)), and a ternary equivalence.

Levels 9/32 and 10/32 have

$$\binom{32}{9} = 28,048,800 \quad \text{and} \quad \binom{32}{10} = 64,512,240 \qquad (5.11)$$

nodes, respectively. Table 5.25 displays Levels 9/32 and 10/32 with the logical forms (i) a × (a + e), a × (h × (a + j)), a × (i × (a + j)), b + (a × (a × e)), and (ii) a × h, b + (a × e), e + (a × g) = (e + e) + (a × e), (a × g) + ((a × e) + (a × e)) = e + (a × e) + (a × e) + (a × e).

Levels 11/32 and 12/32 have

$$\binom{32}{11} = 129,024,480 \quad \text{and} \quad \binom{32}{12} = 225,792,840 \qquad (5.12)$$

Table 5.27. Levels 13/32 and 14/32.

0000 1011 1100 0000 1111 0000 0000 1111	$X_{16}(X_1 + X_4 + X_{13}X_6')$
0000 1111 1111 0000 1101 0000 0000 1001	$X_{16}(X_1' + X_4 + X_6')$
	$\times (X_1' + X_4' + X_{13}' + X_6)$
0100 0101 0010 0010 1010 1010 0101 0101	$X_6(X_1 + X_4') + X_1'X_{16}X_4X_{13}'X_6$
0001 1000 0000 0001 1101 1100 1011 0011	$X_1X_4 + X_4'X_{13}X_6' + X_1'X_{16}'X_4X_{13}X_6$
0011 0011 1100 1100 0100 1100 0010 0011	$X_4(X_1' + X_{13} + X_6')$
0100 0111 0010 0010 1010 1010 0101 0101	$X_6(X_1 + X_4') + X_1'X_{16}X_4X_{13}'$
0000 1011 0000 0001 1101 1100 1011 0011	$X_1X_4 + X_4'X_{13}X_6' + X_1'X_{16}X_4X_{13}'$
0011 0011 1111 0000 0111 0000 0000 1011	$X_1X_{16}(X_{13} + X_6') + X_1'X_{16}X_4$
	$+X_1'X_{16}'X_4X_{13} + X_1'X_{16}X_4'X_{13}'$

Table 5.28. Levels 15/32 and 16/32.

0000 1101 1111 0000 1111 0000 0000 1111	$X_{16}(X_1 + X_4' + X_{13} + X_6)$
0001 1000 0011 0001 1101 1100 1011 0011	$X_1X_4 + X_4'X_{13}X_6'$
	$+X_1'X_{16}X_4'X_{13}' + X_1'X_{16}'X_4X_{13}X_6$
1011 0011 1100 1100 0100 1100 0010 0011	$X_4(X_1' + X_{13} + X_6') + X_1'X_{16}'X_4'X_{13}'X_6'$
0101 0100 0010 0010 1110 1010 0101 0111	$X_6(X_1 + X_4') + X_1X_{16}X_4X_6'$
	$+X_1'X_{16}'X_4X_{13}X_6$
0011 0011 1100 1100 1100 1100 0011 0011	X_4
1100 0011 0011 1100 1111 0000 0000 1111	$X_1X_{16} + X_1'X_{13}'$
0011 0111 1110 1100 1110 0000 0000 0111	$X_{16}X_4 + X_1'X_{16}'X_4 + X_{16}X_4'X_6$
1000 1011 0001 1101 1001 0010 1001 0111	$X_4'X_{13}X_6' + X_1X_{13}'X_6 + X_1'X_{13}'X_6'$
	$+X_1X_{16}X_4X_{13} + X_1'X_4X_{13}'X_6$
1111 0000 0000 1111 1111 0000 0000 1111	$X_1 \sim X_{16}$

nodes, respectively. Table 5.26 displays some nodes and the relative logical forms (i) $a \times (a + f)$, $a \times (c \times (a + j))$, $a \times (j \times (a + i))$, $(b + (a \times e)) + (a \times (a \times e))$, and (ii) f, g, $e + (e + e)$, $(a \times e) + (e + (a \times g))$.

Levels 13/32 and 14/32 have

$$\binom{32}{13} = 347,373,600 \quad \text{and} \quad \binom{32}{14} = 471,435,600 \qquad (5.13)$$

nodes, respectively. Table 5.27 displays the following logical forms: (i) $a \times (a + h)$, $a \times (a + i)$, $f + (a \times (a \times e))$, $g + (a \times (a \times e))$, and (ii) $a \times j$, $f + (a \times e)$, $g + (a \times e)$, $(a \times e) + (a \times f) + (a \times g)$.

Levels 15/32 and 16/32 have

$$\binom{32}{15} = 565,722,720 \quad \text{and} \quad \binom{32}{16} = 601,080,390 \qquad (5.14)$$

	$\mathcal{B}_4$	$\mathcal{B}_5$
1/2	b+(e+e)=cx(jxj)	(e+(e+e))+((axe)+(axe))
15/32		ax(a+j) (axj)+(ax(axe)) (g+(axe))+(ax(axe)) (f+(axe))+(ax(axe))
7/16	axj f+(axe) g+(axe)	(axe)+(axf)+(axg)
13/32		ax(a+i) ax(a+h) f+(ax(axe)) g+(ax(axe))
3/8	e+(e+e)	(axe)+(e+(axg))
11/32		ax(a+f) ax(cx(a+j)) ax(jx(a+i)) (e+(axg))+(ax(axe))
5/16	axh b+(axe) e+(axg)=(e+e)+(axe)	e+((axe)+(axe)+(axe))
9/32		ax(a+e) ax(hx(a+j)) ax(ix(a+j)) b+(ax(axe))
1/4		(axe)+(axe)+(axe)+(axe)
7/32		ax(axj) ax(f+(axe)) (ax(g+(axe))=e+(axe)+(ax(axe))
3/16	axf axg	(axe)+(axe)+(axe)
5/32		ax(axh) ax(b+(axe))=e+(ax(axe)) ax(e+(axg)=(axe)+(axe)+(ax(axe))
1/8		(axe)+(axe)
3/32		ax(axg)=(axe)+(ax(axe)) ax(axf)
1/16	axe	
1/32		ax(axe)

Figure 5.2.   Comparison between new logical forms of $\mathcal{B}_4$ and $\mathcal{B}_5$.

nodes, respectively. Table 5.28 displays the following logical forms: (i) a × (a + j), (g + (a × e)) + (a × (a × e)), (a × j) + (a × (a × e)), (f + (a × e)) + (a × (a × e)), and (ii) a, d, b + (e + e), (e + (e + e)) + ((a × e) + (a × e)), and binary equivalences. In Figure 5.2 there is a summary of all the new forms and a comparison with the new logical forms of $\mathscr{B}_4$.

Chapter 6

Codification

6.1. Codification in General

A cubic representation of the 3D SD variables, as shown in Figure 6.1, can be helpful for visualising some transformations. In such cases, the **0** needs to shift each time from one position to another. This shift is also insightful regarding other aspects, as stressed in the caption. With this in view, let us now apply such structures to some codes.

6.2. Colour Codification

We expect that logical (and possible) relations are sufficiently powerful to ground and frame most of the relations or linear codifications that are instantiated in the real world. It is therefore helpful to explore some examples that are not the result of human industry. An example of two main CAs and 4 ternary SCAs for each one ($\mathscr{B}_3$) is represented by colour codification. The colours of the visible spectrum are arranged as in Figure 6.2. Now, it turns out that humans have developed a colour codification (for diurnal vision) on triadic basis according to whether the wavelength is short (blue), middle (green), or long (red)[1] (see Figure 6.3). It is remarkable that each

---

[1] See [FARAH 2000].

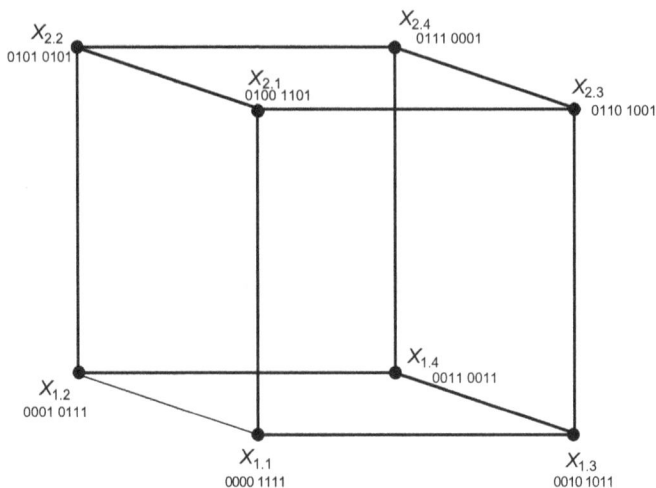

Figure 6.1.   The four variables generated by X_1 are at the bottom, while the four variables generated by X_2 are on the top (they correspond to the four variables above and the four below the horizontal line, respectively, in Table 3.3). The corners may be thought of as terminal points of vectors. The two main 3D CAs (1 and 2) are given by two opposite corners of the bottom square and the other two rotated opposite corners on the top square. All basic variables of a SCA are two steps farther from each other (the distance between informational elements is called Hamming distance). The first SCA (1.a) with $X_{1.1}, X_{1.4}, X_{2.2}$ takes the corner $X_{1.2}$ as origin: this is the sole variable that is one step farther from the previous three. To get the second SCA (1.b) with $X_{1.4}, X_{1.1}, X_{2.3}$, we need to take $X_{1.3}$ as origin, which is a reflection or a rotation of about $180°$ of the bottom square. This explains while in the sets (3.6) $X_{1.4}$ replaces $X_{1.1.}$ being the Cartesian x of the reference frame. Similarly for the other cases. Note that the centre is always one of the elements of another SCA. It may be further noted that opposite corners in the cube (like $X_{1.1}$ and $X_{2.4}$) represent the most distant elements. In $\mathcal{B}_3$, $X_{1.1}$ and $X_{2.4}$ are antipodal (their products gives 0000 0001 as the value) as well as the couples $X_{2.3}$-$X_{1.2}$, $X_{1.4}$-$X_{2.1}$, $X_{2.2}$-$X_{1.3}$.

colour can be generated through a combination of these colours. By partitioning the axes in 256 colours (from 0 to 255), we get the cube displayed in Figure 6.4. Note that, according to our visual system, white is the blend of all colours and therefore can be understood as the intersection of the three basic colours, while black is the absence of colour (and light) and can be therefore understood as the intersection of the complementaries of the three basic colours or, in other words, as the negation of the sum of the three basic colours.

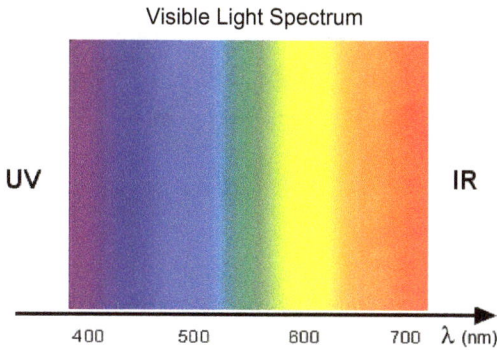

Figure 6.2. The ultraviolet (UV) and the infrared (IR) light spectrums are invisible to a naked eye. The visible spectrum is arranged according to the wavelength of light. From the left we have the shortest wavelength (and the most energetic one), violet, and on the right we have the longest wavelength (and the less energetic one), red. The wavelengths λ are expressed in nanometers (nm), i.e. 10^{-9} m.

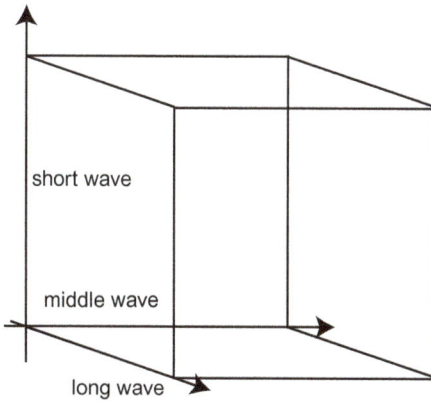

Figure 6.3. The three basic colours of human vision can be arranged as a set of orthogonal vectors. Adapted from [AULETTA 2011, p. 109].

Using the three basic colours is not the only way to codify colours. In fact, printers use another convention, since the sum of all colours cannot be white, but need to be black. In this case, the basic colours are not red–green–blue but yellow–cyan–magenta, as displayed in Figure 6.5. In fact, what happens is that we can chose any triad of orthogonal axes, or 8 SCAs, as displayed in Figure 6.6.

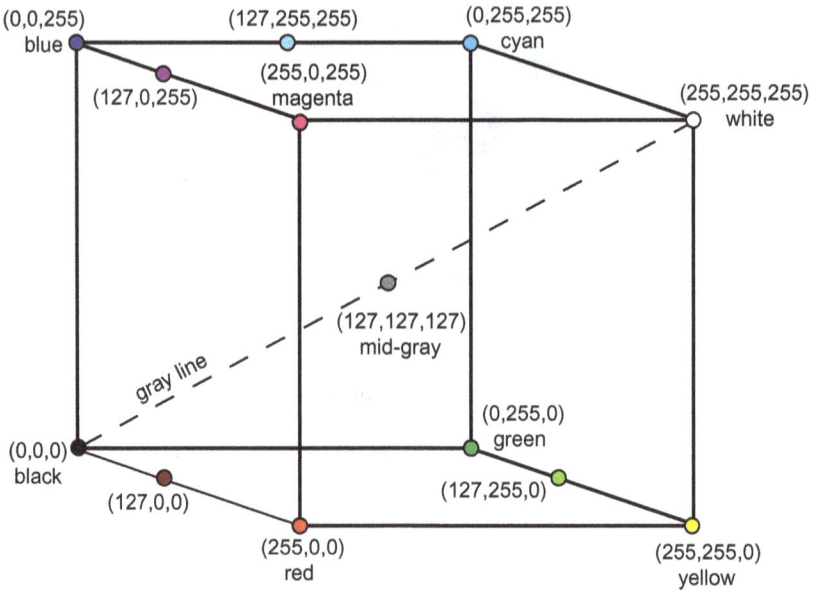

Figure 6.4.   The colour cube. Note the central circle representing the intermediate point of all cross–diagonals and therefore being the mid–grey. Adapted from [AULETTA 2011, p. 71].

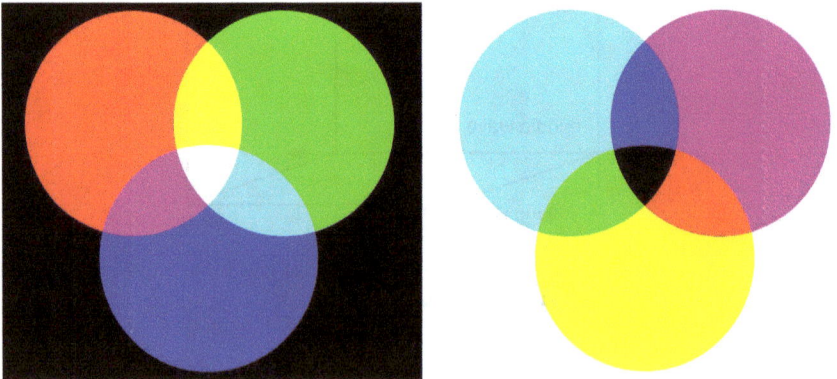

Figure 6.5.   On the left we have the human code for diurnal vision, on the right the printers' codification. Note that we have an inversion black–white when passing from a system to another. Note also that by superposing two colours of the latter code we get the colours of the first code back. See also Figure 3.1.

Figure 6.6. The top row represents the four triadic combinations of white, red, green, and blue, while the bottom row represents their respective antipodal choices, with the four triadic combinations of black, yellow, cyan, and magenta.

Thus, this pattern allows us to use colour codification as an example of codification in $\mathcal{B}_3$. First, let us consider that the counterparts of X_1 and X_2 (for $\mathcal{B}_2$) would be the two–color vision of black–and–white (nocturnal) vision. In this perspective, the two colours generate the other 6 colours by crossing them in two different subsets. In particular, black (B) generates itself as well as all colours of the bottom surface, green (G), red (R), and yellow (Y), as displayed in Figure 6.4, while white (W) generates itself and cyan (C), magenta (M), and blue (Bl), all located in the upper surface. White and blue together with red and green constitute CA1, while black and yellow together with magenta and cyan constitute CA2:

$$\{W, G, Bl, R\}, \ \{C, M, B, Y\}. \tag{6.1}$$

Note that any passage from a triad of CA1 to the antipodal triad of CA2 determines an inversion between overall product and overall neg–sum of those colours that corresponds to the inversion of central lines in Tables 3.7–3.8. In other words, independently of the logical aspects and considering codification only, we expect always a kind of inversion when passing from CA1 to CA2, or vice versa, and permutations of the alphabet when going from a SCA to another in the same main code. In the latter case, we pass from Code R to Code G by exchanging R and G as well as W and Bl, and similarly when passing from Code Bl to Code W, where we need to exchange Bl and W as well as R and G; however, we pass from Code R to Code B by exchanging R–V with Bl–B, and similarly when passing from Code G to Code W. We have the same behaviour for the second row of Figure 6.6. With such a structure, we can produce any node of $\mathscr{B}_3$. The results of the combinations will be sums of specific areas. For instance, the equivalent of $X_{1.1}(X_{1.4}+X_{2.2})$ is $Y+W+M$ (f + g + h).

6.3.   The Genetic Code

Life displays a hierarchical organisation that requires control, and, in turn, control requires codes.[2] The genetic subsystem of the organism is a complex machinery whose goal is the production of proteins, according to its (metabolic) necessities, on the basis of information codified in the DNA. The DNA presents itself as a double-helix structure (see Figure 6.7), whose backbone is constituted by sugars and phosphates, on which the bases (representing the genetic information) stick.[3] The bases are connected across the two strands through so-called hydrogen bonds (Figure 6.8). Note that letters (which are the "negative" of each other) are connected in this way: A (adenine) with T (thymine) as well as G (guanine) with C (cytosine). This pattern determines a complete independence of the basis from the chemical substrate to which it is attached. In fact, any possible sequence of the letters A, C, G, T could occur. This is a necessary requirement for having information. In fact, the information contained in some scientific paper cannot depend on the chemistry of the page (especially since the paper can also be online). This is what establishes a connection between information and possibility (and thus logic). Note that DNA works like an information processor in that it not only guarantee the transmission of genetic information to following generations but also contributes to variety thanks to mutations

---

[2][BARBIERI 2003].
[3][ALBERTS *et al.* 1983, Part II].

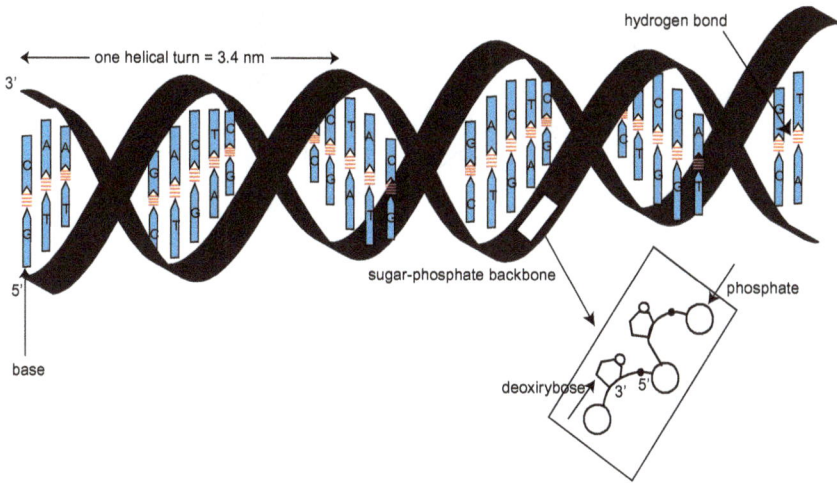

Figure 6.7. DNA's double–helix structure. A helix is a common structural motif (actually, a type of wave) in biological structures. Adapted from [AULETTA 2011, p. 214].

that sometimes occur (although with a low frequency still sufficient to promote evolution of the species) as well as recombination and reshuffling of whole codifying segments. Genetic transmission makes DNA activity a true computation.[4]

Essentially, the process through which a protein is built starting from the genetic information can be cast in three steps[5]:

- First of all, the information contained in the DNA needs to be expressed (better: it needs to be activated). In fact, the DNA is a molecular complex that is chemically inert (and encapsulated in a fibre called chromatin) and therefore dormant from an informational point of view. Such an expression starts from some internal signal (chemical pattern), essentially a request for producing a certain protein. The activation is performed through the so-called RNA polymerase, often in conjunction with other complex molecules. The polymerase cuts the connection between the strands (i.e. the hydrogen bonds), see Figure 6.9(a)–(b).

---

[4][ADLEMAN 1998]. See also [AULETTA 2011, Section 9.7].
[5][AULETTA 2011, Section 7.4].

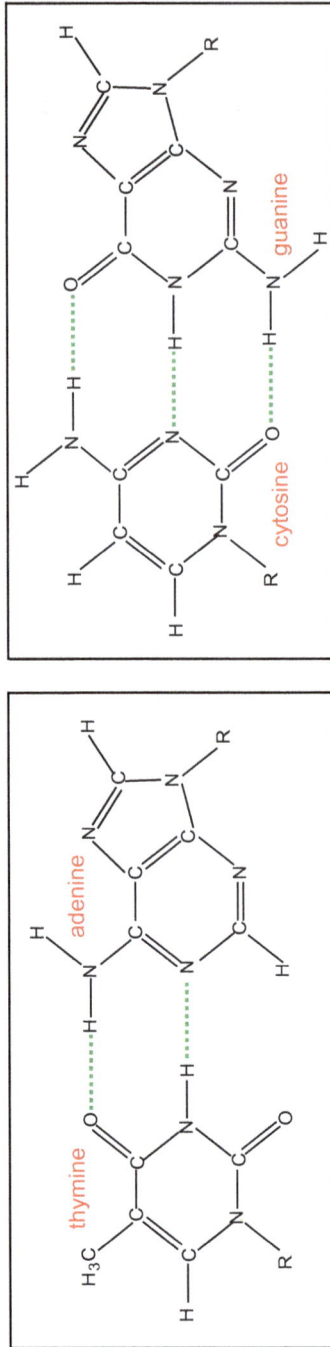

Figure 6.8. The base T establishes hydrogen bonds (shown in dashed line) with A (left) as well as C with G (right). Adapted from [AULETTA 2011, p. 214].

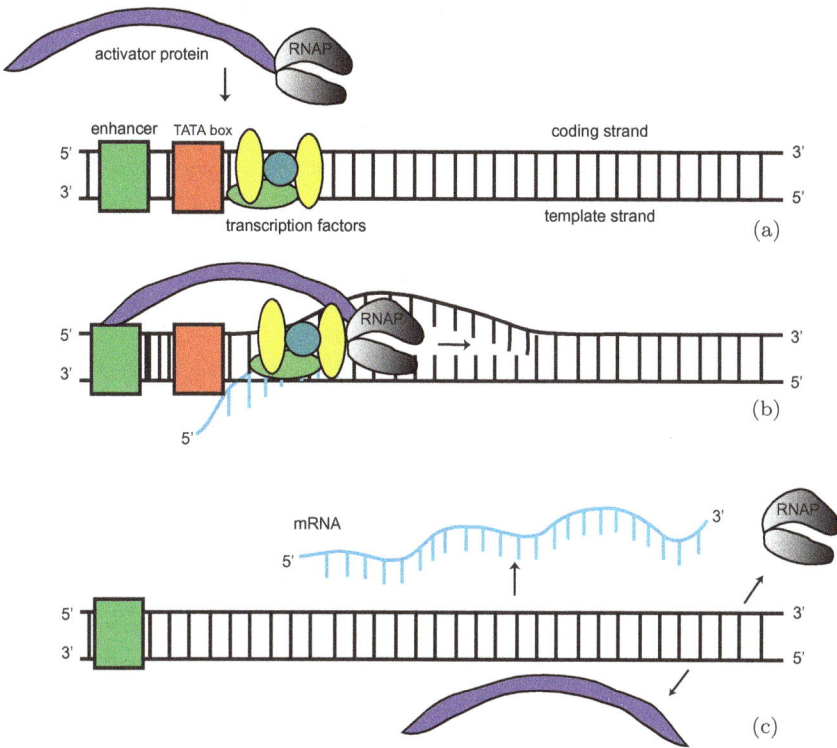

Figure 6.9. (a) RNA polymerase (RNAP) with a protein tail, (b) activating a gene by acting on the region determined (in eukaryotes) by a complex of transcription factors or by a promoter (in bacteria), while attached protein acts as an enzyme on the enhancer region (other eukaryotic elements like mediators, chromatin remodelling complexes, and histone modifying enzymes are not shown). (c) The final resulting mRNA. Only one of the two DNA strands is transcribed. This strand is called the template strand because it provides the template for ordering the sequence of nucleotides in an RNA transcript. The other strand is called the coding strand because its sequence is the same as the newly created RNA transcript, except for U (uracil) replacing T (thymine). The RNA's (genetic) code should be more ancient than DNA's code. Adapted from [AULETTA 2011, p. 223].

- The second step is called transcription (Figure 6.9(c)), where, thanks to the polymerase, the information contained in the DNA template is matched with those of the messenger RNA (mRNA). In such a process, non-coding sequences are eliminated, an operation called splicing. This information is no longer the result of a pure and random combinatorics, but fulfills the requirements of permutation with repetition. In other words, the specific sequence of the letters does matter, so that, for example, AG is different from GA. This information now represents a set of instructions. Thus, the RNA establishes a code (the genetic code) through which different triplets of letters correspond to 20 different amino acids (Table 6.1). All proteins (apart from few exceptions) can be produced through combination of these amino acids.

Table 6.1. The 20 amino acids and their bases constituting the genetic code. Since the possible combinations of bases in the different codons (a triplet of nucleotides) is 4^3, which is an instance of formula n^k for the so-called permutation with repetition (where n is the number of elements and k is the length of the "string"), this is larger than 20. In other words, each amino acid may be coded using different bases (degeneracy). Only 61 codons are employed to codify for amino acids, so that three combinations of bases (UAA, UGA, UAG) are free to provide the stop signal that terminates the translation process. Legend: p = polar, np = non–polar, n = neutral, sb = strongly basic, wb = weakly basic, a = acidic.

Amino acid	Abbr.	Symbol	Codons	Side chain polarity	Side chain acidity or basicity	Hydropathy index
Alanine	Ala	A	GCA, GCC, GCG, GCU	np	n	1.8
Cysteine	Cys	C	UGC, UGU	p	n	2.5
Aspartic acid	Asp	D	GAC, GAU	p	a	~3.5
Glutamic acid	Glu	E	GAA, GAG	p	a	~3.5
Phenylalanine	Phe	F	UUC, UUU	np	n	2.8
Glycine	Gly	G	GGA, GGC, GGG, GGU	np	n	~0.4
Histidine	His	H	CAC, CAU	p	wb	~3.2
Isoleucine	Ile	I	AUA, AUC, AUU	np	n	4.5
Lysine	Lys	K	AAA, AAG	p	b	~3.9
Leucine	Leu	L	UUA, UUG, CUA, CUC, CUG, CUU	np	n	3.8
Methionine	Met	M	AUG	np	n	1.9
Asparagine	Asn	N	AAC, AAU	p	n	~3.5
Proline	Pro	P	CCA, CCC, CCG, CCU	np	n	~1.6
Glutamine	Gln	Q	CAA, CAG	p	n	~3.5
Arginine	Arg	R	AGA, AGG, CGA, CGC, CGG, CGU	p	sb	~4.5
Serine	Ser	S	AGC, AGU, UCA, UCC, UCG, UCU	p	n	~0.8
Threonine	Thr	T	ACA, ACC, ACG, ACU	p	n	~0.7
Valine	Val	V	GUA, GUC, GUG, GUU	np	n	4.2
Tryptophan	Trp	W	UGG	np	n	~0.9
Tyrosine	Tyr	Y	UAC, UAU	p	n	~1.3

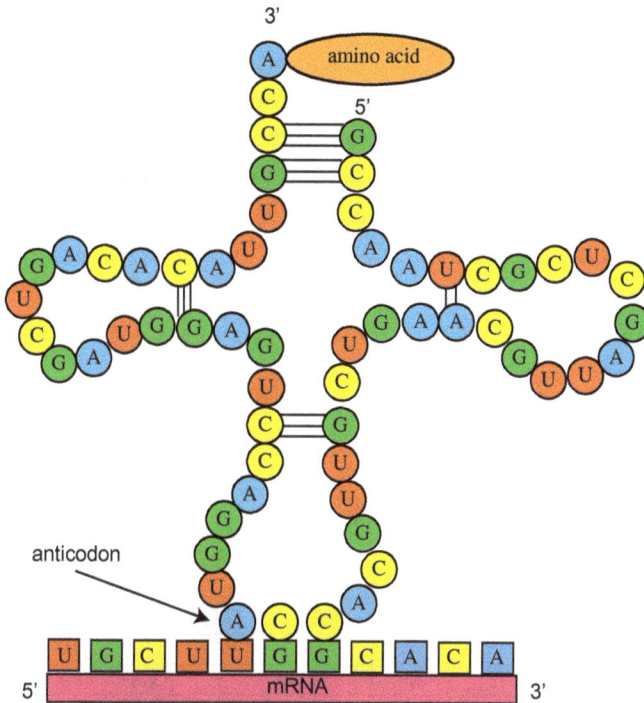

Figure 6.10. A schematic depiction of the tRNA structure. The letters of the tRNA are called anticodons. Also some hydrogen bonds are shown. Adapted from [AULETTA 2011, p. 225].

- Finally, the information is used for selecting a particular amino acid that will be a piece of the sequence constituting a protein. This process, called translation, is realised because another kind of RNA, called transfer RNA (tRNA), sticks to a triplet of the mRNA if its own triplet is the negative of the former one (see Figure 6.10). Since an amino acid is attached to the other end of the tRNA, the latter can enter the sequence of the amino acid, thereby constituting the outgoing protein. This process happens in a particular component of the cell called the ribosome. Note that the mRNA and the tRNA work as coupling devices, while the final sequential, segmented production of the protein can be understood as a sequence of information selections.

It is difficult to understand how this code may have been established.[6] However, it is likely that the code started with two initial different codes (A-G and U-C) that were perhaps used by different RNAs. The two subcodes must have been merged (note that all six codons of both serine and leucine have all codons with the first two letters that are either a subcode or the other), and, although it is not clear in which sequence, the third letter was added. When they merged, several regularities were built. By rearranging the elements of the genetic code as in Table 6.2, we see immediately that all elements of

Table 6.2. All codons that codify for the same amino acid (apart from L, R, S, which are codified by 6 different codons which are split in two parts) share the first two letters. These two letters individuate a single column in Figure 6.11. Note that splits happen when the first two letters are one of the following cases: AA-AG-AU, UA-UG-UU, CA and GA.

Amino acid	Abbr.	Symbol	Codons
Alanine	Ala	A	**GCA, GCC, GCG, GCU**
Cysteine	Cys	C	**UGC, UGU**
Aspartic acid	Asp	D	**GAC, GAU**
Glutamic acid	Glu	E	**GAA, GAG**
Phenylalanine	Phe	F	**UUC, UUU**
Glycine	Gly	G	**GGA, GGC, GGG, GGU**
Histidine	His	H	**CAC, CAU**
Isoleucine	Ile	I	**AUA, AUC, AUU**
Lysine	Lys	K	**AAA, AAG**
Asparagine	Asn	N	**AAC, AAU**
Proline	Pro	P	**CCA, CCC, CCG, CCU**
Glutamine	Gln	Q	**CAA, CAG**
Threonine	Thr	T	**ACA, ACC, ACG, ACU**
Valine	Val	V	**GUA, GUC, GUG, GUU**
Tyrosine	Tyr	Y	**UAC, UAU**
Leucine (1st part)	Leu	L	**UUA, UUG**
Leucine (2nd part)	Leu	L	**CUA, CUC, CUG, CUU**
Arginine (1st part)	Arg	R	**AGA, AGG**
Arginine (2nd part)	Arg	R	**CGA, CGC, CGG, CGU**
Serine (1st part)	Ser	S	**AGC, AGU**
Serine (2nd part)	Ser	S	**UCA, UCC, UCG, UCU**

---

[6][CRICK 1968, OSAWA *et al.* 1992].

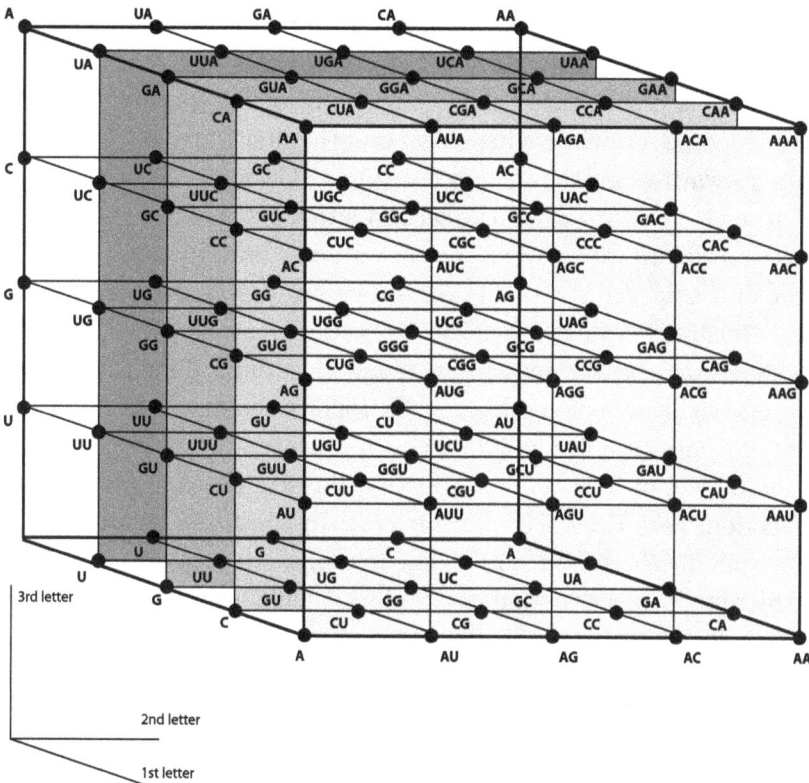

Figure 6.11. A way for representing the genetic code. I have made use of the fact that A can be considered the "negative" of U as well as G of C (it is likely that originally there were two different subcodes: A-G and U-C). The four bases can be understood as vectors of different length. The three axes here represent the positions of the letters (in fact, we deal here with permutation and not simply combinatorics). All triadic and dyadic combinations, as well as monadic letters, are shown (dyadic combinations are in fact codewords): 64+16+4=84 items, where monadic codewords are counted only one time (in the three series). Also dyadic relations are counted one time (they are on the 1st-2nd letter cube face, the 2nd-3rd letter cube face, and 3rd-1st letter cube face). I have used four vertical surfaces (in different grey scales), each containing 16 triadic codewords, three of which are parallel to the front face of the cube (obviously, we can choose any other face of the cube and again consider the parallel surfaces; this leads to the same result). The first surface (the only one with a complete grid) can be called the "A surface", which (counterclockwise [CCW]) has corners A, AA, AAA, and AA. The second surface is the "C surface", which (CCW) has corners C, CA, CAA, and CA. The third surface is the "G surface", which (CCW) has corners G, GA, GAA, and GA. The fourth surface is the "U surface", which (CCW) has corners U, UA, UAA, and UA.

the same line share the first two letters. In such a case, we recover the bottom face of the cube of Figure 6.11. We see that when the first letter is C or G, we have a single split of the quartet (when we have CA or GA), while when A or U is the first letter we have three splits (always when the second letter is A, G, or U). In other words, we never have splits when the second letter is C.

It is difficult to conceive of this process without the establishment and enlargement of the code representing an adaptive advantage "pushing" somehow in this direction (the so-called adaptive theory). Further, the expansion along this third dimension in the case of dyadic sets of codons strictly follows the distinction between the two codes (see Table 6.3). In such a process we have the constitution of sets of four codons, splitting the sequences in two parts of two codons (see Table 6.3; the only cases in which the quadruplet has been split in three and one codon is represented by isoleucine and methionine, respectively), and merging of some sets. Thus, during the code evolution, subsets of codons for precursor amino acids have been reassigned to encode product amino acids (the so-called co-evolution theory). Apart from the already mentioned case of the serine, we have two further cases of amino acids (leucine and argirine) coded

Table 6.3. All couples of codons that codify for the same amino acids have the last letter forming one of the two original subcodes (that is, they are both either purine or pyrimidine).

Amino acid	Abbr.	Symbol	Codons
Cysteine	Cys	C	UGC, UGU
Aspartic acid	Asp	D	GAC, GAU
Glutamic acid	Glu	E	GAA, GAG
Phenylalanine	Phe	F	UUC, UUU
Histidine	His	H	CAC, CAU
Lysine	Lys	K	AAA, AAG
Asparagine	Asn	N	AAC, AAU
Glutamine	Gln	Q	CAA, CAG
Tyrosine	Tyr	Y	UAC, UAU
Leucine (1st part)	Leu	L	UUA, UUG
Arginine (1st part)	Arg	R	AGA, AGG
Serine (1st part)	Ser	S	AGC, AGU

by six triplets. It may be noted that the structure of these two sets is similar and perfectly fits with the informational plot of Figure 6.11. At this stage the informational aspects became prevalent and we can assume that the code was fully operative.[7]

Although the genetic code can be plotted in an abstract information space, displayed by Figure 6.11, when we try to translate it in a Boolean network we encounter some problems. We have codewords of length 3 only, and these give rise to different permutations. This situation means that logical sum and product cannot be used here. However, we have 64 elements, and this situation fits with the Boolean algebra with ID of 6 digits, which in turn is closed only under complementation, reversal, and neg–reversal (Subsection 3.4.2). The same happens for the genetic code, as displayed in Table 6.4. Note

Table 6.4. Permutations of the genetic code according to logical transformations.

	Reverse	Complement	Neg–reverse
AAA	AAA	UUU	UUU
CCC	CCC	GGG	GGG
ACA	ACA	UGU	UGU
AGA	AGA	UCU	UCU
AUA	AUA	UAU	UAU
CAC	CAC	GUG	GUG
CGC	CGC	GCG	GCG
CUC	CUC	GAG	GAG
AAC	CAA	UUG	GUU
AAG	GAA	UUC	CUU
AAU	UAA	UUA	AUU
CCA	ACC	GGU	UGG
CCG	GCC	GGC	CGG
CCU	UCC	GGA	AGG
ACG	GCA	UGC	CGU
ACU	UCA	UGA	AGU
AGC	CGA	UCG	GCU
CAG	GAC	GUC	CUG
CAU	UAC	GUA	AUG
CUA	AUC	GAU	UAG

---

[7][RODIN *et al.* 2011].

that the first eight rows show symmetry (invariance under reversal). However, the first two rows can also be grouped with the third
collection (with two letters equal but with no symmetry) since they
complete the permutation with the first two letters equal. The fourth
collection shows all cases with three different letters.

Then, referring to Table 3.33, we can establish a biunivocal correspondence between the genetic code and that network. Of course,
as a consequence of the previous points, we cannot expect that the
different nodes that result as combinations of SD variables correspond here to a combination of the relative letters. In other words,
such a correspondence has a certain degree of arbitrariness, but the
three involved operations work quite well. There is also an additional
difficulty: complementation is valid for any single letter, so that we
cannot apply it to codewords. As a matter of fact, the complementary form of AAC is UUG and not $U + U + G$, which is forbidden by
the syntax of the code. Thus, the result can be cast as in Table 6.5.
Note that I have used symmetric forms in relation to SD variables
and material equivalences, as well as **0** and **1**. Note also that, as for
the IDs, the genetic codewords in the left column are the negations
of those in the right column, as well as the codewords on the same
level are the reverse of each other.

Having a system giving rise to a code in which (at the opposite to
what happens for colour codification) logical sum and product are not
allowed generates a number of items naturally mapped to an algebra
that is not closed under sum and product may not be a complete
coincidence. More work on different kinds of natural codifications
will tell us more.

6.4.   Classifying Things and Data

6.4.1.   *Basic classifications*

Up to now we have explored examples of codification that already
exist in nature. Let us now develop some simple models that can
help us to understand the recursive power and the plasticity of such a
formalism when we are classifying different kinds of objects. Suppose
we are interested in having some statistics on the population of a

Table 6.5. The genetic code cast in Boolean terms.

000000	CUC	111111	GAG
000001	UCA	111110	AGU
000010	CGA	111101	GCU
000100	CUA	111011	GAU
001000	AUC	110111	UAG
010000	AGC	101111	UCG
100000	ACU	011111	UGA
000011	UUA	111100	AAU
000101	AAC	111010	UUG
000110	AAG	111001	UUC
001001	UGG	110110	ACC
001010	UCC	110101	AGG
001100	CAC	110011	GUG
010001	CCG	101110	GGC
010010	AGA	101101	UCU
100010	GCC	011101	CGG
100001	ACA	011110	UGU
010100	CCU	101011	GGA
100100	GGU	011011	CCA
011000	GAA	100111	CUU
101000	CAA	010111	GUU
110000	AUU	001111	UAA
000111	AAA	111000	UUU
001011	AUA	110100	UAU
001101	UGC	110010	ACG
001110	AUG	110001	UAC
010110	CAG	101001	GUC
011010	GAC	100101	CUG
011100	GUA	100011	CAU
101100	CGU	010011	GCA
010101	CCC	101010	GGG
011001	CGC	100110	GCG

given country. We are content to use a 3D space, which represents a rough classification of data but is nevertheless good for our purposes. We are interested, in particular, in the social–economical situation of people. We could proceed in this way: $X_{1.1}$ represents the property "wealthy" according to some national standard ($X'_{1.1}$ "not wealthy"), $X_{1.4}$ "working" ($X'_{1.4}$ "not working"), while $X_{2.2}$ could be "living in a town" ($X'_{2.2}$ "living in a rural area").

Then, we have the classes of those who are both wealthy and living in town $(X_{1.1}X_{2.2})$ and wealthy and rural $(X_{1.1}X'_{2.2})$. Furthermore, there are the classes of those who are urban and not wealthy $(X'_{1.1}X_{2.2})$, as well as of rural and not wealthy $(X'_{1.1}X'_{2.2})$. We also have the classes of people being wealthy and working $(X_{1.1}X_{1.4})$, not wealthy and working $(X'_{1.1}X_{1.4})$, wealthy and not working $(X_{1.1}X'_{1.4})$, not wealthy and not working $(X'_{1.1}X'_{1.4})$, and similarly the classes of those working and urban $(X_{1.4}X_{2.2})$, working and rural $(X_{1.4}X'_{2.2})$, not working and urban $(X'_{1.4}X_{2.2})$, not working and rural $(X'_{1.4}X'_{2.2})$. Do not forget then that we have the classes $X_{1.1}X_{1.4}X_{2.2}$ or $X_{1.1}X'_{1.4}X_{2.2}$.

$\mathscr{B}_3$ allows far more combinations. For instance, we have the class of people who are wealthy and working or living in town $(X_{1.1}(X_{1.4} + X_{2.2}))$ or of those who are working and are not wealthy or rural $(X_{1.4}(X'_{1.1} + X_{2.2}))$, as well as those who are wealthy and working or not wealthy, not working, and living in a town $(X_{1.1}X_{1.4} + X'_{1.1}X'_{1.4}X_{2.2})$. Furthermore, we have the class of those who are wealthy and working or not wealthy and living in a town $(X_{1.1}X_{1.4} + X'_{1.1}X_{2.2})$. Also there are the classes of those who are wealthy or not working and living in a rural area $(X_{1.1} + X'_{1.4}X'_{2.2})$ or who are not wealthy or working and living in a town $(X'_{1.1} + X_{1.4}X_{2.2})$. Then there is the class collecting all people but those who are wealthy and living in a town $(X'_{1.1} + X'_{2.2})$, or of all people but those who do not work and not live in a rural area $(X_{1.4} + X_{2.2})$. Finally, there are the classes that collect all people but those who are wealthy, work, and live in a town $(X'_{1.1} + X'_{1.4} + X'_{2.2})$, and similar ones.

Of course, we can use more SCAs with different properties in order to explore some alternative grids. There will be some nodes that are shared by two or more SCAs, but there is no ambiguity here since they mean something in one code and something different in another code. To this purpose, let us introduce another triplet of properties: to be a couple $(X_{1.4})$ or not $(X'_{1.4})$, to have children $(X_{1.1})$ or not $(X'_{1.1})$, to be heterosexual $(X_{2.3})$ or not $(X'_{2.3})$. Of course, we repeat for the triplet all the forms already explored. We either use a SCA or the other and never together. We can have up to four different SCAs for each CA, and we can use the four triplets of the other CA to this purpose, which can make up to eight alternative codes. We can further analyse all of the four SCAs of CA1 as a way to explore what the combination of three basic properties gives by changing one code each time. In such a case, we can understand which triplets allow the best social classification.

Furthermore, we can enlarge our triplets of properties to a quadruplet, so that to the properties (i) be wealthy, (ii) employed, (iii) living in a town, we add (iv) be a member of a couple. If we are only interested in exploring binary or ternary products of such properties (and their complementary sums), $\mathscr{B}_3$ could suffice. If, at the opposite, we like to explore more relations, then we need to resort to $\mathscr{B}_4$.

We can also use alternative CAs as systems for codifying different classes of objects. Here, the same node can stand for different objects according to the code used. When dealing with such alternatives, it is better to use different CAs that guarantee fully independent classification, because the use of different SCAs as alternatives can give rise to confusion. In such a case, the four properties of CA1 and the four properties of CA2 can pertain to totally different domains, for instance, people, books, colours, and so on. They can also respond to different ways to classify a broad subject area. For instance, for classifying people we can use the grid wealthy–employed–living in a town–living in a couple (always employing CA1), but we can also use the grid (employing CA2) culture (low or high according to some standard), education (low or high), scientific–technical knowledge (low or high), participation in religious rites (low or high). This second grid is not about the social–economical situation of people, but about their background knowledge and cultural interests.

6.4.2. *Hierarchies*

Now, suppose we need more refined ways to classify people (where the start is represented by the grid wealthy–employed–living in a town–living in a couple). In such a case, the recursive character of Boolean algebra will help us in a natural way. Thus, as mentioned, we may desire to pass to a 4D network (I use here the same convention as in Section 4.3). Our aim about the class of wealthy people may be to distinguish those who invest their money and those who have ensured incomes like loans. So, we introduce the classes X_{1a} and X_{1b}, respectively. Moreover, we may wish to further determine those employed workers (X_{4a}) and directors or managers (X_{4b}), those who live in town that reside stably in a town (X_{2a}) and those who often change their address (X_{2b}), and similarly among couples, those who are married or legally united (X_{3a}) and those who live together (X_{3b}). Of course, we can also have people who reside stably in a town for a certain period or who lived together and subsequently married, or even those who are married and do not live together. Now, should we need all such specifications, the number of combinations would increase considerably. Remaining confined to products, we have for instance the class of those who are investors, workers (e.g. managers), and are married $(X_{1a}X_{4a}X_{3a})$, as well as the class of those who live off their assets but are not investors and live in smaller towns $(X'_{1a}X_{1b}X_{2b})$. The possibilities are more interesting and also more numerous binary products, as in the following examples:

- Both investor and receiving a fixed income $(X_{1a}X_{1b})$, with all variations of the theme $X'_{1a}X_{1b}$, $X_{1a}X'_{1b}$, $X'_{1a}X'_{1b}$.
- Both investor and worker $(X_{1a}X_{4a})$, with all variations $X'_{1a}X_{4a}$, $X_{1a}X'_{4a}$, $X'_{1a}X'_{4a}$.
- Both investor and manager $(X_{1a}X_{4b})$, with variations $X'_{1a}X_{4b}$, $X_{1a}X'_{4b}$, $X'_{1a}X'_{4b}$.
- Both investor and reside in a town $(X_{1a}X_{2a})$, with variations $X'_{1a}X_{2a}$, $X_{1a}X'_{2a}$, $X'_{1a}X'_{2a}$.
- Both investor and married $(X_{1a}X_{3a})$, with variations $X'_{1a}X_{3a}$, $X_{1a}X'_{3a}$, $X'_{1a}X'_{3a}$.
- Both investor and living with someone else $(X_{1a}X_{3b})$, with variations $X'_{1a}X_{3b}$, $X_{1a}X'_{3b}$, $X'_{1a}X'_{3b}$.

And so on. Of course, we could decide from the start to make use of a 4D algebra. In any case, it is important to observe the different rules for getting the proper SCAs.

6.4.3. *Playing with networks*

An interesting game is the following: consider the paths leading from **0** to other nodes as a flux of information (but even of energy). What happens if we drop or annihilate a node? Will this affect other nodes? Let us start with $\mathscr{B}_2$ for simplicity. Suppose that we drop a product, such as $X_1 X_2$. As it is easy to see, nothing happens to the other nodes, as any of the nodes of Level 2/4 can be reached from the other three nodes of Level 1/4. The network is robust to such a failure. However, suppose that we drop both $X_1 X_2$ and $X_1 X_2'$. As is evident, the node represented by X_1 can no longer exist, although $X_1 + X_2$ does exist since X_2 exists, and this whatever X_1 means or is. However, if we also drop a third node, e.g. $X_1' X_2$, then neither X_2 exists, nor $X_1 \sim X_2$, and we no longer have $X_1 + X_2$. The opposite consideration is also interesting: if we turn on e.g. the product $X_1 X_2$ (that is, it is true), automatically all other three products are turned off (they are false).

Why is this game interesting? Because to drop a node is in fact a selection. Now, there are several kinds of networks (for instance, our brain) that temporarily or permanently do inhibit some nodes (neurons).

Let us now play another game. As I have mentioned, each Boolean network contains several Boolean subnetworks of lower dimension (Subsection 3.1.2). For instance, in $\mathscr{B}_3$ we have 6+6 2D subnetworks, and one of them is shown in Table 6.6. Of course, the number of subnetworks grows exponentially with the number of dimensions; for

Table 6.6. Example of 2D subnetwork in $\mathscr{B}_3$.

0000 0000	**0**	**1**	1111 1111
0000 0101	$X_{1.1} X_{2.2}$	$X_{1.1}' + X_{2.2}'$	1111 1010
0000 1010	$X_{1.1} X_{2.2}'$	$X_{1.1}' + X_{2.2}$	1111 0101
0101 0000	$X_{1.1}' X_{2.2}$	$X_{1.1} + X_{2.2}'$	1010 1111
1010 0000	$X_{1.1}' X_{2.2}'$	$X_{1.1} + X_{2.2}$	0101 1111
0000 1111	$X_{1.1}$	$X_{1.1}'$	1111 0000
0101 0101	$X_{2.2}$	$X_{2.2}'$	1010 1010
1010 0101	$X_{1.1} \sim X_{2.2}$	$X_{1.1} \not\sim X_{2.2}$	0101 1010

example, in $\mathscr{B}_4$ we have 896 3D subalgebras. Now, some of these subalgebras could be subnetworks that work independently and in parallel (again the brain with its areas is a good example, but also many social networks). In principle, all of these subnetworks share the supremum and the infimum of $\mathscr{B}_4$. However, it could be more interesting to take two other nodes for each autonomous subnetwork and let them be its **0** and **1** (of course, in other codes these two nodes keep their original meaning). In this way, **0** could be the gateway for accessing the network, while **1** its exit gate, so that such a subnetwork can from time to time be recruited for a more general task involving other nodes or simply receive some (e.g. inhibitory or excitatory) input from other subnetworks and in turn contribute to other networks. Of course, we can can also have hierarchies of containment. For instance, a 3D subnetwork contains three 2D subnetworks, and we can reiterate or vary previous procedures. Such possibilities are very interesting, especially when combined with the previous game. Indeed, not only can single nodes inside the subnetwork be silenced but, as said, the whole subnetwork itself can be shut down if we drop the gateway node.

Another possibility is to make use of partial networks inside a complete network (Section 3.4). For instance, we could have two 6–digit subnetworks embedded in a 3D network, and, making use of the fact that each of them can display four SD variables (Table 3.33), we could extend considerably possibilities. In such a case, the remaining 130 nodes could be shared by both subnetworks or also contain other two 6–digit subnetworks. In both cases, this could considerably extend our capabilities to reproduce structures of the real word, like neural architectures.

6.4.4. *Properties*

Boolean networks are logically built through combinations of properties. When we use these resources for classifying data or for solving problems, we deal with the problem of how to single out the specific properties we are dealing with. The problem is that some of what appears as properties could be in fact classes of objects that display some combinations of properties that are more fundamental. We have already met such a problem, without taking notice, when we enlarged our 3D space to a 4D space and introduced properties like "wealthy people who invest their money". We may in fact raise the question whether this is not a composed property. Let us make another example: suppose that, at a given stage of knowledge, we take *mathematical* as a basic property referred to a given collection of items (likely Plato's original point of view). However, with the progress of knowledge we understand that *mathematical* is in fact an intersection or product of *formal* and *abstract* (of course, an intersection of these two properties may also be considered for other classes of items, such as music too). How should we proceed in such a case?

First, allow me to say that we can never take for granted that a given property (like *abstract*) will remain basic forever. The fundamental reason for that seems to be that in the real world the properties that we ascribe to different kinds of objects should be considered as extrapolations of relations those objects have with other objects.[8] This reason is strictly related to the fact that a single classification for everything (as understood by Aristotle) does not exist, and in many cases not even for specific fields. Thus, there is no way to solve the problem on a pure theoretical basis. The only viable alternative is to deal with this issue in pragmatic terms. We can always take a property as basic (even our "wealthy people who invest their money") as far as it does not determine inconsistencies or conflicts with needs of some kind. If there is such a situation, we need to map our original network into one of higher dimension in which to locate the new basic properties. Since what we have discovered to be a class is no longer a property, it needs to be mapped to a node that is no longer characterised by a SD ID. Although many relations with the previous nodes will be maintained, this demands a partial reset of the new subnetwork. Of course, whenever we need to come back to a compound property, we only need to make a kind of coarse graining of the network.

---

[8][AULETTA 2019].

6.4.5. *Individuals*

Although I have explained that Boolean algebras deal with schemes of classes, when we apply these resources to concrete cases, we need to also consider the involved individual items. This consideration is already evident for the previous examples. Thus, a certain class will contain a certain number of individuals, and we could attach a label to each of them. Now, the interesting issue is that individual items like people are classified according to different sets of basic properties (or codifications, like financial condition, social status, culture, hobbies, and so on), and therefore they pertain to different classes, thus cumulating a number of characters that is potentially indefinite, depending only on the size of the network and the level of resolution in attributing properties. Something similar happens for the classes themselves. For instance, humans pertain to the classes of mammals, bipeds, sexual organisms, animals, cultural actors, and so on. The list of classes could be indefinitely long.

Another interesting point is that individuals can themselves be part of networks (like social networks, companies, and so on) as individuals. In this case, we have individual–networks grafted on Boolean networks that are not necessarily Boolean themselves but contribute to the specification of the individuals.

6.4.6. *Some additional words*

Some additional considerations seem opportune here. The different nodes of a Boolean network codify for information. However, the whole information contained in the network is bigger than at first sight. There are relations due to implications, connections among nodes due to the De Morgan's laws, and the fact that half of the nodes are complementary relative to the other half and the same for reversal, and, by adding self neg–reversal, also for neg–reversal. There is also additional information in the recurrent and combinatorial patterns that go through levels of an algebra, as well as across different–dimensional algebras. Some of them have already been explored, especially for $\mathcal{B}_4$ and $\mathcal{B}_5$. Moreover, as we see in the following there are linear transformations contributing to additional information.

Chapter 7

Logical Spaces and Quantum Information

7.1. From IDs to Vectors

Traditionally, Boolean algebra is understood as connecting logic and set theory. Such a connection is ensured by category theory and its dealing with the general notion of class.[1] Here, I show that a connection with vector spaces can be found. In particular, I show that Boolean algebra can be mapped into a portion of the associated vector space. In fact, in the finite case, we can associate any algebra $\mathscr{B}_n$ with a vector space $\mathscr{V}^n$ (with $n \in \mathbb{N}$). In the infinite case, we associate $\mathscr{B}_\infty$ with a a vector space $\mathscr{V}^\infty$. Indeed, the ID can be put in a kind of vectorial form as, for instance, a column or a row of numbers. Since we do not deal with complex numbers, I sometimes use the row convention. This convention gives the *direction* of the logical vector (LV) representing the element of the algebra.

The purpose is to build a connection with physical theories like quantum mechanics and in particular with quantum information. In such a context, we can admit vectors of only maximal length **1**. This means that we cannot prolong any vector by any length. Thus, the best way to represent a logical vector space (LVS) is by means of

---

[1][LEINSTER 2014, SPIVAK 2013].

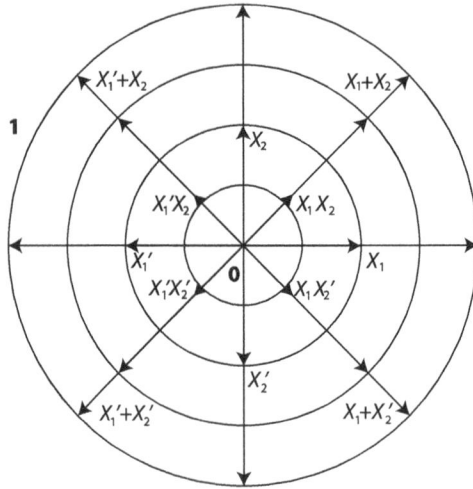

Figure 7.1. The way in which we can represent the 2D logical vector space (LVS). Note that we have here five circles (one of them represented by the **0** point) corresponding to five levels (determined by the number of 1s in the ID). Note the vectors of the kind $X_1 X_2$ have length $1/4$, vectors of the kind X_1, X_2 have length $1/2$, and vectors of the kind $X_1 + X_2$ have length $3/4$ across any $\mathscr{B}_n$.

concentric hyperspheres centred on **0** whose rays are given by

$$\frac{0}{2^n}, \frac{1}{2^n}, \frac{2}{2^n}, \ldots, \frac{2^n}{2^n},$$

and with external hypersurface corresponding to **1**, as displayed in Figure 7.1. In other words, any LV, when prolonged to length 1, is in fact **1**. For this reason, the latter can be called the universal LV (as well as the LV corresponding to **0** as the null LV). While the LV corresponding to **1** has length 1, the LV corresponding to **0** has length 0. Another way to consider the problem is the following: for each nD algebra, half of the nodes are logically contravalent relative to the other half. This means that the sum of the relative logical expressions is **1** (and their product **0**). If we denote the logical contravalence by $\neq$, we have, for instance, $X + Y \neq X'Y'$, that is, $(X + Y) + X'Y' = \mathbf{1}$. Thus, we can start from any node and, using the logical rule of addition, reach the surface of length 1 so that any vector can be extended up to length 1.

Thus, by also attributing *length* to all LVs, we can build a LVS understood as a unitary hypersphere in the relative vector space. A nD LVS (with $n \in \mathbb{N}$), covering a collection of vectors $\in \mathscr{V}^n$ consists of LVs $x, y, z, \ldots$ such that, $\forall X \in \mathscr{B}_n$, $\exists x \in \mathscr{V}^n$ is defined as

$$x = \frac{j}{2^n}(X), \tag{7.1}$$

where $j \in \mathbb{N}$ is determined by the sum of 1s present in the ID of X. In other words, the length of a LV depends on the ID of the corresponding logical expression. Thus, the mapping $X \longmapsto x$ is injective. I stress that the length of the vectors has a geometric significance (length determines the structure of the LVS) and not a logical one. It should be further noted that the length of any LV is invariant across LVSs of different dimensions; for example, all LVs representing SD variables have length of $1/2$ in any nD LVS.

Moreover, the LVS is endowed with a vector sum. The *vector sum* is defined as follows: $\forall n \in \mathbb{N}$ and $\forall x_j, x_k \in \mathscr{V}^n$ is given by the logical sum of the IDs of the corresponding $X_j, X_k \in \mathscr{B}_n$ times the number $j/2^n \in \mathbb{Q}$ given by Equation (7.1). Thus, (for 3D)

$$x_{1.1} = \frac{4}{8}(X_{1.1}) = \frac{4}{8}(00001111) \quad \text{and} \quad x_{1.4} = \frac{4}{8}(X_{1.4}) = \frac{4}{8}(00110011)$$

we have

$$x_{1.1} + x_{1.4} = \frac{6}{8}(00111111). \tag{7.2}$$

Of course, the 1s in the logical IDs follow the convention of logical sum. The number $6/8$ (and any other rational number denoting the length) is easily obtained by summing the 1s in the ID over the total number of its components.

We say that a set collects *linear independent* vectors when no vector of the set can be written as a linear combination (sum) of other vectors of the set. Since any set of normal linearly independent vectors in the geometric (ordinary) vector space can be written as having a 1 in only one slot, linear independence can be expressed in this case as affirming that no such vectors can have a 1 in the same slot (have orthogonal directions). However, in the logical space, the 0s do not denote "absence" (no component in that direction) but express truth values (as well as the 1s). Thus, the crucial notion for judging whether two or more LVs are linearly independent is how many truth values are shared (how many 0s and 1s are in the same slots). In particular, linearly independent LVs need to share the minimal possible number of truth values. The minimal number is clearly one half of the possible assignation. So, linearly independent LVs need to share exactly one half of the truth values, no more, no less.

A subset S of a LVS is said to *span* the latter if any element of LVS can be written as a linear combination of elements of S. A subset S of LVS is said to be a *basis* for LVS if S spans $\mathscr{V}$ and is a collection of linearly independent vectors. Note that only LVs corresponding to SD variables (in short, SD LVs) can represent a logical basis (they are the only sets of LVs with n elements for any $\mathscr{B}_n$ to share the half of the truth values).

Thus, when dealing with geometrical properties of the LVS it would be better to conceive the two numbers occurring here as -1 (for 0) and $+1$ (for 1) or simply as $-,+$. This allows us to define the *scalar product* as, $\forall x, y$ with

$$x_j = \frac{j}{2^n}(X_j) \quad \text{and} \quad x_k = \frac{k}{2^n}(X_k), \tag{7.3}$$

given by

$$(x_j, x_k) = \sum_i \frac{m}{2^n} X_j^i Y_k^i, \tag{7.4}$$

where the X_j^i, X_k^i are the ith truth value occurring in the logical part of the LVs x_j, x_k, respectively, and the number $m/2^n \in \mathbb{Q}$ given by Equation (7.1). Note that for a 3D LVS we can have 0, 2, and 4 as possible numerators. In general, for any nD LVS ($n \in \mathbb{N}$), we have the series $0, 2, 4, \ldots 2^{n-1}$. It is easy to verify that in such a case the scalar product between linearly independent vectors is in fact zero:

consider the LVs

$$x_{1.1} = \frac{1}{2}(----+++ +), \quad x_{1.4} = \frac{1}{2}(--++--++),$$

$$x_{2.2} = \frac{1}{2}(-+-+-+-+).$$

It is clear that the sum of all products column by column of any pair of these LV will yield 0. One can also verify that the product of any LV $x = j/2^n(X)$ by itself gives $j/2^n$. This is also true for LVs $x = j/2^n(X)$ and $x' = j/2^n(X')$ (where x and x' have opposite orientations), which shows that the scalar product and the logical product (see Equation (2.1)) are not the same.

Thus, we are distinguishing between the notion of geometric orthogonality and logical orthogonality. Geometrically speaking, a basis of orthogonal vectors could be represented by the eight vectors of Tables 3.7 or 3.8. However, the logical product is not zero here. On the other hand, $\forall X$, X and X' are orthogonal in a certain sense (given that their product is zero). However, such an orthogonality concerns the mutual exclusiveness of the aspects or objects (or even events) represented by them, and is not relevant here.

The fact that vectors with opposite orientation are antipodal on the hypersphere and do not make a right angle can be considered an anomaly. However, on the (hyper-)sphere of the quantum density matrices (or on the Poincaré sphere), we represent mutually exclusive events in the same way. Thus, in the logical space two opposite options still have such a property. In fact, it turns out that the LV spaces can be mapped into the quantum Poincaré hypersphere or the hypersphere of density matrices. In particular, we can map SD variables or LVs like $X, X', Y, \ldots$ to projectors $\hat{P}_x, \hat{P}_x^{\perp}, \hat{P}_y, \ldots$, i.e. into the *component states* of a state $\hat{P}_\psi$ in that space, while statements of the form $X + Y$ map to hybrid states that we deal with below. In general, amplitude and phase cannot be represented in the logical space, and in fact depend on specific physical conditions and not on logical structures. Expressions like XY could also represent coincident (classes of) possible events $(\hat{P}_x \hat{P}_y)$, while material equivalence $(X \sim Y)$ and contravalence $(X \nsim Y)$ to (parallel and antiparallel, respectively) entangled states. Note that tautology **1** represents a pure state and is in fact a scalar covering the whole surface of the unitary sphere. This state corresponds to the fundamental property of all quantum systems to occupy as much "space" as possible and thus, when free, to cover all possibilities. In the following we see the consequences of this.

7.2. Practical Representation

It can be helpful to take advantage of Figure 6.1 and try to consider which kind of rotation changes of SCA and CA will induce (see also Table 3.9). Of course, in such schemes, not all vectors appear to be linearly independent. Thus, such a representation has only a practical value. We can pass from SCA1.a to SCA1.b by rotating the cube horizontally by $180°$ so that the new origin $(X_{1.3})$ is now situated in what in the previous reference frame was $X_{1.2}$. We accomplish the change from SCA1.a to SCA1.c by rotating the frame vertically and horizontally by $90°$ so that the bottom square now has corners $X_{1.1}$, $X_{1.2}$, $X_{2.1}$, and $X_{2.2}$, and the origin is now in the old $X_{2.1}$. Finally, we pass from SCA1.a to SCA1.d by applying a vertical rotation of $180°$ in order that the bottom square now has corners $X_{2.1}$, $X_{2.3}$, $X_{2.2}$, and $X_{2.4}$.

We can perform similar transformations for SCA2.x ($x =$ a,b,c,d). In fact, we pass from SCA2.a to SCA2.b by rotating the cube horizontally by $180°$ so that the origin is now $X_{1.4}$. In order to obtain the transformation from SCA2.a to SCA2.c, we rotate the cube vertically by $90°$ so that the origin becomes $X_{2.2}$. Finally, for obtaining the transformation from SC2.a to SCA2.d, we rotate vertically $90°$ and horizontally $180°$ so that the origin is now $X_{2.3}$.

7.3. Quantum Networks

Some of the most interesting developments in the physical sciences are in quantum computation.[2] The use of Boolean algebra

---

[2] A good textbook is [NIELSEN/CHUANG 2000].

in quantum computation is interesting not only for the robustness provided by its codes but also for the likelihood that in the next decades classical and quantum computation will be integrated, the latter used for problems of considerable complexity or difficulty, the former for performing rather ordinary calculations and operations. It is also likely that such an integration will be pushed towards single machines combining these two forms of computation. It is therefore mandatory to explore ways for adopting a single formalism for both approaches.

7.3.1. *Superposition state*

When trying to translate quantum states in the language of logic, one of the first problems that we meet is the significance of the minus sign. We can assume that the minus sign is logically an AND NOT (Table 2.1 and Section 2.5). What is crucial is that we cannot have an isolated term like $-X$, which is deprived of logical meaning. Let us first deal with a superposition of the kind $X' - X$, where, in the binary case, we logically represent the state vector $|0\rangle$ by X' and the state vector $|1\rangle$ by X. The term X can be expanded as $1 - X'$. Assuming that $X' - X = 1$ (where the expression on the left-hand side needs to be taken as an irreducible unit), we have $(1 + X) - (1 - X')$, which clearly is $= 1$. But also the sum of the two terms is equal to 1, so that we have

$$(1 + X) \pm (1 - X') = 1. \tag{7.5}$$

As a consequence, the logical respective of the Hadamard gate acts according to

$$\mathrm{H}X' = X' + X, \quad \mathrm{H}(X' + X) = X', \tag{7.6a}$$

$$\mathrm{H}X = X' - X, \quad \mathrm{H}(X' - X) = X. \tag{7.6b}$$

This shows that a Hadamard gate produces tautologies, and given the generality of this transformation, as a consequence, any superposition needs to be interpreted as a tautology. In particular, if the superposition with the plus sign could be interpreted as giving raise to fringes, that with the minus sign to antifringes.[3]

A historical remark is opportune here. It is quite common to assume that the algebra or the logic underlying quantum mechanics is not Boolean because it is believed that distribution laws (1.6), that is

$$X(Y + Z) = XY + XZ \quad \text{and} \quad X + YZ = (X + Y)(X + Z),$$

$$(7.7)$$

are not satisfied.[4] The whole trouble here is represented by the fact that, often, it is understood that a superposition state must be represented by a contingent logical sum of the kind $X + Y$. Now, a logical sum includes the cases in which *only* X or *only* Y (that could stand for paths in an interferometer) is true, while superposition truly means that *both "ways" are occupied* (if both are open, the quantum system will certainly go both ways), although in a weaker sense than that represented by a conjunction XY that expresses joint real or happened events. Now, once we interpret a superposition state as a sum like $Y + Z$, it is quite natural to consider experimental arrangements for which $X(Y + Z) \neq XY + XZ$. This interpretation is believed to stem from a famous contribution of the American mathematician George Birkhoff (1884–1944) and J. von Neumann,[5] although their original paper is much more careful about this point. In fact, in their paper the authors speak about possible measurement outcomes, and therefore of *property* ascription, and not about the nature of the superposition *state* (and therefore not about the structure of quantum vector [Hilbert] spaces). The reason for the presumed violation of distributivity is due to the incompatibility of experimental arrangements in measuring two non–commuting observables. However, such an incompatibility only concerns the way

---

[3] See e.g. [ENGLERT *et al.* 1995].
[4] See [BELTRAMETTI/CASSINELLI 1981, Chap. 12]. Unfortunately, in the past I have also followed this approach ([AULETTA *et al.* 2009, Sec. 2.4]). Moreover, one finds this statement still today ([COECKE/PAQUETTE 2011, pp. 246–47]).
[5] [BIRKHOFF/VON NEUMANN 1936]. See also [LANDSMAN 2017, Sec. 2.10].

in which we sample the possible events in classes. But these two different sets of events are otherwise fully classical.

In fact, the two possibilities, for example, going on one path in an interferometer and going in the other, are mutually exclusive. Facing a binary and classical choice (at the final detectors), if X' represents the statement that "$|0\rangle$ is observed", it is clear that the statement that "$|1\rangle$ is observed" needs to be represented by X. Thus, as said, beam splitting (and the Hadamard gate) need to be considered as a kind of gate for producing tautologies. I think that part of the trouble here derives from our dealing with projectors. A *projector*, as density matrices, is a mathematical tool for describing *states* (pure states are described by a single projector while mixed states as a weighted sum of projectors), and quantum mechanics is essentially a theory of states and their evolution. As seen, Boolean algebra can be well-mapped to the hypersphere of density matrices. However, the notion of *projection* (often confused with the previous one) is far more tangled. In von Neumann's view it should catch that of property or even the logical notion of proposition or statement (from here the so–called projection postulate). However, property ascription is in general (apart from trivial cases) forbidden in quantum mechanics, if not in dependence of an event (like the click of a detector). Moreover, the use of the projection postulate became even superfluous in the last decades, and in fact most physicists no longer use such a postulate.[6]

Tautology shares two fundamental characters with the superposition state:

- If such a state is described by a state vector $|\psi\rangle$, there is a probability 1 that the system is in such a state (we also know any of its future or past states into which and from which the same probability evolves unitarily), so that we have a certitude. Of course, tautologies also represent certitude as far as they cannot be false and thus are necessarily true.
- However, if we try to get a specific answer (like about the position of the system), we receive none, so that there is incertitude about

---

[6][D'ARIANO *et al.* 2017].

the specific behaviour of the system. Similarly, from a tautology $X + X'$ we cannot know which of the two terms is true.

Thus, we have a tight connection between these two concepts.

7.3.2.   Bell states

For convenience, let us adopt the vector form of representing states in applying the previous resources to quantum computation. First, let us remark that the usual computational basis for two quantum systems

$$|00\rangle := \begin{pmatrix} 1 \\ 0 \\ 0 \\ 0 \end{pmatrix}, \quad |01\rangle := \begin{pmatrix} 0 \\ 1 \\ 0 \\ 0 \end{pmatrix}, \quad |10\rangle := \begin{pmatrix} 0 \\ 0 \\ 1 \\ 0 \end{pmatrix}, \quad |11\rangle := \begin{pmatrix} 0 \\ 0 \\ 0 \\ 1 \end{pmatrix}$$

$$(7.8)$$

perfectly corresponds to the four products of $\mathscr{B}_2$:

$$X_1'X_2' = 1000, \quad X_1'X_2 = 0100, \quad X_1X_2' = 0010, \quad X_1X_2 = 0001,$$

$$(7.9)$$

so that it suffices to put the ID vertically. Let us now consider the transformation of the input states (7.8) into the Bell states[7]

$$|\Phi^+\rangle = \frac{1}{\sqrt{2}} \begin{pmatrix} 1 \\ 0 \\ 0 \\ 1 \end{pmatrix}, \quad |\Psi^+\rangle = \frac{1}{\sqrt{2}} \begin{pmatrix} 0 \\ 1 \\ 1 \\ 0 \end{pmatrix},$$

$$|\Phi^-\rangle = \frac{1}{\sqrt{2}} \begin{pmatrix} 1 \\ 0 \\ 0 \\ -1 \end{pmatrix}, \quad |\Psi^-\rangle = \frac{1}{\sqrt{2}} \begin{pmatrix} 0 \\ 1 \\ -1 \\ 0 \end{pmatrix}, \quad (7.10)$$

respectively. For LVs, we should replace the coefficient $1/\sqrt{2}$ by $1/2$ (although a pure logical approach is sufficient here). Since both

---

[7][BRAUNSTEIN *et al.* 1992].

material equivalence and contravalence individuate two opposite points on the half–unitary sphere, we can represent all four vectors in this way. Here, equivalence stands for correlations ($|\Phi^{\pm}\rangle$ states) and contravalence for anticorrelations ($|\Psi^{\pm}\rangle$ states).

Now, in order to the deal with the minus sign, consider that $X_1 X_2'$ is not the complement of $X_1' X_2$, nor $X_1 X_2$ of $X_1' X_2'$. Thus, in order to assign nodes to the expressions with the minus sign, we need to adopt another strategy. We map $X_1' X_2' + X_1 X_2$ and $X_1' X_2' - X_1 X_2$, as well as the two anticorrelations, into nodes of the 3D algebra, making use of the fact that each 2D SD variable generates 3D doublets. In particular, we associate $X_1' X_2' + X_1 X_2$ with $X_{1.1} \sim X_{2.2}$ (or $X_{1.4} \sim X_{2.3}$; ID: 1010 0101), $X_1' X_2' - X_1 X_2$ with $X_{1.1} \sim X_{2.3}$ (or $X_{1.4} \sim X_{2.2}$; ID: 1001 1001), $X_1' X_2 + X_1 X_2'$ with $X_{1.1} \not\sim X_{2.2}$ (or $X_{1.4} \not\sim X_{2.3}$; ID: 0101 1010), and $X_1' X_2 - X_1 X_2'$ with $X_{1.1} \not\sim X_{2.3}$ (or $X_{1.4} \not\sim X_{2.2}$; ID: 0110 0110). In other words, we are embedding our network in a larger (e.g. 3D) network and adding to the subnetwork two additional nodes. Of course, we do not enter the significance of the minus term, and the two nodes can still be expressed in different ways in alternative codifications. What is important is to back-translate these two expressions into the original ones when performing calculations. Indeed, the correctness of the above formulations is proved by the correspondences

$$\frac{1}{\sqrt{2}}\left(|\Phi^+\rangle + |\Phi^-\rangle\right) = |00\rangle, \quad X_1' X_2' + X_1 X_2 + X_1' X_2' - X_1 X_2 = X_1' X_2',$$

$$(7.11a)$$

$$\frac{1}{\sqrt{2}}\left(|\Psi^+\rangle + |\Psi^-\rangle\right) = |01\rangle, \quad X_1' X_2 + X_1 X_2' + X_1' X_2 - X_1 X_2' = X_1' X_2,$$

$$(7.11b)$$

$$\frac{1}{\sqrt{2}}\left(|\Psi^+\rangle - |\Psi^-\rangle\right) = |10\rangle, \quad X_1' X_2 + X_1 X_2' - X_1' X_2 + X_1 X_2' = X_1 X_2',$$

$$(7.11c)$$

$$\frac{1}{\sqrt{2}}\left(|\Phi^+\rangle - |\Phi^-\rangle\right) = |11\rangle, \quad X_1' X_2' + X_1 X_2 - X_1' X_2' + X_1 X_2 = X_1 X_2.$$

$$(7.11d)$$

Of course, in an entanglement like, say $|\Phi^+\rangle$, we expect that measuring one of the two subsystems and getting a certain result, the other subsystem will be in the same state. The logical counterpart is easily provided by a product, namely:

$$X(XY + X'Y') = XY. \tag{7.12}$$

Since we expect the same result if we measure the other system, we can write the equality

$$X(XY + X'Y') = Y(XY + X'Y'). \tag{7.13}$$

The equation is similar for the other forms of entanglement.

Now, the production of the Bell states (7.10) is given by the circuit

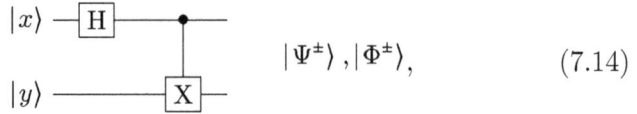

$$|\Psi^\pm\rangle, |\Phi^\pm\rangle, \tag{7.14}$$

where $x, y = 0, 1$. The first gate is the Hadamard gate, while the second the CNOT gate. We can use the matrix product

$$\text{CNOT H} = \begin{bmatrix} 1 & 0 & 0 & 0 \\ 0 & 1 & 0 & 0 \\ 0 & 0 & 0 & 1 \\ 0 & 0 & 1 & 0 \end{bmatrix} \begin{bmatrix} 1 & 0 & 1 & 0 \\ 0 & 1 & 0 & 1 \\ 1 & 0 & -1 & 0 \\ 0 & 1 & 0 & -1 \end{bmatrix} = \begin{bmatrix} 1 & 0 & 1 & 0 \\ 0 & 1 & 0 & 1 \\ 0 & 1 & 0 & -1 \\ 1 & 0 & -1 & 0 \end{bmatrix}$$

$$\tag{7.15}$$

for generating the whole transformation.

Thus, the two steps of the circuit (7.14) can be resumed as

$$X_1' X_2' \longmapsto (X_1' + X_1)X_2' \longmapsto X_1' X_2' + X_1 X_2, \quad |00\rangle \longmapsto |\Phi^+\rangle, \tag{7.16a}$$

$$X_1' X_2 \longmapsto (X_1' + X_1)X_2 \longmapsto X_1' X_2 + X_1 X_2', \quad |01\rangle \longmapsto |\Psi^+\rangle, \tag{7.16b}$$

$$X_1 X_2' \longmapsto (X_1' - X_1) X_2' \longmapsto X_1' X_2' - X_1 X_2, \quad |10\rangle \longmapsto |\Phi^-\rangle ,$$

$$(7.16c)$$

$$X_1 X_2 \longmapsto (X_1' - X_1) X_2 \longmapsto X_1' X_2 - X_1 X_2', \quad |11\rangle \longmapsto |\Psi^-\rangle .$$

$$(7.16d)$$

By taking the inverse (transpose) transformation

$$
\text{H CNOT} =
\begin{bmatrix}
1 & 0 & 1 & 0 \\
0 & 1 & 0 & 1 \\
1 & 0 & -1 & 0 \\
0 & 1 & 0 & -1
\end{bmatrix}
\begin{bmatrix}
1 & 0 & 0 & 0 \\
0 & 1 & 0 & 0 \\
0 & 0 & 0 & 1 \\
0 & 0 & 1 & 0
\end{bmatrix}
=
\begin{bmatrix}
1 & 0 & 0 & 1 \\
0 & 1 & 1 & 0 \\
1 & 0 & 0 & -1 \\
0 & 1 & -1 & 0
\end{bmatrix},
$$

$$(7.17)$$

that is, inverting the order of the Hadamard and the CNOT gate, we can obtain the computational basis back:

$$X_1' X_2' + X_1 X_2 \longmapsto (X_1' + X_1) X_2' \longmapsto X_1' X_2', \quad |\Phi^+\rangle \longmapsto |00\rangle ,$$

$$(7.18a)$$

$$X_1' X_2 + X_1 X_2' \longmapsto (X_1' + X_1) X_2 \longmapsto X_1' X_2, \quad |\Psi^+\rangle \longmapsto |01\rangle ,$$

$$(7.18b)$$

$$X_1' X_2' - X_1 X_2 \longmapsto (X_1' - X_1) X_2' \longmapsto X_1 X_2', \quad |\Phi^-\rangle \longmapsto |10\rangle ,$$

$$(7.18c)$$

$$X_1' X_2 - X_1 X_2' \longmapsto (X_1' - X_1) X_2 \longmapsto X_1 X_2, \quad |\Psi^-\rangle \longmapsto |11\rangle ,$$

$$(7.18d)$$

which of course means that the product of H CNOT and CNOT H, in whatever order, gives the identity matrix (i.e. the whole transformation is unitary or reversible).

7.3.3. *A worry*

A possible source of problems is how to interpret in quantum–mechanical vectorial terms a logical expression like $X_1 + X_2$. From a logical point of view we have the identity

$$X_1 + X_2 = X_1 X_2 + X_1 X_2' + X_1' X_2, \tag{7.19}$$

so that we can interpret such expression as

$$\frac{1}{\sqrt{3}}(|11\rangle + |01\rangle + |10\rangle) = \frac{1}{\sqrt{3}}|11\rangle + \sqrt{\frac{2}{3}}|\Psi^+\rangle. \tag{7.20}$$

It would be perhaps more appropriate to interpret such an expression as a hybrid state like[8]

$$|\Psi_H\rangle = \frac{1}{\sqrt{3}}(|ha\rangle + |va\rangle + |hb\rangle), \tag{7.21}$$

where h, v are horizontal and vertical polarisations, respectively, for photon 1, and a, b are the two paths of an interferometer, respectively, for photon 2. Nevertheless, for the present discussion the previous expression can go. Since the term $|00\rangle$ is absent, the vector (7.20) tells us that when the state of either system 1 or system 2 is $|0\rangle$, the state of the other system is $|1\rangle$: we thus have a kind of half-entanglement. As a matter of fact, the logical counterparts are halfway between entangled states (represented by equivalences and contravalences) and tautology. We reach tautology when we add the term $X_1' X_2'$. In fact, in such a case, if we correspondently add the term $|00\rangle$ to the state (7.20), we get the product state

$$\frac{1}{2}(|00\rangle + |01\rangle + |10\rangle + |11\rangle) = \frac{1}{2}(|0\rangle + |1\rangle)_1 (|0\rangle + |1\rangle)_2, \tag{7.22}$$

which is fully parallel to

$$X_1' X_2' + X_1' X_2 + X_1 X_2' + X_1 X_2 = (X_1' + X_1)(X_2' + X_2). \tag{7.23}$$

For a full understanding of all this, consider that entanglements set a constraint on the relatively and otherwise unconstrained product states and as such also encompass a classical component.[9]

---

[8][MA *et al.* 2013].
[9][AULETTA 2019, Subsec. 5.1.2].

7.3.4. *Toffoli gate*

The Toffoli gate is a way to make a logical sum or product reversible, which are otherwise irreversible operations (from the output you cannot know the value of the inputs apart from the sum being **0** or the product being **1**).[10] This can be quantum–mechanically implemented thanks to the circuit

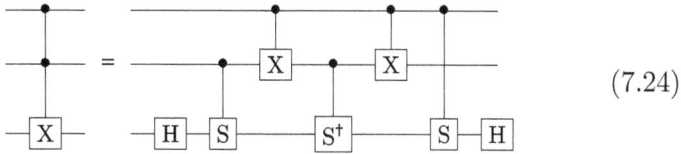

$$(7.24)$$

Note that the half-phase matrix S is defined as

$$S := \begin{bmatrix} 1 & 0 \\ 0 & i \end{bmatrix}, \tag{7.25}$$

so that the square of S is the phase shift

$$Z = S^2 = \begin{bmatrix} 1 & 0 \\ 0 & i \end{bmatrix} \begin{bmatrix} 1 & 0 \\ 0 & i \end{bmatrix} = \begin{bmatrix} 1 & 0 \\ 0 & -1 \end{bmatrix}. \tag{7.26}$$

Thanks to the notorious Euler's identity

$$e^{\pi i} + 1 = 0, \tag{7.27}$$

we can interpret $e^{\pi i}$ as -1 as well as $e^{2\pi i}$ as **1**. Note that the Z operator in logical terms acts as follows:

$$\hat{Z}(X' + X) = X' + (-X), \tag{7.28}$$

where I have used a hat for avoiding confusion with the SD variable Z. However, it is more difficult to interpret S with its $e^{\frac{\pi}{2}i} = i$. The best choice is to leave it as it stands. This is no worry since, in general, in quantum computing protocols the imaginary units are usually dropped in the last step. Moreover, a global phase factor can be dropped every time.

---

[10][TOFFOLI 1980].

What is relevant is that circuit (7.24) leaves all states but $|110\rangle$ and $|111\rangle$ unchanged and commutes the latter two. In other words, it flips the third qubit when the first two are in the state $|1\rangle$. Let us therefore consider what happens to these two inputs:

$$|110\rangle \mapsto \frac{1}{\sqrt{2}}(|110\rangle + |111\rangle) \mapsto \frac{1}{\sqrt{2}}(|110\rangle + |11\rangle\, i\,|1\rangle)$$

$$\mapsto \frac{1}{\sqrt{2}}(|100\rangle + |10\rangle\, i\,|1\rangle) \mapsto \frac{1}{\sqrt{2}}(|100\rangle + |10\rangle\, i\,|1\rangle)$$

$$\mapsto \frac{1}{\sqrt{2}}(|110\rangle + |11\rangle\, i\,|1\rangle) \mapsto \frac{1}{\sqrt{2}}(|110\rangle + |11\rangle\, (-|1\rangle))$$

$$\mapsto \frac{1}{2}(|110\rangle + |111\rangle - |110\rangle + |111\rangle) = |111\rangle, \qquad (7.29a)$$

$$|111\rangle \mapsto \frac{1}{\sqrt{2}}(|110\rangle - |111\rangle) \mapsto \frac{1}{\sqrt{2}}(|110\rangle - |11\rangle\, i\,|1\rangle)$$

$$\mapsto \frac{1}{\sqrt{2}}(|100\rangle - |10\rangle\, i\,|1\rangle) \mapsto \frac{1}{\sqrt{2}}(|100\rangle - |10\rangle\, i\,|1\rangle)$$

$$\mapsto \frac{1}{\sqrt{2}}(|110\rangle - |11\rangle\, i\,|1\rangle) \mapsto \frac{1}{\sqrt{2}}(|110\rangle - |11\rangle\, (-|1\rangle))$$

$$\mapsto \frac{1}{2}(|110\rangle + |111\rangle + |110\rangle - |111\rangle) = |110\rangle. \qquad (7.29b)$$

This can put in logical terms as follows (having previously proved the correspondence with logical expressions of $\mathscr{B}_2$, we can now use general symbols valid for any dimensions $n \geq 2$):

$$XYZ' \mapsto XYZ' + XYZ \mapsto \hat{S}_3(XYZ' + XYZ)$$

$$\mapsto \hat{S}_3(XY'Z' + XY'Z) \mapsto \hat{S}_3(XY'Z' + XY'Z)$$

$$\mapsto \hat{S}_3(XYZ' + XYZ) \mapsto XYZ' - XYZ$$

$$\mapsto XYZ' + XYZ - XYZ' + XYZ = XYZ, \qquad (7.30a)$$

$$XYZ \mapsto XYZ' - XYZ \mapsto \hat{S}_3(XYZ' - XYZ)$$

$$\mapsto \hat{S}_3(XY'Z' - XY'Z) \mapsto \hat{S}_3(XY'Z' - XY'Z)$$

$$\mapsto \hat{S}_3(XYZ' - XYZ) \mapsto XYZ' + XYZ$$

$$\mapsto XYZ' + XYZ + XYZ' - XYZ = XYZ'. \qquad (7.30b)$$

7.3.5. *Deutsch's problem*

By using such Boolean operations, it is possible to solve quantum–mechanically Deutsch's problem. Suppose a device that can evaluate the operation O and that it is allowed to run only once. The so-called *Deutsch's problem* is to ask whether, under these conditions, it is possible to determine if the operation O is constant or balanced (it gives always the same output or half the time 0 and half the time 1),[11] which cannot be solved classically. Quantum–mechanically we can set the problem as follows:

$$|x\rangle\,|y\rangle \overset{\hat{U}}{\longmapsto} |x\rangle\,|O(x)\oplus y\rangle\,, \tag{7.31}$$

where $|x\rangle\,,|y\rangle = |0\rangle\,,|1\rangle$ and $\oplus$ is the modulo-2 addition (2.15). In other words, O acts on x and is then modulo-2 added to y. Quantum mechanics allows any initial superposition of the computational basis, so that the previous equation can be generalised to

$$\sum_j c_j\,|x_j\rangle\,|y\rangle \overset{\hat{U}}{\longmapsto} \sum_j c_j\,|x_j\rangle\,|O(x_j)\oplus y\rangle\,, \tag{7.32}$$

where $|x_j\rangle$ are arbitrary qubit states and c_j's $\in \mathbb{C}$ are coefficients satisfying the normalisation condition. Let us now run the following circuit:

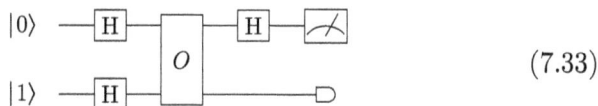

$$\tag{7.33}$$

The two qubits are first transformed separately by the two Hadamard gates

$$|01\rangle \overset{\mathrm{H\otimes H}}{\longmapsto} \mathrm{H}\,|0\rangle\,\mathrm{H}\,|1\rangle = \frac{1}{2}\left(|0\rangle + |1\rangle\right)\left(|0\rangle - |1\rangle\right). \tag{7.34}$$

For later convenience, the resultant state of Equation (7.34) can be rewritten as

$$|\Psi\rangle = \frac{1}{2}\left(|00\rangle + |10\rangle - |01\rangle - |11\rangle\right). \tag{7.35}$$

---

[11][DEUTSCH 1985a, DEUTSCH 1985b].

The state $|\Psi\rangle$ is then processed by the Boolean gate, leading to four possible results, depending on the nature of the operation O. If O is a constant operation, then we obtain

$$
\hat{U}(0,0) : |\Psi\rangle \longmapsto \frac{1}{2}(|0\rangle \, |0 \oplus 0\rangle + |1\rangle \, |0 \oplus 0\rangle
$$
$$
- |0\rangle \, |0 \oplus 1\rangle - |1\rangle \, |0 \oplus 1\rangle)
$$
$$
= \frac{1}{2}[(|0\rangle + |1\rangle) \, |0\rangle - (|0\rangle + |1\rangle) \, |1\rangle]
$$
$$
= |+\rangle \, |-\rangle , \tag{7.36a}
$$

$$
\hat{U}(1,1) : |\Psi\rangle \longmapsto \frac{1}{2}(|0\rangle \, |1 \oplus 0\rangle + |1\rangle \, |1 \oplus 0\rangle
$$
$$
- |0\rangle \, |1 \oplus 1\rangle - |1\rangle \, |1 \oplus 1\rangle)
$$
$$
= \frac{1}{2}(|01\rangle + |11\rangle - |00\rangle - |10\rangle)
$$
$$
= -|+\rangle \, |-\rangle , \tag{7.36b}
$$

where

$$
|+\rangle = \frac{1}{\sqrt{2}}(|0\rangle + |1\rangle), \quad |-\rangle = \frac{1}{\sqrt{2}}(|0\rangle - |1\rangle). \tag{7.36c}
$$

The expression $\hat{U}(0,0)$ means $O(0) = O(1) = 0$, while $\hat{U}(1,1)$ means $O(0) = O(1) = 1$ (note that the two outputs only differ for a global phase factor). On the other hand, for the balanced operations we have

$$
\hat{U}(0,1) : |\Psi\rangle \longmapsto \frac{1}{2}(|0\rangle \, |0 \oplus 0\rangle + |1\rangle \, |1 \oplus 0\rangle
$$
$$
- |0\rangle \, |0 \oplus 1\rangle - |1\rangle \, |1 \oplus 1\rangle)
$$
$$
= \frac{1}{2}(|00\rangle + |11\rangle - |01\rangle - |10\rangle)
$$
$$
= |-\rangle \, |-\rangle , \tag{7.36d}
$$

$$
\hat{U}(1,0) : |\Psi\rangle \longmapsto \frac{1}{2}(|0\rangle \, |1 \oplus 0\rangle + |1\rangle \, |0 \oplus 0\rangle - |0\rangle \, |1 \oplus 1\rangle
$$
$$
- |1\rangle \, |0 \oplus 1\rangle)
$$
$$
= \frac{1}{2}(|01\rangle + |10\rangle - |00\rangle - |11\rangle)
$$
$$
= -|-\rangle \, |-\rangle , \tag{7.36e}
$$

where again the two outputs differ for a global phase factor. Each of the above four results (two for the constant case and two for the balanced case) is further processed by the final Hadamard gate, leading to the results

$$\pm|+\rangle\,|-\rangle \xrightarrow{\mathrm{H}\otimes\hat{I}} \pm|0\rangle\,|-\rangle \quad \text{if } O \text{ is constant,} \tag{7.37a}$$

$$\pm|-\rangle\,|-\rangle \xrightarrow{\mathrm{H}\otimes\hat{I}} \pm|1\rangle\,|-\rangle \quad \text{if } O \text{ is balanced.} \tag{7.37b}$$

In order to formulate this in logical terms, let us have a closer look at the four properties (2.1). It is evident that the two properties on the left always give the same result (either $\mathbf{0}$ or $\mathbf{1}$) independently from the value of the expression (there X) to which they are connected, while the two properties on the right give a truth value that completely depends on the value of that expression. In other words, these four cases can be considered as a set of Boolean operations ($\mathbf{0}$ times, $\mathbf{1}$ plus, $\mathbf{0}$ plus, and $\mathbf{1}$ times, respectively) such that the first two are constant (always give the same result), while the other two are balanced, since they give in the mean half-time 0 and half-time 1.

Now, we can translate the previous computation in logical terms. First, let us apply the Hadamard transformations (7.6) to the initial state $|01\rangle$:

$$(\mathrm{H}X')(\mathrm{H}Y) = X'Y' - X'Y + XY' - XY. \tag{7.38}$$

When the systems pass the Boolean gate they undergo the following transformations:

$$\mathbf{0}\times: \quad X'Y' - X'Y + XY' - XY, \tag{7.39}$$

$$\mathbf{1}+: \quad X'Y - X'Y' + XY - XY', \tag{7.40}$$

$$\mathbf{0}+: \quad X'Y' - X'Y + XY - XY', \tag{7.41}$$

$$\mathbf{1}\times: \quad X'Y' - X'Y + XY - XY'. \tag{7.42}$$

Note that the last two outputs are identical; a global minus sign will change nothing. Finally, by using the Hadamard gate again on X, we get $\pm X'$ if O is constant and $\pm X$ if balanced.

7.3.6.   *Different superpositions*

Up to now we have dealt with the computational basis ($|0\rangle$, $|1\rangle$) and symmetric forms of superposition ($|+\rangle$, $|-\rangle$). However, when we deal with quantum computation, very often we deal with (asymmetric) superpositions with different coefficients. Moreover, often we have two different superpositions in the same expression, like the following product state of two systems

$$(c_0\,|0\rangle + c_1\,|1\rangle)(d_0\,|0\rangle + d_1\,|1\rangle), \tag{7.43}$$

with all coefficients different and $\in \mathbb{C}$. In many cases we need to distinguish between these two states (e.g. for teleportation). Logic is the combinatorics of the possibilities and, therefore, also possible events or experimental outcomes, but says nothing about the probabilities to get those events (computed as square moduli of the relative coefficients), since it is two-valued. We can vary the coefficients of a superposition state as we like (with the exclusion of 0 and 1), but the possible outcomes remain always the same. In other words, a superposition, independently of the value of the coefficients (apart from 0 an 1), needs to be expressed always as a tautology. Then, how can we distinguish, say, between the left and the right superposition in the previous equation?

It turns out that, by varying the coefficients (and the relative phase) of a superposition, we effect a change of basis, i.e. of codification, so that the coefficient themselves (and the relative phase of the superposition) represent the rules according to which we *combine* the two computational basis states $|0\rangle$ and $|1\rangle$, and these combinations can be understood as *codewords*. Thus, when we change the coefficients, we are logically expressing a given quantum state in a different code alphabet.

Consider Figure 7.2. Let us call γ the angle between $|\psi\rangle$ and $|x_2\rangle$ and β the angle between $|\psi'\rangle$ and $|x_1'\rangle$. We can write the initial state of the system under consideration as

$$|\psi\rangle = \cos(2\alpha + \beta)\,|x_1\rangle + \sin\gamma\,|x_2\rangle. \tag{7.44}$$

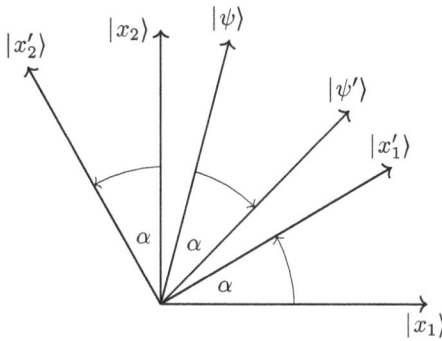

Figure 7.2. A transformation may be considered from two equivalent viewpoints: from the *active* point of view, the state vector $|\psi\rangle$ is transformed (here represented by a clockwise rotation by an angle α) into the state vector $|\psi'\rangle$ while the basis vectors $|x_1\rangle$ and $|x_2\rangle$ are kept fixed; from the *passive* point of view, the basis vectors $|x_1\rangle$ and $|x_2\rangle$ are transformed in a reverse manner (here represented by a counterclockwise rotation by the same angle α) to the basis vectors $|x_1'\rangle$ and $|x_2'\rangle$, respectively. Note that the original state $|\psi\rangle$ in the new basis $\{|x_1'\rangle, |x_2'\rangle\}$ is equivalent to the transformed state $|\psi'\rangle$ in the old basis $\{|x_1\rangle, |x_2\rangle\}$.

Now we can let perform a rotation of the state vector and get a new state vector

$$|\psi'\rangle = \cos(\alpha + \beta)|x_1\rangle + \sin(\alpha + \gamma)|x_2\rangle, \qquad (7.45)$$

which can be also expressed in terms of another basis:

$$|\psi\rangle = \cos(\alpha + \beta)|x_1'\rangle + \sin(\alpha + \gamma)|x_2'\rangle, \qquad (7.46)$$

which shows that the state vector $|\psi\rangle$ in the basis $\{|x_1'\rangle, |x_2'\rangle\}$ is equivalent to the state vector $|\psi'\rangle$ in the basis $\{|x_1\rangle, |x_2\rangle\}$.

Thus, to know in a quantum information protocol what is the basis used is not irrelevant. Therefore, we need to specify the basis or the alphabet code in which a given superposition is expressed. This can be done with a simple label. For instance, if we express the product state (7.43) as $\psi_A\psi_B$, we can rephrase it in logical terms as 1_A1_B. This allows us to formulate, for example, the teleportation protocol without problems. We can teleport a system 1 in the usual state[12]

$$|\psi\rangle_1 = c_0 |0\rangle + c_1 |1\rangle , \tag{7.47}$$

while the initially entangled systems 2 and 3 are in the state

$$|\Phi^+\rangle = \frac{1}{\sqrt{2}} (|00\rangle + |11\rangle). \tag{7.48}$$

Then, we can build the following circuit:

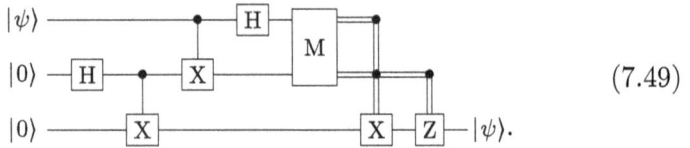

$$\tag{7.49}$$

The first two gates on the left allow the preparation of the state (7.48). The box M denotes the joint measurement on systems 1–2. Note that double lines carry classical information (from Alice to Bob). The classical output of M on the first row is x, while that of the second row is y.

Then, we proceed in explicit computational form as follows:

$$|\psi\rangle |\Phi^+\rangle \overset{\text{CNOT}_{12}}{\longmapsto} \frac{1}{\sqrt{2}} [c_0 |0\rangle (|00\rangle + |11\rangle) + c_1 |1\rangle (|10\rangle + |01\rangle)]$$

$$\overset{H_1}{\longmapsto} \frac{1}{2} [|00\rangle (c_0 |0\rangle + c_1 |1\rangle) + |01\rangle (c_0 |1\rangle + c_1 |0\rangle)$$

$$+ |10\rangle (c_0 |0\rangle - c_1 |1\rangle) + |11\rangle (c_0 |1\rangle - c_1 |0\rangle)]. \tag{7.50}$$

Now, a measurement performed not on the Bell states but using the basis (7.8) for systems 1–2, whose results xy are random, leaves the

---

[12][BENNETT/WIESNER 1992, BENNETT *et al.* 1993, GOTTESMAN/CHUANG 1999].

qubit belonging to Bob in the state $\hat{U}_{xy}\,|\psi'\rangle$, where we have

$$\hat{U}_{00} = \hat{I}, \qquad i.e.\ |\psi'\rangle \longmapsto |\psi'\rangle, \tag{7.51}$$

$$\hat{U}_{01} = X, \qquad i.e.\ |\psi'\rangle \longmapsto |\psi' \oplus 1\rangle, \tag{7.52}$$

$$\hat{U}_{10} = Z, \qquad i.e.\ |\psi'\rangle \longmapsto (-1)^{\psi'}\,|\psi'\rangle, \tag{7.53}$$

$$\hat{U}_{11} = iY, \qquad i.e.\ |\psi'\rangle \longmapsto i(-1)^{\psi'}\,|\psi' \oplus 1\rangle. \tag{7.54}$$

The modulo-2 addition and exponentiation need to be performed on the two components of the qubit separately. Bob needs only to apply the inverse $\hat{U}_{xy}^{\dagger} = \hat{U}_{xy}$ of one of the previous transformations (according to Alice's classical communication) in order to recover $|\psi\rangle$.

Practically, the chosen basis for systems 1–2 works as a two–bit code. The final unitary operation of Bob (third row) is a controlled operation: if Alice gets $|01\rangle$, it is the second qubit to control the Z operation; if the Alice's result is $|10\rangle$, it is the first qubit to control the X gate; if it is $|11\rangle$, both qubits control the gates ($|00\rangle$ is mapped to the identity transformation). In other words, such operations can be performed without any human intervention (apart from the measurement step).

We have already met the Pauli matrix Z (Equation (7.28)). The matrix X is a qubit flipper, so we can write in logical terms

$$\hat{X}(X + X') = (X)' + (X')', \tag{7.55}$$

where the order is important when different coefficients are involved. Of course, $\hat{Y} = i\hat{X}\hat{Z}$. It is evident that two applications of the logical counterparts of Pauli matrices give the original expression. Thus, we can write in synthetic terms

$$\mathbf{1}_1(YZ + Y'Z') \mapsto Y'Z'(\mathbf{1}_1) + Y'Z(\hat{X}\mathbf{1}_1) + YZ'(\hat{Z}\mathbf{1}_1) + YZ(\hat{X}\hat{Z}\mathbf{1}_1). \tag{7.56}$$

Logically speaking, it is a map $\mathbf{1}_1 \mapsto \mathbf{1}_1$, and Bob's needed operations are then evident.

In conclusion, let us consider some powerful combination of classical and quantum information. Consider two 2D Boolean networks. The two ordinary SD variables and their complements for each network need to be mapped to 4 different particles (e.g. electrons trapped in artificial atoms or photons trapped in microcavities). Each node tells us how many photons or electrons are trapped, if any. Of course, all possible combinations are displayed. Note that if one of the contingent nodes is true, some others are also true (or turned on), and similarly for nodes turned off. Now, suppose that each of these particles is entangled with one of the twins particles in the other network. Then, depending on the kind of entanglement, if we perform a determinative measurement (or premeasurement) one of these particles (or even two or more jointly) in one network in order to get e.g. 1 as a result (i.e. the node is true), we automatically now a lot of different things about the other network. The IDs of the nodes of the other network that are turned on could represent strings of information and constituted therefore a code. Another way to consider the model is that, with a single operation on a Boolean network we can change a lot of things in another Boolean network. Of course, increasing the dimensions allows far more possibilities.

Chapter 8

Application of Representation Theory

8.1. Linear Transformations on the Same "Surface"

According to representation theory, a representation is a vector space with relative linear transformations into which we can map an algebraic structure. Having shown that we can build a collection of vectors in a vector space that is isomorphic to any $\mathscr{B}_n$, we now show that there are also linear transformations (matrices) that bring transformation from one vector to another. In particular, we find a single general matrix at least for each class or subclass of transformations. Having shown that all nodes can be represented either as the negation of each other or the neg–reversal of each other or of themselves, we run into an immediate difficulty: we cannot provide a general matrix for negation. However, what we can do is use a transformation that exchanges half of each chunk with the other half. Thus, for $\mathscr{B}_2$, we transform 0001 into 0100 and vice versa, 0010 into 1000 and vice versa, 0111 into 1101 and vice versa, 1011 into 1110 and vice versa, 0110 into 1001 and vice versa, and 0011 into 1100 and vice versa, while 0101 and 1010 are invariant. Of course, we cannot adopt such a procedure for 0000 and 1111, which nevertheless are the null vector and the surface encircling the space. The general matrix for this algebra is

$$\epsilon_2 = \begin{bmatrix} 0 & 0 & 1 & 0 \\ 0 & 0 & 0 & 1 \\ 1 & 0 & 0 & 0 \\ 0 & 1 & 0 & 0 \end{bmatrix}. \tag{8.1}$$

For building the matrices for higher dimensional algebras, we simply follow the reiteration

$$\epsilon_n = \begin{bmatrix} \epsilon_{n-1} & 0 \\ 0 & \epsilon_{n-1} \end{bmatrix}, \tag{8.2}$$

where $n > 3$ and 0 denotes a block matrix. Thus, for $\mathscr{B}_3$ we obtain

$$\epsilon_3 = \begin{bmatrix} 0 & 0 & 1 & 0 & 0 & 0 & 0 & 0 \\ 0 & 0 & 0 & 1 & 0 & 0 & 0 & 0 \\ 1 & 0 & 0 & 0 & 0 & 0 & 0 & 0 \\ 0 & 1 & 0 & 0 & 0 & 0 & 0 & 0 \\ 0 & 0 & 0 & 0 & 0 & 0 & 1 & 0 \\ 0 & 0 & 0 & 0 & 0 & 0 & 0 & 1 \\ 0 & 0 & 0 & 0 & 1 & 0 & 0 & 0 \\ 0 & 0 & 0 & 0 & 0 & 1 & 0 & 0 \end{bmatrix}. \tag{8.3}$$

Note that all of the above matrices are the inverse of themselves (and therefore they are all Hermitian). Of course, there are other possible solutions to this problem, but this is one of the easiest ones.

8.2. Raising and Lowering Operators

The previous formalism is correct, but with a substantial limitation: it transforms vectors of a given length (with a fixed number of 1s) into vectors of the same length (same number of 1s). We may be interested, on the other hand, in transformations that correspond to those occurring among different levels of the algebra. In this case, we introduce raising and lowering operators, in particular, a set of operators that allows us to transition from any level to any next higher (or lower) level. Such a formalism parallels that of quantum mechanics. There is also a limitation here: if each row of a columnar vector represents a state (e.g. an energy level), this formalism tells us whether such a state is occupied or not and not how many systems (in the case of bosons) are in that state. Moreover, we do not introduce transformations from and into **0** and **1**, although in the latter case are obvious and in the first case, when dealing with raising operators, we need to insert an additional constant 1 representing, for instance, the environment.

Let us first consider the 2D case. The set of raising operators is the following:

$$\begin{bmatrix} 1 & 0 & 0 & 0 \\ 1 & 1 & 0 & 0 \\ 0 & 0 & 1 & 0 \\ 0 & 0 & 0 & 1 \end{bmatrix}, \begin{bmatrix} 1 & 0 & 0 & 0 \\ 0 & 1 & 0 & 0 \\ 1 & 0 & 1 & 0 \\ 0 & 0 & 0 & 1 \end{bmatrix}, \begin{bmatrix} 1 & 0 & 0 & 0 \\ 0 & 1 & 0 & 0 \\ 0 & 0 & 1 & 0 \\ 1 & 0 & 0 & 1 \end{bmatrix}, \quad (8.4a)$$

$$\begin{bmatrix} 1 & 0 & 0 & 0 \\ 0 & 1 & 0 & 0 \\ 0 & 1 & 1 & 0 \\ 0 & 0 & 0 & 1 \end{bmatrix}, \begin{bmatrix} 1 & 0 & 0 & 0 \\ 0 & 1 & 0 & 0 \\ 0 & 0 & 1 & 0 \\ 0 & 1 & 0 & 1 \end{bmatrix}, \quad (8.4b)$$

$$\begin{bmatrix} 1 & 0 & 0 & 0 \\ 0 & 1 & 0 & 0 \\ 0 & 0 & 1 & 0 \\ 0 & 0 & 1 & 1 \end{bmatrix}. \quad (8.4c)$$

The reader may verify that when the 6 operators are applied to vectors of Level (length) 1/4 they give rise to the four vectors of Level (length) 1/2, and when applied to those of Level (length) 1/2, they give rise to vectors of Level (length) 3/4 (each repeated 3 times). Of course, other solutions are possible, but the one proposed here is relatively simple and can be reiterated for higher dimensional spaces. Now, let us consider a set of lowering operators bringing back vectors of any length to the next lower length:

$$
\begin{bmatrix} 0 & 0 & 0 & 0 \\ 0 & 1 & 0 & 0 \\ 0 & 0 & 1 & 0 \\ 0 & 0 & 0 & 1 \end{bmatrix},\quad
\begin{bmatrix} 1 & 0 & 0 & 0 \\ 0 & 0 & 0 & 0 \\ 0 & 0 & 1 & 0 \\ 0 & 0 & 0 & 1 \end{bmatrix},\quad
\begin{bmatrix} 1 & 0 & 0 & 0 \\ 0 & 1 & 0 & 0 \\ 0 & 0 & 0 & 0 \\ 0 & 0 & 0 & 1 \end{bmatrix},\quad
\begin{bmatrix} 1 & 0 & 0 & 0 \\ 0 & 1 & 0 & 0 \\ 0 & 0 & 1 & 0 \\ 0 & 0 & 0 & 0 \end{bmatrix}.
$$

$$(8.5)$$

As said, such a procedure is recursive. In fact, for $\mathscr{B}_3$ we can build a set of 28 raising operators allowing the passage from any length or level to the next length or level:

$$
\begin{bmatrix}
1&0&0&0&0&0&0&0\\
1&1&0&0&0&0&0&0\\
0&0&1&0&0&0&0&0\\
0&0&0&1&0&0&0&0\\
0&0&0&0&1&0&0&0\\
0&0&0&0&0&1&0&0\\
0&0&0&0&0&0&1&0\\
0&0&0&0&0&0&0&1
\end{bmatrix},\quad
\begin{bmatrix}
1&0&0&0&0&0&0&0\\
0&1&0&0&0&0&0&0\\
1&0&1&0&0&0&0&0\\
0&0&0&1&0&0&0&0\\
0&0&0&0&1&0&0&0\\
0&0&0&0&0&1&0&0\\
0&0&0&0&0&0&1&0\\
0&0&0&0&0&0&0&1
\end{bmatrix},\quad
\begin{bmatrix}
1&0&0&0&0&0&0&0\\
0&1&0&0&0&0&0&0\\
0&0&1&0&0&0&0&0\\
1&0&0&1&0&0&0&0\\
0&0&0&0&1&0&0&0\\
0&0&0&0&0&1&0&0\\
0&0&0&0&0&0&1&0\\
0&0&0&0&0&0&0&1
\end{bmatrix},
$$

$$
\begin{bmatrix}
1&0&0&0&0&0&0&0\\
0&1&0&0&0&0&0&0\\
0&0&1&0&0&0&0&0\\
0&0&0&1&0&0&0&0\\
1&0&0&0&1&0&0&0\\
0&0&0&0&0&1&0&0\\
0&0&0&0&0&0&1&0\\
0&0&0&0&0&0&0&1
\end{bmatrix},\quad
\begin{bmatrix}
1&0&0&0&0&0&0&0\\
0&1&0&0&0&0&0&0\\
0&0&1&0&0&0&0&0\\
0&0&0&1&0&0&0&0\\
0&0&0&0&1&0&0&0\\
1&0&0&0&0&1&0&0\\
0&0&0&0&0&0&1&0\\
0&0&0&0&0&0&0&1
\end{bmatrix},\quad
\begin{bmatrix}
1&0&0&0&0&0&0&0\\
0&1&0&0&0&0&0&0\\
0&0&1&0&0&0&0&0\\
0&0&0&1&0&0&0&0\\
0&0&0&0&1&0&0&0\\
0&0&0&0&0&1&0&0\\
1&0&0&0&0&0&1&0\\
0&0&0&0&0&0&0&1
\end{bmatrix},
$$

$$
\begin{bmatrix}
1&0&0&0&0&0&0&0\\
0&1&0&0&0&0&0&0\\
0&0&1&0&0&0&0&0\\
0&0&0&1&0&0&0&0\\
0&0&0&0&1&0&0&0\\
0&0&0&0&0&1&0&0\\
0&0&0&0&0&0&1&0\\
1&0&0&0&0&0&0&1
\end{bmatrix},
$$

$$
\begin{bmatrix}
1&0&0&0&0&0&0&0\\
0&1&0&0&0&0&0&0\\
0&1&1&0&0&0&0&0\\
0&0&0&1&0&0&0&0\\
0&0&0&0&1&0&0&0\\
0&0&0&0&0&1&0&0\\
0&0&0&0&0&0&1&0\\
0&0&0&0&0&0&0&1
\end{bmatrix},\quad
\begin{bmatrix}
1&0&0&0&0&0&0&0\\
0&1&0&0&0&0&0&0\\
0&0&1&0&0&0&0&0\\
0&1&0&1&0&0&0&0\\
0&0&0&0&1&0&0&0\\
0&0&0&0&0&1&0&0\\
0&0&0&0&0&0&1&0\\
0&0&0&0&0&0&0&1
\end{bmatrix},\quad
\begin{bmatrix}
1&0&0&0&0&0&0&0\\
0&1&0&0&0&0&0&0\\
0&0&1&0&0&0&0&0\\
0&0&0&1&0&0&0&0\\
0&1&0&0&1&0&0&0\\
0&0&0&0&0&1&0&0\\
0&0&0&0&0&0&1&0\\
0&0&0&0&0&0&0&1
\end{bmatrix},
$$

$$
\begin{bmatrix}
1&0&0&0&0&0&0&0\\
0&1&0&0&0&0&0&0\\
0&0&1&0&0&0&0&0\\
0&0&0&1&0&0&0&0\\
0&0&0&0&1&0&0&0\\
0&1&0&0&0&1&0&0\\
0&0&0&0&0&0&1&0\\
0&0&0&0&0&0&0&1
\end{bmatrix},
\begin{bmatrix}
1&0&0&0&0&0&0&0\\
0&1&0&0&0&0&0&0\\
0&0&1&0&0&0&0&0\\
0&0&0&1&0&0&0&0\\
0&0&0&0&1&0&0&0\\
0&0&0&0&0&1&0&0\\
0&1&0&0&0&0&1&0\\
0&0&0&0&0&0&0&1
\end{bmatrix},
\begin{bmatrix}
1&0&0&0&0&0&0&0\\
0&1&0&0&0&0&0&0\\
0&0&1&0&0&0&0&0\\
0&0&0&1&0&0&0&0\\
0&0&0&0&1&0&0&0\\
0&0&0&0&0&1&0&0\\
0&0&0&0&0&0&1&0\\
0&1&0&0&0&0&0&1
\end{bmatrix},
$$

$$
\begin{bmatrix}
1&0&0&0&0&0&0&0\\
0&1&0&0&0&0&0&0\\
0&0&1&0&0&0&0&0\\
0&0&1&1&0&0&0&0\\
0&0&0&0&1&0&0&0\\
0&0&0&0&0&1&0&0\\
0&0&0&0&0&0&1&0\\
0&0&0&0&0&0&0&1
\end{bmatrix},
\begin{bmatrix}
1&0&0&0&0&0&0&0\\
0&1&0&0&0&0&0&0\\
0&0&1&0&0&0&0&0\\
0&0&0&1&0&0&0&0\\
0&0&1&0&1&0&0&0\\
0&0&0&0&0&1&0&0\\
0&0&0&0&0&0&1&0\\
0&0&0&0&0&0&0&1
\end{bmatrix},
\begin{bmatrix}
1&0&0&0&0&0&0&0\\
0&1&0&0&0&0&0&0\\
0&0&1&0&0&0&0&0\\
0&0&0&1&0&0&0&0\\
0&0&0&0&1&0&0&0\\
0&0&1&0&0&1&0&0\\
0&0&0&0&0&0&1&0\\
0&0&0&0&0&0&0&1
\end{bmatrix},
$$

$$
\begin{bmatrix}
1&0&0&0&0&0&0&0\\
0&1&0&0&0&0&0&0\\
0&0&1&0&0&0&0&0\\
0&0&0&1&0&0&0&0\\
0&0&0&0&1&0&0&0\\
0&0&0&0&0&1&0&0\\
0&0&1&0&0&0&1&0\\
0&0&0&0&0&0&0&1
\end{bmatrix},
\begin{bmatrix}
1&0&0&0&0&0&0&0\\
0&1&0&0&0&0&0&0\\
0&0&1&0&0&0&0&0\\
0&0&0&1&0&0&0&0\\
0&0&0&0&1&0&0&0\\
0&0&0&0&0&1&0&0\\
0&0&0&0&0&0&1&0\\
0&0&1&0&0&0&0&1
\end{bmatrix},
$$

$$
\begin{bmatrix}
1&0&0&0&0&0&0&0\\
0&1&0&0&0&0&0&0\\
0&0&1&0&0&0&0&0\\
0&0&0&1&0&0&0&0\\
0&0&0&1&1&0&0&0\\
0&0&0&0&0&1&0&0\\
0&0&0&0&0&0&1&0\\
0&0&0&0&0&0&0&1
\end{bmatrix},
\begin{bmatrix}
1&0&0&0&0&0&0&0\\
0&1&0&0&0&0&0&0\\
0&0&1&0&0&0&0&0\\
0&0&0&1&0&0&0&0\\
0&0&0&0&1&0&0&0\\
0&0&0&1&0&1&0&0\\
0&0&0&0&0&0&1&0\\
0&0&0&0&0&0&0&1
\end{bmatrix},
\begin{bmatrix}
1&0&0&0&0&0&0&0\\
0&1&0&0&0&0&0&0\\
0&0&1&0&0&0&0&0\\
0&0&0&1&0&0&0&0\\
0&0&0&0&1&0&0&0\\
0&0&0&0&0&1&0&0\\
0&0&0&1&0&0&1&0\\
0&0&0&0&0&0&0&1
\end{bmatrix},
$$

$$
\begin{bmatrix}
1&0&0&0&0&0&0&0\\
0&1&0&0&0&0&0&0\\
0&0&1&0&0&0&0&0\\
0&0&0&1&0&0&0&0\\
0&0&0&0&1&0&0&0\\
0&0&0&0&0&1&0&0\\
0&0&0&0&0&0&1&0\\
0&0&0&1&0&0&0&1
\end{bmatrix},
$$

$$
\begin{bmatrix}
1&0&0&0&0&0&0&0\\
0&1&0&0&0&0&0&0\\
0&0&1&0&0&0&0&0\\
0&0&0&1&0&0&0&0\\
0&0&0&0&1&0&0&0\\
0&0&0&0&1&1&0&0\\
0&0&0&0&0&0&1&0\\
0&0&0&0&0&0&0&1
\end{bmatrix},
\begin{bmatrix}
1&0&0&0&0&0&0&0\\
0&1&0&0&0&0&0&0\\
0&0&1&0&0&0&0&0\\
0&0&0&1&0&0&0&0\\
0&0&0&0&1&0&0&0\\
0&0&0&0&0&1&0&0\\
0&0&0&0&1&0&1&0\\
0&0&0&0&0&0&0&1
\end{bmatrix},
\begin{bmatrix}
1&0&0&0&0&0&0&0\\
0&1&0&0&0&0&0&0\\
0&0&1&0&0&0&0&0\\
0&0&0&1&0&0&0&0\\
0&0&0&0&1&0&0&0\\
0&0&0&0&0&1&0&0\\
0&0&0&0&0&0&1&0\\
0&0&0&0&1&0&0&1
\end{bmatrix},
$$

$$\begin{bmatrix} 1&0&0&0&0&0&0&0\\ 0&1&0&0&0&0&0&0\\ 0&0&1&0&0&0&0&0\\ 0&0&0&1&0&0&0&0\\ 0&0&0&0&1&0&0&0\\ 0&0&0&0&0&1&0&0\\ 0&0&0&0&0&1&1&0\\ 0&0&0&0&0&0&0&1 \end{bmatrix}, \begin{bmatrix} 1&0&0&0&0&0&0&0\\ 0&1&0&0&0&0&0&0\\ 0&0&1&0&0&0&0&0\\ 0&0&0&1&0&0&0&0\\ 0&0&0&0&1&0&0&0\\ 0&0&0&0&0&1&0&0\\ 0&0&0&0&0&0&1&0\\ 0&0&0&0&0&1&0&1 \end{bmatrix},$$

$$\begin{bmatrix} 1&0&0&0&0&0&0&0\\ 0&1&0&0&0&0&0&0\\ 0&0&1&0&0&0&0&0\\ 0&0&0&1&0&0&0&0\\ 0&0&0&0&1&0&0&0\\ 0&0&0&0&0&1&0&0\\ 0&0&0&0&0&0&1&0\\ 0&0&0&0&0&0&1&1 \end{bmatrix}. \tag{8.6}$$

For instance, making reference to Table 3.10, the first seven raising operators generate vectors represented in rows 1, 2, 3, 4, 5, 6, and 7, respectively, and so on. The lowering operators are also built in a recursive way:

$$\begin{bmatrix} 0&0&0&0&0&0&0&0\\ 0&1&0&0&0&0&0&0\\ 0&0&1&0&0&0&0&0\\ 0&0&0&1&0&0&0&0\\ 0&0&0&0&1&0&0&0\\ 0&0&0&0&0&1&0&0\\ 0&0&0&0&0&0&1&0\\ 0&0&0&0&0&0&0&1 \end{bmatrix}, \begin{bmatrix} 1&0&0&0&0&0&0&0\\ 0&0&0&0&0&0&0&0\\ 0&0&1&0&0&0&0&0\\ 0&0&0&1&0&0&0&0\\ 0&0&0&0&1&0&0&0\\ 0&0&0&0&0&1&0&0\\ 0&0&0&0&0&0&1&0\\ 0&0&0&0&0&0&0&1 \end{bmatrix}, \begin{bmatrix} 1&0&0&0&0&0&0&0\\ 0&1&0&0&0&0&0&0\\ 0&0&0&0&0&0&0&0\\ 0&0&0&1&0&0&0&0\\ 0&0&0&0&1&0&0&0\\ 0&0&0&0&0&1&0&0\\ 0&0&0&0&0&0&1&0\\ 0&0&0&0&0&0&0&1 \end{bmatrix},$$

$$\begin{bmatrix} 1&0&0&0&0&0&0&0\\ 0&1&0&0&0&0&0&0\\ 0&0&1&0&0&0&0&0\\ 0&0&0&0&0&0&0&0\\ 0&0&0&0&1&0&0&0\\ 0&0&0&0&0&1&0&0\\ 0&0&0&0&0&0&1&0\\ 0&0&0&0&0&0&0&1 \end{bmatrix}, \begin{bmatrix} 1&0&0&0&0&0&0&0\\ 0&1&0&0&0&0&0&0\\ 0&0&1&0&0&0&0&0\\ 0&0&0&1&0&0&0&0\\ 0&0&0&0&0&0&0&0\\ 0&0&0&0&0&1&0&0\\ 0&0&0&0&0&0&1&0\\ 0&0&0&0&0&0&0&1 \end{bmatrix}, \begin{bmatrix} 1&0&0&0&0&0&0&0\\ 0&1&0&0&0&0&0&0\\ 0&0&1&0&0&0&0&0\\ 0&0&0&1&0&0&0&0\\ 0&0&0&0&1&0&0&0\\ 0&0&0&0&0&0&0&0\\ 0&0&0&0&0&0&1&0\\ 0&0&0&0&0&0&0&1 \end{bmatrix},$$

$$\begin{bmatrix} 1&0&0&0&0&0&0&0\\ 0&1&0&0&0&0&0&0\\ 0&0&1&0&0&0&0&0\\ 0&0&0&1&0&0&0&0\\ 0&0&0&0&1&0&0&0\\ 0&0&0&0&0&1&0&0\\ 0&0&0&0&0&0&0&0\\ 0&0&0&0&0&0&0&1 \end{bmatrix}, \begin{bmatrix} 1&0&0&0&0&0&0&0\\ 0&1&0&0&0&0&0&0\\ 0&0&1&0&0&0&0&0\\ 0&0&0&1&0&0&0&0\\ 0&0&0&0&1&0&0&0\\ 0&0&0&0&0&1&0&0\\ 0&0&0&0&0&0&1&0\\ 0&0&0&0&0&0&0&0 \end{bmatrix}.$$

We could ask about the reason for 6 raising and 4 lowering operators for $\mathcal{B}_2$ and 28 raising and 8 lowering operators for $\mathcal{B}_3$. This number of operators is simply due to the chosen convention to always have a couple of 1s in a single row per matrix in the case of a raising operators and all 0s in a single row per matrix for lowering operators. Those numbers quite naturally correspond to the nodes of Levels 2/4 and 3/4 and of Levels 2/8 and 7/8, respectively. These are the lowest levels (apart from $\mathbf{0}$) and the highest levels (apart from $\mathbf{1}$) for the two algebras, respectively.

Generalisation to higher dimensional spaces is obvious. For both raising and lowering operators, we start with the identity matrix. For raising operators, we add first a 1 in the position a_{21}, then in the position a_{31}, and so on up to a_{m1}, with $m =$ to the number of digits of the ID; then we restart by adding a 1 in the position a_{32}, then in the position a_{42}, and so on up to a_{m2}. We repeat the procedure up to $a_{m,m-1}$. For the lowering operators, we simply replace a 1 with a 0 along the diagonal starting by a_{11} and going on up to a_{mm}.

Thus, the number of raising (and of lowering) operators decreases relatively to the total number of vectors when the dimension of the space increases. In fact, we have the series $6/16 = 3/8$, $28/256 = 7/64$, $120/65536 = 15/8192$, and so on, according to

$$\lim_{m \to \infty} \frac{\binom{m}{2}}{\sum_{x=0}^{m} \binom{m}{x}} = 0. \tag{8.7}$$

8.3. Basis Vectors in the 3D Case

The previous formalism transforms whatever vector of a LVS associated with the Boolean algebra $\mathcal{B}_n$ in another one and vice versa. However, the nodes of the algebra are also built through combinations of SD variables that for $n \geq 3$ pertain to a CA or a SCA and this corresponds to a choice of a vector basis spanning the space. In other words, the logical expressions provide for a finer analysis of representation, so that we need to explore the linear transformations from one basis to another. Let us focus on $\mathscr{B}_3$. First, as explained in Subsection 3.1.2, we pass from CA1 to CA2 and vice versa by exchanging columns (here rows) d and e. Thus, the relative transformation matrix is

$$C^3 = \begin{bmatrix} 1 & 0 & 0 & 0 & 0 & 0 & 0 & 0 \\ 0 & 1 & 0 & 0 & 0 & 0 & 0 & 0 \\ 0 & 0 & 1 & 0 & 0 & 0 & 0 & 0 \\ 0 & 0 & 0 & 0 & 1 & 0 & 0 & 0 \\ 0 & 0 & 0 & 1 & 0 & 0 & 0 & 0 \\ 0 & 0 & 0 & 0 & 0 & 1 & 0 & 0 \\ 0 & 0 & 0 & 0 & 0 & 0 & 1 & 0 \\ 0 & 0 & 0 & 0 & 0 & 0 & 0 & 1 \end{bmatrix}. \tag{8.8}$$

For instance, let us apply this matrix to $X_{1.1}$:

$$\begin{bmatrix} 1 & 0 & 0 & 0 & 0 & 0 & 0 & 0 \\ 0 & 1 & 0 & 0 & 0 & 0 & 0 & 0 \\ 0 & 0 & 1 & 0 & 0 & 0 & 0 & 0 \\ 0 & 0 & 0 & 0 & 1 & 0 & 0 & 0 \\ 0 & 0 & 0 & 1 & 0 & 0 & 0 & 0 \\ 0 & 0 & 0 & 0 & 0 & 1 & 0 & 0 \\ 0 & 0 & 0 & 0 & 0 & 0 & 1 & 0 \\ 0 & 0 & 0 & 0 & 0 & 0 & 0 & 1 \end{bmatrix} \begin{pmatrix} 0 \\ 0 \\ 0 \\ 0 \\ 1 \\ 1 \\ 1 \\ 1 \end{pmatrix} = \begin{pmatrix} 0 \\ 0 \\ 0 \\ 1 \\ 0 \\ 1 \\ 1 \\ 1 \end{pmatrix}, \tag{8.9}$$

which is $X_{1.2}$. Now, I consider here only the transformations pertaining to CA1 because those for CA2 behave similarly. Using Table 3.6, we perform the transformations from SCA1.a to SCA1.b and vice versa (exchange of c-d with f-e) thanks to the matrix

$$S_1^3 = \begin{bmatrix} 1 & 0 & 0 & 0 & 0 & 0 & 0 & 0 \\ 0 & 1 & 0 & 0 & 0 & 0 & 0 & 0 \\ 0 & 0 & 0 & 0 & 0 & 1 & 0 & 0 \\ 0 & 0 & 0 & 0 & 1 & 0 & 0 & 0 \\ 0 & 0 & 0 & 1 & 0 & 0 & 0 & 0 \\ 0 & 0 & 1 & 0 & 0 & 0 & 0 & 0 \\ 0 & 0 & 0 & 0 & 0 & 0 & 1 & 0 \\ 0 & 0 & 0 & 0 & 0 & 0 & 0 & 1 \end{bmatrix}, \tag{8.10}$$

the transformations from SCA1.a into SCA1.c and vice versa (exchange of b-d with g-e) thanks to the matrix

$$S_2^3 = \begin{bmatrix} 1 & 0 & 0 & 0 & 0 & 0 & 0 & 0 \\ 0 & 0 & 0 & 0 & 0 & 0 & 1 & 0 \\ 0 & 0 & 1 & 0 & 0 & 0 & 0 & 0 \\ 0 & 0 & 0 & 0 & 1 & 0 & 0 & 0 \\ 0 & 0 & 0 & 1 & 0 & 0 & 0 & 0 \\ 0 & 0 & 0 & 0 & 0 & 1 & 0 & 0 \\ 0 & 1 & 0 & 0 & 0 & 0 & 0 & 0 \\ 0 & 0 & 0 & 0 & 0 & 0 & 0 & 1 \end{bmatrix}, \tag{8.11}$$

and finally from SCA1.a into SCA1.d and vice versa (exchange of b-c with g-f) thanks to the matrix

$$S_3^3 = \begin{bmatrix} 1 & 0 & 0 & 0 & 0 & 0 & 0 & 0 \\ 0 & 0 & 0 & 0 & 0 & 0 & 1 & 0 \\ 0 & 0 & 0 & 0 & 0 & 1 & 0 & 0 \\ 0 & 0 & 0 & 1 & 0 & 0 & 0 & 0 \\ 0 & 0 & 0 & 0 & 1 & 0 & 0 & 0 \\ 0 & 0 & 1 & 0 & 0 & 0 & 0 & 0 \\ 0 & 1 & 0 & 0 & 0 & 0 & 0 & 0 \\ 0 & 0 & 0 & 0 & 0 & 0 & 0 & 1 \end{bmatrix}. \tag{8.12}$$

If we want the transformation from SCA1.b into SCA1.c, we first apply S_1^3 to the vectors pertaining to SCA1.b and then S_2^3 to the result. Of course, we can use any of those matrices as alternatives to the method explored in the previous section.

8.4. Basis Vectors in the 4D Case

In this section, I show the transformations from a CA1.x into a CA1.y ($1 \leq x, y \leq 8$) vector for $\mathscr{B}_4$ (the situation is analogous for the $CA2.x$, $1 \leq x \leq 8$). Using the results of Section 4.1, we get the following transformation from CA1.1 into CA.1.2 and vice versa

(exchange of columns, here rows, h and i)

$$C_1^4 = \begin{bmatrix}
1 & 0 & 0 & 0 & 0 & 0 & 0 & 0 & 0 & 0 & 0 & 0 & 0 & 0 & 0 & 0 \\
0 & 1 & 0 & 0 & 0 & 0 & 0 & 0 & 0 & 0 & 0 & 0 & 0 & 0 & 0 & 0 \\
0 & 0 & 1 & 0 & 0 & 0 & 0 & 0 & 0 & 0 & 0 & 0 & 0 & 0 & 0 & 0 \\
0 & 0 & 0 & 1 & 0 & 0 & 0 & 0 & 0 & 0 & 0 & 0 & 0 & 0 & 0 & 0 \\
0 & 0 & 0 & 0 & 1 & 0 & 0 & 0 & 0 & 0 & 0 & 0 & 0 & 0 & 0 & 0 \\
0 & 0 & 0 & 0 & 0 & 1 & 0 & 0 & 0 & 0 & 0 & 0 & 0 & 0 & 0 & 0 \\
0 & 0 & 0 & 0 & 0 & 0 & 1 & 0 & 0 & 0 & 0 & 0 & 0 & 0 & 0 & 0 \\
0 & 0 & 0 & 0 & 0 & 0 & 0 & 0 & 1 & 0 & 0 & 0 & 0 & 0 & 0 & 0 \\
0 & 0 & 0 & 0 & 0 & 0 & 0 & 1 & 0 & 0 & 0 & 0 & 0 & 0 & 0 & 0 \\
0 & 0 & 0 & 0 & 0 & 0 & 0 & 0 & 0 & 1 & 0 & 0 & 0 & 0 & 0 & 0 \\
0 & 0 & 0 & 0 & 0 & 0 & 0 & 0 & 0 & 0 & 1 & 0 & 0 & 0 & 0 & 0 \\
0 & 0 & 0 & 0 & 0 & 0 & 0 & 0 & 0 & 0 & 0 & 1 & 0 & 0 & 0 & 0 \\
0 & 0 & 0 & 0 & 0 & 0 & 0 & 0 & 0 & 0 & 0 & 0 & 1 & 0 & 0 & 0 \\
0 & 0 & 0 & 0 & 0 & 0 & 0 & 0 & 0 & 0 & 0 & 0 & 0 & 1 & 0 & 0 \\
0 & 0 & 0 & 0 & 0 & 0 & 0 & 0 & 0 & 0 & 0 & 0 & 0 & 0 & 1 & 0 \\
0 & 0 & 0 & 0 & 0 & 0 & 0 & 0 & 0 & 0 & 0 & 0 & 0 & 0 & 0 & 1
\end{bmatrix},$$

$$(8.13)$$

from CA.2 into CA.3 (exchange of columns, here rows, g-h and j-i)

$$C_2^4 = \begin{bmatrix}
1 & 0 & 0 & 0 & 0 & 0 & 0 & 0 & 0 & 0 & 0 & 0 & 0 & 0 & 0 & 0 \\
0 & 1 & 0 & 0 & 0 & 0 & 0 & 0 & 0 & 0 & 0 & 0 & 0 & 0 & 0 & 0 \\
0 & 0 & 1 & 0 & 0 & 0 & 0 & 0 & 0 & 0 & 0 & 0 & 0 & 0 & 0 & 0 \\
0 & 0 & 0 & 1 & 0 & 0 & 0 & 0 & 0 & 0 & 0 & 0 & 0 & 0 & 0 & 0 \\
0 & 0 & 0 & 0 & 1 & 0 & 0 & 0 & 0 & 0 & 0 & 0 & 0 & 0 & 0 & 0 \\
0 & 0 & 0 & 0 & 0 & 1 & 0 & 0 & 0 & 0 & 0 & 0 & 0 & 0 & 0 & 0 \\
0 & 0 & 0 & 0 & 0 & 0 & 0 & 0 & 0 & 1 & 0 & 0 & 0 & 0 & 0 & 0 \\
0 & 0 & 0 & 0 & 0 & 0 & 0 & 0 & 1 & 0 & 0 & 0 & 0 & 0 & 0 & 0 \\
0 & 0 & 0 & 0 & 0 & 0 & 0 & 1 & 0 & 0 & 0 & 0 & 0 & 0 & 0 & 0 \\
0 & 0 & 0 & 0 & 0 & 0 & 1 & 0 & 0 & 0 & 0 & 0 & 0 & 0 & 0 & 0 \\
0 & 0 & 0 & 0 & 0 & 0 & 0 & 0 & 0 & 0 & 1 & 0 & 0 & 0 & 0 & 0 \\
0 & 0 & 0 & 0 & 0 & 0 & 0 & 0 & 0 & 0 & 0 & 1 & 0 & 0 & 0 & 0 \\
0 & 0 & 0 & 0 & 0 & 0 & 0 & 0 & 0 & 0 & 0 & 0 & 1 & 0 & 0 & 0 \\
0 & 0 & 0 & 0 & 0 & 0 & 0 & 0 & 0 & 0 & 0 & 0 & 0 & 1 & 0 & 0 \\
0 & 0 & 0 & 0 & 0 & 0 & 0 & 0 & 0 & 0 & 0 & 0 & 0 & 0 & 1 & 0 \\
0 & 0 & 0 & 0 & 0 & 0 & 0 & 0 & 0 & 0 & 0 & 0 & 0 & 0 & 0 & 1
\end{bmatrix}.$$

$$(8.14)$$

It is, in fact, more convenient to consider the series of transformations. For the transformation from CA.3 into CA.4 and vice versa we have that $C_1^4 = C_3^4$. For the transformation of CA.14 into CA1.5 and vice versa we have (exchange of f-g-h with k-j-i)

$$
C_4^4 =
\begin{bmatrix}
1 & 0 & 0 & 0 & 0 & 0 & 0 & 0 & 0 & 0 & 0 & 0 & 0 & 0 & 0 & 0 \\
0 & 1 & 0 & 0 & 0 & 0 & 0 & 0 & 0 & 0 & 0 & 0 & 0 & 0 & 0 & 0 \\
0 & 0 & 1 & 0 & 0 & 0 & 0 & 0 & 0 & 0 & 0 & 0 & 0 & 0 & 0 & 0 \\
0 & 0 & 0 & 1 & 0 & 0 & 0 & 0 & 0 & 0 & 0 & 0 & 0 & 0 & 0 & 0 \\
0 & 0 & 0 & 0 & 1 & 0 & 0 & 0 & 0 & 0 & 0 & 0 & 0 & 0 & 0 & 0 \\
0 & 0 & 0 & 0 & 0 & 0 & 0 & 0 & 0 & 0 & 1 & 0 & 0 & 0 & 0 & 0 \\
0 & 0 & 0 & 0 & 0 & 0 & 0 & 0 & 0 & 1 & 0 & 0 & 0 & 0 & 0 & 0 \\
0 & 0 & 0 & 0 & 0 & 0 & 0 & 0 & 1 & 0 & 0 & 0 & 0 & 0 & 0 & 0 \\
0 & 0 & 0 & 0 & 0 & 0 & 0 & 1 & 0 & 0 & 0 & 0 & 0 & 0 & 0 & 0 \\
0 & 0 & 0 & 0 & 0 & 0 & 1 & 0 & 0 & 0 & 0 & 0 & 0 & 0 & 0 & 0 \\
0 & 0 & 0 & 0 & 0 & 1 & 0 & 0 & 0 & 0 & 0 & 0 & 0 & 0 & 0 & 0 \\
0 & 0 & 0 & 0 & 0 & 0 & 0 & 0 & 0 & 0 & 0 & 1 & 0 & 0 & 0 & 0 \\
0 & 0 & 0 & 0 & 0 & 0 & 0 & 0 & 0 & 0 & 0 & 0 & 1 & 0 & 0 & 0 \\
0 & 0 & 0 & 0 & 0 & 0 & 0 & 0 & 0 & 0 & 0 & 0 & 0 & 1 & 0 & 0 \\
0 & 0 & 0 & 0 & 0 & 0 & 0 & 0 & 0 & 0 & 0 & 0 & 0 & 0 & 1 & 0 \\
0 & 0 & 0 & 0 & 0 & 0 & 0 & 0 & 0 & 0 & 0 & 0 & 0 & 0 & 0 & 1 \\
\end{bmatrix},
\tag{8.15}
$$

while for CA1.5 into CA1.6 and vice versa we have $C_5^4 = C_1^4$, from CA1.6 into CA1.7 and vice versa $C_6^4 = C_2^4$, and for CA1.7 into CA1.8 and vice versa $C_7^4 = C_1^4$.

Now, I consider some transformations from SCAs into SCAs. Given the huge number of possibilities for each CA, I focus on the eight bases represented in Equation (4.8), using the same symbolism:

$$\{X_{1a}, X_{4a}, X_{2a}, X_{3b}\}, \quad \{X_{1b}, X_{4b}, X_{2b}, X_{3a}\}, \tag{8.16a}$$

$$\{X_{4a}, X_{1a}, X_{3a}, X_{2b}\}, \quad \{X_{4b}, X_{1b}, X_{3b}, X_{2a}\}, \tag{8.16b}$$

$$\{X_{2a}, X_{3a}, X_{1a}, X_{4b}\}, \quad \{X_{2b}, X_{3b}, X_{1b}, X_{4a}\}, \tag{8.16c}$$

$$\{X_{3a}, X_{2a}, X_{4a}, X_{1b}\}, \quad \{X_{3b}, X_{2b}, X_{4b}, X_{1a}\}. \tag{8.16d}$$

In particular, I perform the transformation from $\{X_{1a}, X_{4a}, X_{2a}, X_{3b}\}$ into $\{X_{4a}, X_{1a}, X_{3a}, X_{2b}\}$ and vice versa (exchange of c-d with n-m and of g-h with j-i) thanks to

$$
S_1^4 = \begin{bmatrix}
1 & 0 & 0 & 0 & 0 & 0 & 0 & 0 & 0 & 0 & 0 & 0 & 0 & 0 & 0 & 0 \\
0 & 1 & 0 & 0 & 0 & 0 & 0 & 0 & 0 & 0 & 0 & 0 & 0 & 0 & 0 & 0 \\
0 & 0 & 0 & 0 & 0 & 0 & 0 & 0 & 0 & 0 & 0 & 0 & 0 & 1 & 0 & 0 \\
0 & 0 & 0 & 0 & 0 & 0 & 0 & 0 & 0 & 0 & 0 & 0 & 1 & 0 & 0 & 0 \\
0 & 0 & 0 & 0 & 1 & 0 & 0 & 0 & 0 & 0 & 0 & 0 & 0 & 0 & 0 & 0 \\
0 & 0 & 0 & 0 & 0 & 1 & 0 & 0 & 0 & 0 & 0 & 0 & 0 & 0 & 0 & 0 \\
0 & 0 & 0 & 0 & 0 & 0 & 0 & 0 & 0 & 1 & 0 & 0 & 0 & 0 & 0 & 0 \\
0 & 0 & 0 & 0 & 0 & 0 & 0 & 0 & 1 & 0 & 0 & 0 & 0 & 0 & 0 & 0 \\
0 & 0 & 0 & 0 & 0 & 0 & 0 & 1 & 0 & 0 & 0 & 0 & 0 & 0 & 0 & 0 \\
0 & 0 & 0 & 0 & 0 & 0 & 1 & 0 & 0 & 0 & 0 & 0 & 0 & 0 & 0 & 0 \\
0 & 0 & 0 & 0 & 0 & 0 & 0 & 0 & 0 & 0 & 1 & 0 & 0 & 0 & 0 & 0 \\
0 & 0 & 0 & 0 & 0 & 0 & 0 & 0 & 0 & 0 & 0 & 1 & 0 & 0 & 0 & 0 \\
0 & 0 & 0 & 1 & 0 & 0 & 0 & 0 & 0 & 0 & 0 & 0 & 0 & 0 & 0 & 0 \\
0 & 0 & 1 & 0 & 0 & 0 & 0 & 0 & 0 & 0 & 0 & 0 & 0 & 0 & 0 & 0 \\
0 & 0 & 0 & 0 & 0 & 0 & 0 & 0 & 0 & 0 & 0 & 0 & 0 & 0 & 1 & 0 \\
0 & 0 & 0 & 0 & 0 & 0 & 0 & 0 & 0 & 0 & 0 & 0 & 0 & 0 & 0 & 1
\end{bmatrix}, \quad (8.17)
$$

from $\{X_{4a}, X_{1a}, X_{3a}, X_{2b}\}$ into $\{X_{2a}, X_{3a}, X_{1a}, X_{4b}\}$ and vice versa (exchange of b-c with o-n and of f-g with k-j)

$$
S_2^4 = \begin{bmatrix}
1 & 0 & 0 & 0 & 0 & 0 & 0 & 0 & 0 & 0 & 0 & 0 & 0 & 0 & 0 & 0 \\
0 & 0 & 0 & 0 & 0 & 0 & 0 & 0 & 0 & 0 & 0 & 0 & 0 & 0 & 1 & 0 \\
0 & 0 & 0 & 0 & 0 & 0 & 0 & 0 & 0 & 0 & 0 & 0 & 0 & 1 & 0 & 0 \\
0 & 0 & 0 & 1 & 0 & 0 & 0 & 0 & 0 & 0 & 0 & 0 & 0 & 0 & 0 & 0 \\
0 & 0 & 0 & 0 & 1 & 0 & 0 & 0 & 0 & 0 & 0 & 0 & 0 & 0 & 0 & 0 \\
0 & 0 & 0 & 0 & 0 & 0 & 0 & 0 & 0 & 0 & 1 & 0 & 0 & 0 & 0 & 0 \\
0 & 0 & 0 & 0 & 0 & 0 & 0 & 0 & 0 & 1 & 0 & 0 & 0 & 0 & 0 & 0 \\
0 & 0 & 0 & 0 & 0 & 0 & 0 & 1 & 0 & 0 & 0 & 0 & 0 & 0 & 0 & 0 \\
0 & 0 & 0 & 0 & 0 & 0 & 0 & 0 & 1 & 0 & 0 & 0 & 0 & 0 & 0 & 0 \\
0 & 0 & 0 & 0 & 0 & 0 & 1 & 0 & 0 & 0 & 0 & 0 & 0 & 0 & 0 & 0 \\
0 & 0 & 0 & 0 & 0 & 1 & 0 & 0 & 0 & 0 & 0 & 0 & 0 & 0 & 0 & 0 \\
0 & 0 & 0 & 0 & 0 & 0 & 0 & 0 & 0 & 0 & 0 & 1 & 0 & 0 & 0 & 0 \\
0 & 0 & 0 & 0 & 0 & 0 & 0 & 0 & 0 & 0 & 0 & 0 & 1 & 0 & 0 & 0 \\
0 & 0 & 1 & 0 & 0 & 0 & 0 & 0 & 0 & 0 & 0 & 0 & 0 & 0 & 0 & 0 \\
0 & 1 & 0 & 0 & 0 & 0 & 0 & 0 & 0 & 0 & 0 & 0 & 0 & 0 & 0 & 0 \\
0 & 0 & 0 & 0 & 0 & 0 & 0 & 0 & 0 & 0 & 0 & 0 & 0 & 0 & 0 & 1
\end{bmatrix}, \quad (8.18)
$$

while for the transformation of $\{X_{2a}, X_{3a}, X_{1a}, X_{4b}\}$ into $\{X_{3a}, X_{2a}, X_{4a}, X_{1b}\}$ and vice versa I have that $S_3^4 = S_1^4$. I perform the transformation from $\{X_{3a}, X_{2a}, X_{4a}, X_{1b}\}$ into $\{X_{1b}, X_{4b}, X_{2b}, X_{3a}\}$ and vice versa (exchange of b-c with o-n and of e-h with l-i) thanks to

$$S_4^4 = \begin{bmatrix}
1 & 0 & 0 & 0 & 0 & 0 & 0 & 0 & 0 & 0 & 0 & 0 & 0 & 0 & 0 & 0 \\
0 & 0 & 0 & 0 & 0 & 0 & 0 & 0 & 0 & 0 & 0 & 0 & 0 & 0 & 1 & 0 \\
0 & 0 & 0 & 0 & 0 & 0 & 0 & 0 & 0 & 0 & 0 & 0 & 0 & 1 & 0 & 0 \\
0 & 0 & 0 & 1 & 0 & 0 & 0 & 0 & 0 & 0 & 0 & 0 & 0 & 0 & 0 & 0 \\
0 & 0 & 0 & 0 & 0 & 0 & 0 & 0 & 0 & 0 & 0 & 1 & 0 & 0 & 0 & 0 \\
0 & 0 & 0 & 0 & 1 & 0 & 0 & 0 & 0 & 0 & 0 & 0 & 0 & 0 & 0 & 0 \\
0 & 0 & 0 & 0 & 0 & 1 & 0 & 0 & 0 & 0 & 0 & 0 & 0 & 0 & 0 & 0 \\
0 & 0 & 0 & 0 & 0 & 0 & 0 & 0 & 1 & 0 & 0 & 0 & 0 & 0 & 0 & 0 \\
0 & 0 & 0 & 0 & 0 & 0 & 0 & 1 & 0 & 0 & 0 & 0 & 0 & 0 & 0 & 0 \\
0 & 0 & 0 & 0 & 0 & 0 & 0 & 0 & 0 & 1 & 0 & 0 & 0 & 0 & 0 & 0 \\
0 & 0 & 0 & 0 & 0 & 0 & 0 & 0 & 0 & 0 & 1 & 0 & 0 & 0 & 0 & 0 \\
0 & 0 & 0 & 0 & 1 & 0 & 0 & 0 & 0 & 0 & 0 & 0 & 0 & 0 & 0 & 0 \\
0 & 0 & 0 & 0 & 0 & 0 & 0 & 0 & 0 & 0 & 0 & 0 & 1 & 0 & 0 & 0 \\
0 & 0 & 1 & 0 & 0 & 0 & 0 & 0 & 0 & 0 & 0 & 0 & 0 & 0 & 0 & 0 \\
0 & 1 & 0 & 0 & 0 & 0 & 0 & 0 & 0 & 0 & 0 & 0 & 0 & 0 & 0 & 0 \\
0 & 0 & 0 & 0 & 0 & 0 & 0 & 0 & 0 & 0 & 0 & 0 & 0 & 0 & 0 & 1
\end{bmatrix}. \quad (8.19)$$

From here onwards the remaining three transformations of the right column of bases (4.8) go parallel to those in the left column, i.e. $S_5^4 = S_1^4$, $S_6^4 = S_2^4$, and $S_7^4 = S_1^4$.

Chapter 9

A Few Concluding Words

9.1. Infinity

I already mentioned (in Subsection 3.1.2) that each $\mathscr{B}_n$ contains several complete subalgebras $\mathscr{B}_m$, with $m < n$. For instance, $\mathscr{B}_2$ contains two $\mathscr{B}_1$ complete subalgebras; $\mathscr{B}_3$ contains eight $\mathscr{B}_1$ complete subalgebras and twelve $\mathscr{B}_2$ complete subalgebras (six couples for each CA displayed in Table 3.3); and so on. Note that all of the $\mathscr{B}_m$ so defined are ideal subalgebras, where *ideal* is meant[1] a subset $\mathscr{B}_m$ of any algebra $\mathscr{B}_n$ such that

- $\mathbf{0} \subset \mathscr{B}_m$,
- If $X \subset \mathscr{B}_m$ and $Y \subset \mathscr{B}_m$, then $X + Y \subset \mathscr{B}_m$,
- If $X \subset \mathscr{B}_m$ and $Y \subset \mathscr{B}_m$, then $XY \subset \mathscr{B}_m$.

Of course, we can also have other kinds of subalgebras, as explained in Section 3.4.

When the number n of the dimensions of the space grows tending towards infinity (and so the length of the ID), the number $2^{\frac{m}{2}}$ (with $m = 2^n$) of SD variables (representing sets) relatively shrinks tending towards a set of zero measure, according to the series

$$\frac{1}{2^{\frac{m}{2}}} = \frac{1}{2^{2^{n-1}}} \to 0 \quad \text{for } n \to \infty. \tag{9.1}$$

---

[1] See [GIVANT/HALMOS 2009, pp. 149–50].

For instance, for a 3D algebra, the SD variables (including their complements) are 1/16 of all k objects; for a 4D algebra, these variables are 1/256 of all k objects; for a 5D algebra, the SD variables are 1/65,536 of all k objects, and so on. Thus, $\mathscr{B}_\infty$ should contain in itself any possible relation giving rise to any thinkable class of objects. We can see the growing series of the different $\mathscr{B}_n$ networks (with $n \to \infty$) as a further and finer analysis of the universal set. In other words, any nD algebra is in fact an extrapolation out of $\mathscr{B}_\infty$.

Traditionally, it is believed that we can build logical expressions *ad libitum*. Of course, when we think about $\mathscr{B}_\infty$, it is evident that there is an infinite number of both contingent classes and logical propositions. Nevertheless, although we cannot exhaust such a number, we are ensured by the reiterative building of Boolean algebras that such classes and propositions are not arbitrary, but follow very precise patterns so that for any finite dimension we are able, at least in principle, to tell which form will have both classes and logical propositions.

9.2. Completeness

Therefore this system satisfies completeness since any truth can be found. However, whatever mathematical or formal system we are able to build is necessarily *incomplete*. There are two different meanings of this term. According to the first significance, a system is incomplete if it acknowledges some truths that it cannot prove. This is the current mathematical understanding of this problem, as dealt with for the first time by K. Gödel.[2] The second meaning is the logical one, with a root in Aristotle's work: since each science assumes several truths as foundations that it cannot prove, there are many possible truths that are clearly excluded.

According to the first view, for any truth that we assume we should be able to derive from it other truths. However, a look at Boolean algebra clearly shows that the only way to do that is to

---

[2][GÖDEL 1931].

find nodes of lower levels that imply the nodes whose expression we like to prove. This search leads us to an endless regress, with two consequences: the implying nodes are less and less general, and we approach more and more the contradiction.

According to the second view, many possible truths cannot be included in the object formal system for the simple reason that they could lead to inconsistencies. Note that in any $\mathscr{B}_n$, each set X of objects has possible relations not only with whatever other set of objects Y but also with its complement Y'. For instance, we have both XY and XY' and both $X + Y$ and $X + Y'$. Now, whatever formal system we like to build, we need to choose *certain* relations among classes and not others. These relations are determined by given assumptions. Thus, the class X has only one of the two possible relations with another class Y, either with Y or its complement Y'. In this way discontinuity is generated. This is Hilbert's requirement of consistency: we cannot have both Y and Y'; one of the two must be demonstrable.[3] In other words, the incompleteness of a formal system (in this second sense) is a natural consequence of selecting out of a logical space a formal system that is consistent. Thus, we not only are not able to prove our assumed truths but also cut the access to many others.

This examination shows another important point. If no specific assumptions (and therefore selections of subspaces) are made, we cannot consider the classes of $\mathscr{B}_\infty$ as collecting individual elements of any kind, since no specific relations among those classes have been selected. Therefore, as anticipated, at this level they should rather be considered as abstract *schemes* of classes exhibiting all possible kinds (this means also incompatible ones) among sets.

I finally observe that the IDs provide a quite natural way to solve Gödel's problem of finding a number for each logical expression. This can simplify a lot formal treatments in which such numbers play a role.

---

[3] [CHAITIN 1998, Sec. 1.1].

9.3. Summary

Boolean algebra has fundamental characteristics. Any algebra $\mathscr{B}_n$ is of course an algebra, but also

- The universal set (of all classes),
- The POSet of all POSets (all POSets are embedded in it),
- The directed graph of all directed graphs (all directed graphs are embedded in it),
- The network of all networks (all networks are embedded in it),
- The codification containing all possible codes,
- A portion of vectorial space.

Moreover, Boolean algebra contains fundamental mathematical notions:

- The notion of combinatorics,
- The notion of numerable infinity (and therefore of natural numbers),
- The notions of function (with its values) and map,
- The algebraic notions of sum and product,
- The notion of digits.

Glossary

Basic logical form: a logical expression such that the 1s present in its ID show no superposition of columns. They are present in odd levels only. In the algebra one dimension higher, basic forms give rise to common forms.

Chunk: the array of 4 digits building an ID.

Class: a collection of items sharing different properties (constituted through some combination of self–dual variables).

Common logical form: a logical expression to which all logical expressions of a given level of algebra can lead. Often it is a generalisation of some logical form of a less dimensional algebra.

Contravalence: the negation of equivalence (whether material or logical). It is symmetric but not reflexive and alternate transitive.

Dimension (of the algebra): the minimal number of self–dual variables capable of generating all nodes of the algebra.

Endogenous pairing: pairing of parts (chunks or 2–chunks, and so on) of IDs of a lower-level algebra that pertain to the same CA.

Exogenous pairing: pairing of parts (chunks or 2–chunks, and so on) of IDs of a lower-level algebra that pertain to different CAs.

Equivalence (logical): two expressions that express the same node.

Equivalence (material): equivalence between two (or more) different logical expressions and which satisfies the properties of reflexivity, symmetry, and transitivity.

ID: the string of truth values that univocally individuates a node of the algebra.

Implication: a logical implication, such as a tautology, that is the arrow connecting different nodes of the algebra pertaining to different levels (or a single node with itself). It constitutes the Boolean algebra as a graph.

Irreducible logical form: a logical expression that cannot be made simpler by dropping some term or some product or sum.

Level (of the algebra): the collection of all nodes with IDs presenting the same number of 1s (or 0s).

Main code alphabet (CA): for dimensions ≥ 3, the consistent collection of SD variables allowing the picking up of the different subalphabets that generate every node of the algebra. It fulfils several combinatorial criteria.

Neg–reversal: the operation that combines complementation and reversal.

Node: object defined through the ID and logically expressed by either a self–dual variable (a basic set) or through a combination of variables (a class). It is contingent (apart from the *infimum* and *supremum*).

Proposition: in this book, a tautology is built through implication or combinations of implications.

Reduced logical form: a logical expression derived from another one by simply dropping some terms or some sums or products.

Reversal: the operation that reverses the ID.

Root (**SD**) variable: a SD variable $\in \mathscr{B}_n$ that generates duplets of SD variables in a given main code alphabet $\in \mathscr{B}_{n+1}$; in a wider sense, the root of all SD variables $\in \mathscr{B}_{n+1}$ sharing the first and final parts of the IDs.

Self–dual variables (or basic sets): the variables invariant under neg–reversal that span the whole Boolean algebra (giving rise to all classes of the algebra). They express a property or aspect. For dimensions ≥ 3, there are alternative choices of sets of those variables.

Self–product (or **self–sum**): when two SD variables generated by the same root variable of less dimension give rise to a product (or sum).

Subcode alphabet (**SCA**): any of the choices of SD variables in a main code alphabet that allows the generation of all nodes of the algebra. The SCA need to fulfil combinatorial criteria.

Bibliography

[ADLEMAN 1998] Adleman, Leonard M., "Computing with DNA", *Scientific American* **279** (2): 54–61; doi:10.1038/sj.embor.embor719.

[ALBERTS *et al.* 1983] Alberts, B., Bray, D., Lewis, J., Raff, M., Roberts, K., and Watson, J. D., *The Molecular Biology of the Cell*, New York, Garland P., 1983.

[AULETTA 2011] Auletta, Gennaro, *Cognitive Biology: Dealing with Information from Bacteria to Minds*, Oxford, University Press.

[AULETTA 2013a] —, "Aristotle's Syllogistic Revisited", *Epistemologia* **36** (2): 233–61; doi:10.3280/EPIS2013-002004.

[AULETTA 2013b] —, *Mechanical Logic in Three-Dimensional Space*, Singapore, Pan Stanford Pub.

[AULETTA 2015] —, "Irreducible Statements and Bases for Finite-Dimensional Logical Spaces", *Indian Ocean Review of Science and Technology* **2015**: Article 2.

[AULETTA 2017] —, "A Critical Examination of Peirce's Theory of Natural Inferences", *Revista Portuguesa de Filosofia* **73** (3–4): 1053–1094; doi:10.17990/RPF/2017–73–3–1053.

[AULETTA 2019] —, *The Quantum Mechanics Conundrum: Interpretation and Foundations*, Cham, Springer.

[AULETTA *et al.* 2009] Auletta, G., Fortunato, M., and Parisi, G., *Quantum Mechanics*, Cambridge, University Press, 2009; rev. paperback ed. 2013.

[BARBIERI 2003] Barbieri, Marcello, *The Organic Codes: An Introduction to Semantic Biology*, Cambridge, University Press.

[BELTRAMETTI/CASSINELLI 1981] Beltrametti, E. and Cassinelli, G., *The Logic of Quantum Mechanics*, Redwood City, Addison-Wesley.

[BENNETT/WIESNER 1992] Bennett, C. H. and Wiesner, S. J., "Communication via One- and Two-Particle Operators on EPR States", *Physical Review Letters* **69** (20): 2881–84; doi:10.1103/PhysRevLett.69.2881.

[BENNETT *et al.* 1993] Bennett, C. H., Brassard, G., Crepeau, C., Jozsa, R., Peres, A., and Wootters, W. K., "Teleporting an Unknown Quantum State via Dual Classical and EPR Channels", *Physical Review Letters* **70** (13): 1895–99; doi:10.1103/PhysRevLett.70.1895.

[BIRKHOFF/VON NEUMANN 1936] Birkhoff, G. D. and von Neumann, J., "The Logic of Quantum Mechanics", *Annals of Mathematics* **37**: 823–43.

[BOOLE 1854] Boole, George, *An Investigation of the Laws of Thought, on Which Are Founded the Mathematical Theories of Logic and Probabilities*, London, Walton and Maberly, 1854.

[BRAUNSTEIN *et al.* 1992] Braunstein, S. L., Mann, A., and Revzen, M., "Maximal Violation of Bell Inequalities for Mixed States", *Physical Review Letters* **68** (22): 3259–61; doi:10.1103/PhysRevLett.68.3259.

[CHAITIN 1998] Chaitin, Gregory J., *The Limits of Mathematics: A Course on Information Theory and Limits of Formal Reasoning*, 2nd ed., Singapore, Springer.

[COECKE 2011] Coecke, Bob (Ed.), *New Structures for Physics*, Berlin, Springer.

[COECKE/PAQUETTE 2011] Coecke, B. and Paquette, É. O., "Categories for the Practising Physicist", in [COECKE 2011, 173–286].

[CRICK 1968] Crick, Francis H. C., "The Origin of the Genetic Code", *Journal of Molecular Biology* **38** (3): 367–79; doi:10.1016/0022-2836(68)90392-6.

[D'ARIANO *et al.* 2017] D'Ariano, M., Chiribella, G., and Perinotti, P., *Quantum Theory from First Principles: An Informational Approach*, Cambridge, University Press.

[DEUTSCH 1985a] Deutsch, David, "Quantum Theory as a Universal Physical Theory", *International Journal of Theoretical Physics* **24** (1): 1–41; doi:10.1007/BF00670071.

[DEUTSCH 1985b] Deutsch, David, "Quantum Theory, the Church–Turing Principle and the Universal Quantum Computer", *Proceedings of the Royal Society of London* **A400** (1818): 97–117; doi:10.1098/rspa.1985.0070.

[ENGLERT *et al.* 1995] Englert, B.-G., Scully, M. O., and Walther, H., "Complementarity and Uncertainty", *Nature* **375**: 367–68; doi:10.1038/375367b0.

[FARAH 2000] Farah, Martha J., *The Cognitive Neuroscience of Vision*, Oxford, Blackwell.

[GIVANT/HALMOS 2009] Givant, S. and Halmos, P., *Introduction to Boolean Algebras*, New York, Springer.

[GÖDEL 1931] Gödel, Kurt, "Über formal unentscheidbar Sätze der *Principia Mathematica* und verwandter Systeme. I", *Monatshefte für Mathematik und Physik* **38**: 173–98; rep. in [GÖDEL CW, I, 144–94].

[GÖDEL 1931] —, "Über unentscheidbare Sätze", in [GÖDEL CW, III, 30–34].

[GÖDEL CW] —, *Collected Works*, Oxford, University Press, 1995.

[GOTTESMAN/CHUANG 1999] Gottesman, D. and Chuang, I. L., "Quantum Teleportation is a Universal Computational Primitive", *Nature* **402**: 390–93; doi:10.1038/46503.

[LANDSMAN 2017] Landsman, Klaas, *Foundations of Quantum Theory: From Classical Concepts to Operator Algebras*, Springer.

[LEINSTER 2014] Leinster, Tom, *Basic Category Theory*, Cambridge, University Press.

[MA *et al.* 2013] Ma, X.-S., Kofler, J., Qarry, A., Tetik, N., Scheidl, T., Ursin, R., Ramelow, S., Herbst, T., Ratschbacher, L., Fedrizzi, A., Jennewein, T., and Zeilinger, A., "Quantum Erasure with Causally Disconnected Choice", *Proceedings of the National Academy of Sciences USA* **110**: 1221–26; doi:10.1073/pnas.1213201110.

[NIELSEN/CHUANG 2000] Nielsen, M. A. and Chuang, I. L., *Quantum Computation and Quantum Information*, Cambridge, University Press, 2000, 2002, 2011.

[OSAWA *et al.* 1992] Osawa, S., Jukes, T. H., Watanabe, K., and Muto, A., "Recent Evidence for Evolution of the Genetic Code", *Microbiological Reviews* **56** (1): 229–64.

[RODIN *et al.* 2011] Rodin, A. S., Szathmáry, E., and Rodin, S. N., "On origin of genetic code and tRNA before translation", *Biology Direct* **6**: 14; doi:10.1186/1745-6150-6-14.

[SPIVAK 2013] Spivak, David I., *Category Theory for Scientists*, Cambridge, MA, MIT Press.

[TOFFOLI 1980] Toffoli, Tommaso, "Reversible Computing", *MIT-Laboratory for Computing Science* **Memo 1**: http://pm1.bu.edu/~tt/publ/revcomp-rep.pdf.

[WHITEHEAD/RUSSELL 1910] Whitehead, A. N. and Russell, B., *Principia Mathematica*, Cambridge, University Press, 1910, 2nd ed. 1927.

[WITTGENSTEIN 1914–16] Wittgenstein, Ludwig, *Tagebücher 1914–1916*; in [WITTGENSTEIN W, I, 87–187].

[WITTGENSTEIN 1921] —, *Tractatus Logico–Philosophicus*, 1921; II ed.: London, 1922; III ed.: 1933; London, 1971; in [WITTGENSTEIN W, I, 7–85].

[WITTGENSTEIN W] Wittgenstein, Ludwig, *Werkausgabe*, Frankfurt a. M., Suhrkamp, 1984.